Bayesian Inverse Problems

Fundamentals and Engineering Applications

Editors

Juan Chiachío-Ruano
University of Granada
Spain

Manuel Chiachío-Ruano
University of Granada
Spain

Shankar Sankararaman
Intuit Inc.
USA

CRC Press
Taylor & Francis Group
Boca Raton London New York

CRC Press is an imprint of the
Taylor & Francis Group, an **informa** business
A SCIENCE PUBLISHERS BOOK

Cover credit: Cover image by Dr Elmar Zander (chapter author). It is original and has not been taken from any copyrighted source.

First edition published 2022
by CRC Press
6000 Broken Sound Parkway NW, Suite 300, Boca Raton, FL 33487-2742

and by CRC Press
2 Park Square, Milton Park, Abingdon, Oxon, OX14 4RN

Library of Congress Cataloging-in-Publication Data

Names: Chiachío-Ruano, Juan, 1983- editor.
Title: Bayesian inverse problems : fundamentals and engineering applications / editors, Juan Chiachío-Ruano, University of Granada, Spain, Manuel Chiachío-Ruano, University of Granada, Spain, Shankar Sankararaman, Intuit Inc., USA.
Description: First edition. | Boca Raton : CRC Press, 2021. | Includes bibliographical references and index.
Identifiers: LCCN 2021016121 | ISBN 9781138035850 (hardcover)
Subjects: LCSH: Inverse problems (Differential equations) | Bayesian statistical decision theory.
Classification: LCC QA371 .B3655 2021 | DDC 515/.357--dc23
LC record available at https://lccn.loc.gov/2021016121

ISBN: 978-1-138-03585-0 (hbk)
ISBN: 978-1-032-11217-6 (pbk)
ISBN: 978-1-315-23297-3 (ebk)

DOI: 10.1201/b22018

Typeset in Times New Roman
by Radiant Productions

To the students of any scientific or engineering discipline, who may find in this book an opportunity to learn how mathematics possesses not only truth and supreme beauty, but also an extraordinary power to address many of society's most challenging problems.

Preface

We live in the digital era. As developed societies, we are facing the onset of a new industrial revolution due to the rapid development of technologies including artificial intelligence, internet of things, and soft robotics. As a result, the amount of information and data coming from remotely monitored infrastructures, buildings, vehicles, industrial plants, etc. will increase exponentially over the next few decades. At the same time, our fundamental knowledge about Nature and engineered systems has experienced a rampant increase since the last century, and the computational power available today to process such information has seen a revolutionary transformation. This intersection between empirical (data-driven) and fundamental (physics-based) knowledge has led to the rise of new research topics for knowledge discovery, out of which the Bayesian methods and stochastic simulation are prominent. Just as an example, the increased availability of information coming from *digital twins* models and the grown ability by intelligent algorithms of fusing such an information with real-time data is leading to intelligent *cyber-physical systems* such as autonomous cars, smart buildings, etc. This engineering "revolution" enabled by digital technologies and increasing fundamental knowledge is changing the way the 21st century's engineered assets are designed, built, and operated.

This book is devoted to the so-called "Bayesian methods" and how this class of methods can be useful to rigorously address a range of engineering problems where empirical data and fundamental knowledge come into play. These methods comprise not only the Bayesian formulation of inverse and forward engineering problems, but also the associated stochastic simulation algorithms needed to solve them. All the authors contributing to this book are renowned experts in this field and share the same perception about the importance and relevance of this topic in the upcoming challenges and opportunities brought by the digital revolution in modern engineering.

The Editors

Contents

List of Figures

List of Tables

Contributors

David E. Acuña-Ureta
Pontificia Universidad Católica de Chile
Chile

James L. Beck
California Institute of Technology
USA

Sergio Cantero-Chinchilla
University of Nottingham
UK

Juan Chiachío-Ruano
University of Granada
Spain

Manuel Chiachío-Ruano
University of Granada
Spain

Dimitrios Chronopoulos
University of Nottingham
UK

Noémi Friedman
SZTAKI (Institute for Computer Science and Control, Budapest)
Hungary

Yong Huang
Harbin Institute of Technology
China

María L. Jalón
University of Granada
Spain

Lambros Katafygiotis
Hong Kong University of Science and Technology
China

Hermann G. Matthies
Technische Universität Braunschweig
Germany

Marcos E. Orchard
University of Chile
Chile

Costas Papadimitriou
University of Thessaly
Greece

Shankar Sankararaman
Intuit Inc.
USA

Wang-Ji Yan
University of Macau
China

Elmar Zander
Technische Universität Braunschweig
Germany

Claudia Zoccarato
University of Padova
Italy

Part I

Fundamentals of Bayesian Methods

1

Introduction to Bayesian Inverse Problems

Juan Chiachío-Ruano,[1,*] *Manuel Chiachío-Ruano*[1] and *Shankar Sankararaman*[2]

This chapter formally introduces the concept of uncertainty and explains the impact of uncertainty quantification on an important class of engineering problems such as the *inverse problems*. The treatment of uncertainty can be facilitated through various mathematical methods, though probability has been predominantly used in engineering. A simple introduction to probability theory is presented as the foundation of the Bayesian approach to inverse problems. The interpretation of this Bayesian approach to inverse problems and its practical implications using relevant engineering examples constitute the backbone of this textbook.

1.1 Introduction

Research in the area of uncertainty quantification and the application of stochastic methods to the study of engineering systems has gained considerable attention during the past thirty years. This can be attributed to the necessity and desire to design engineering systems with increasingly complex architectures and new materials. These systems can be multi-level, multi-scale, and multi-disciplinary in nature, and may need sophisticated and complex computational models to facilitate their analysis and design. The development and implementation of computational models is not only increasingly sophisticated and expensive, but also based on physics which is often not well-understood. Furthermore, this complexity is exacerbated when considering the limited availability of full-scale response data and the measurement errors.

This leads to one of the most commonly encountered problems in science and engineering, that is, the identification of a mathematical model or missing parts of a model based on noisy observations (e.g. measurements). This is referred to as the *inverse problem* or *system identification* problem in the literature. The goal of the inverse problem is to use the observed response of a system to "improve" a single or a set of models that idealise that system, so that they make more accurate predictions of the system response to a prescribed, or uncertain, excitation [154].

Different model parameterisations or even model hypotheses representing different physics can be formulated to idealise a single system, yielding a set of different *model classes* [22]. Following the probabilistic formulation of the inverse problem [181], the solution is not a single-valued set

[1] University of Granada, Spain.
[2] Intuit Inc., USA.
* Corresponding author: jchiachio@ugr.es

of model parameters nor a single model class. On the contrary, a range of plausible values for model parameters and a set of candidate model classes constitute a more complete, rigorous and principled solution to the system identification problem. The plausibility of the various possibilities is expressed through probabilities which measure the relative *degree of belief* of the candidate solutions conditional to the available information (e.g. data). This interpretation of probability is not well known in the engineering community where there is a widespread belief that probability only applies to *aleatory uncertainty* (e.g. inherent randomness) and not to *epistemic uncertainty* (missing information). E.T. Jaynes [100], who wrote extensively about Bayesian techniques and probability theory, noted that the assumption of inherent randomness is an example of what he called the *Mind-Projection Fallacy*:

> "Our uncertainty is ascribed to an inherent property of nature, or, more generally, our models of reality are confused with reality."

The goal of this chapter is to provide an introductory overview of the fundamentals of the Bayesian inverse problem and its associate stochastic simulation and uncertainty quantification problems.

1.2 Sources of uncertainty

As introduced before, two types of uncertainty are typically considered in science and engineering. The first type, aleatory uncertainty, is regarded as inherent randomness in nature. If the outcome of an experiment differs each time the experiment is run, then this is an example of aleatory uncertainty. This type of uncertainty is irreducible. The second type, epistemic uncertainty, is regarded as a *lack of knowledge* in relation to a particular quantity and/or physical phenomenon. This type of uncertainty can be reduced (and sometimes eliminated) when new knowledge is available, for example, after some research. Epistemic uncertainty can be present either in the data collected from engineering systems or in the models used to represent the behaviour of these engineering systems.

Therefore, the various sources of uncertainty can be enumerated as follows:

1. **Physical Variability**: As mentioned earlier, this type of uncertainty is referred to as aleatory uncertainty. The inputs to the engineering system may be random (e.g. variability in traffic loads for a bridge deck), or the parameters governing a physical system may present spatial variability (e.g. the elastic modulus of a large bridge structure), and this leads to an uncertain output. It is common to represent such random variables using probability distributions.

2. **Data Uncertainty**: Data uncertainty is a very broad term and can be of different types. The most commonly considered type of data uncertainty is measurement errors (both at the input and output levels). Note that it may be argued that measurement errors occur due to natural variability and hence, must be classified as a type of aleatory uncertainty. The second type of data uncertainty occurs during the characterisation of variability; a large amount of data may be necessary to accurately characterise such a variability and that amount of data may not be available in practice. Sometimes, the available data may be sparse and sometimes, even in the form of intervals. This leads to epistemic uncertainty (i.e. uncertainty reducible in the light of new information). Another type of data uncertainty is related to system-level knowledge; for example, the operational conditions of the system may be partially unknown and therefore, the inputs to the system model will automatically become uncertain.

3. **Model Uncertainty**: The engineering system under study is represented using an idealised mathematical model, and the corresponding mathematical equations are numerically solved using computer codes. This modelling process is an instance of epistemic uncertainty and comprises three different types of errors/uncertainty. First, the intended mathematical equation is solved using a computer code which leads to rounding off errors, solution approximation errors, and coding errors. Second, some model parameters may not be readily known, and field data may be needed in order to *update* them. Third, the mathematical model itself is an idealisation of the physical reality, which leads to prediction errors. The combined effect of solution approximation errors, model prediction errors, and model parameter uncertainty is referred to as *model uncertainty*.

There are several mathematical frameworks that provide varied measures of uncertainty for the purpose of uncertainty representation and quantification. These methods differ not only in the level of granularity and detail, but also in how uncertainty is interpreted. Such methods are based on probability theory [85, 157], possibility theory [61], fuzzy set theory [203], Dempster Shafer's evidence theory [18, 148], interval analysis [119], etc. Amongst these theories, probability theory has received significant attention in the engineering community. As a result, this book will focus only on probabilistic methods and not delve into the aforementioned non-probabilistic approaches.

1.3 Formal definition of probability

The fundamental theory of probability is well-established in the literature, including many textbooks and journal articles. The roots of probability lie in the analysis of games of chance by Gerolamo Cardano in the sixteenth century, and by Pierre de Fermat and Blaise Pascal in the seventeenth century. In earlier days, researchers were interested in Boolean and discrete events only (e.g. win or not win), and therefore on *discrete probabilities*. With the advent of mathematical analysis, the importance of *continuous probabilities* steadily increased. This led to a significant change in the understanding and formal definition of probability. The classical definition of probability was based on counting the number of favorable outcomes, and it was understood that this definition cannot be applied to continuous probabilities (refer to Bertrand's Paradox [101]). Hence, the modern definition of probability, which is based on Set Theory [69] and functional mapping, is more commonly used in recent times.

Discrete probability deals with events where the sample space is discrete and countable. Consider the sample space (Ω), which is equal to the set of all possible outcomes (e.g. $\Omega = \{1, 2, \ldots, 6\}$ for the dice roll experiment). The modern definition of probability maps every element $x \in \Omega$ to a "probability value" $p(x)$ such that:

1. $p(x) \in [0, 1] \; \forall \; x \in \Omega$
2. $\sum\limits_{x \in \Omega} p(x) = 1$

Any event E (e.g. getting a value $x \leqslant 3$ in the dice roll experiment) can be expressed as a subset of the sample space Ω ($E \in \Omega$), and the probability of the event E defined as:

$$P(E) = \sum_{x \in E} p(x) \tag{1.1}$$

Hence, the function $p(x)$ is a mapping from a point x in the sample space to a probability value, and is referred to as the *probability mass function* (PMF).

Continuous probability theory deals with cases where the sample space is continuous and hence uncountable; for example consider the case where the set of outcomes of a random experiment is equal to the set of real numbers ($\mathbb{R}$). In this case, the modern definition of probability introduces the concept of *cumulative distribution function* (CDF), defined as $F_X(x) = P(X \leq x)$, that is, the CDF of the random variable X (e.g. the human lifespan) evaluated at x is equal to the probability that the random variable X can take a value less than or equal to x. This CDF necessarily contains the following properties:

1. $F_X(x)$ is monotonically non-decreasing, and right continuous.
2. $\underset{x \to -\infty}{Lim} \; F_X(x) = 0$
3. $\underset{x \to \infty}{Lim} \; F_X(x) = 1$

If the function $F_X(x)$ is absolutely continuous and differentiable, then the derivative of the CDF is denoted as the *probability density function* (PDF) $p_X(x)$. Therefore,

$$p_X(x) = \frac{dF_X(x)}{dx} \tag{1.2}$$

For any set $E \subseteq \mathbb{R}$ (e.g. lifespan longer than eighty years), the probability of the random variable X belonging to E can be written as

$$P(X \in E) = \int_{x \in E} dF_X(x) \tag{1.3}$$

If the PDF exists, then

$$P(X \in E) = \int_{x \in E} p_X(x)dx \tag{1.4}$$

Note that the PDF exists only for continuous random variables, whereas the CDF exists for all random variables (including discrete variables) whose realisations belong to $\mathbb{R}$. A PDF or CDF is said to be valid if and only if it satisfies all of the above properties. The above discussion can be easily be extended to multiple dimensions by considering the space $\mathbb{R}^n$.

The basic principles of probability theory presented here are not only fundamental to this chapter, but will be repeatedly used along the rest of this book.

1.4 Interpretations of probability

The previous section formally defined probability using the concepts of the cumulative distribution function and the probability density function. What is the meaning of this probability? Although the different concepts of probability are well-established in the literature, there is considerable disagreement among researchers on the interpretation of probability. There are two major interpretations based on *physical* and *subjective* probabilities, respectively. It is essential to understand the difference between these two interpretations before delving deeper into this book.

1.4.1 Physical probability

Physical probabilities [180], also referred to as *objective* or *frequentist* probabilities, are related to random physical systems such as rolling a dice, tossing coins, roulette wheels, etc. Each trial of the experiment leads to an event (which is a subset of the sample space), and in the long run of repeated trials, each event tends to occur at a persistent rate, and this rate is referred to as the "relative frequency". These relative frequencies are expressed and explained in terms of physical probabilities. Thus, physical probabilities are defined *only* in the context of random experiments. The theory of classical statistics is based on physical probabilities. Within the realm of physical probabilities, there are two types of interpretations: von Mises' frequentist [190] and Popper's propensity [146]; the former is more easily understood and widely used.

In the context of physical probabilities, the mean of a random variable, sometimes referred to as the population mean, is deterministic (i.e. a single value). It is meaningless to talk about the uncertainty of this mean. In fact, for any type of parameter estimation, the underlying parameter is assumed to be deterministic. The uncertainty in the parameter estimate is addressed through *confidence intervals.* The interpretation of confidence intervals is sometimes confusing and misleading, and the uncertainty in the parameter estimate cannot be used for further uncertainty quantification. For example, if the uncertainty in the elastic modulus of a material was estimated using several axial loading tests over several beam specimens, this uncertainty cannot be used for quantifying the response of a plate made of the same material. This is a serious limitation, since it is not possible to propagate uncertainty after parameter estimation, which is often necessary in engineering modelling problems. Another disadvantage of this approach is that, when a quantity is not random, but unknown, then the well-known tools of frequentist statistics cannot be used to represent this type of uncertainty (epistemic). The second interpretation of probability, that is, the subjective interpretation, overcomes these limitations.

1.4.2 Subjective probability

Subjective probabilities [54] can be assigned to any "statement". It is not necessary that this statement is related to an event which is a possible outcome of a random experiment. In fact, subjective probabilities can be assigned even in the absence of random experiments. Subjective probabilities are interpreted as *degrees of belief* of the statement, and quantify the extent to which such a statement is supported by existing knowledge and available evidence. The Bayesian methodology to which this book is devoted is based on subjective probabilities. Calvetti and Somersalo [30] explain that "randomness" in the context of physical probabilities is equivalent to a "lack of information" in the context of subjective probabilities.

Using this interpretation of probability, even deterministic quantities can be represented using probability distributions which reflect the subjective degree of the analyst's belief regarding such quantities. As a result, probability distributions can be assigned to parameters that need to be estimated, and therefore, this interpretation facilitates uncertainty propagation after parameter estimation; this is helpful for uncertainty integration across multiple models and scales.

For example, consider the case where a variable is assumed to be normally distributed and the estimation of the mean and the standard deviation based on available data is desired. If sufficient data were available, then it is possible to uniquely estimate these distribution parameters (mean and standard deviation). However, in some cases, data may be sparse and therefore, it may be necessary to quantify the uncertainty in these distribution parameters. Note that this uncertainty reflects our epistemic uncertainty; the quantities may be estimated deterministically with enough data. The former philosophy based on physical probabilities inherently assumes that these distribution parameters are deterministic and expresses the uncertainty through confidence intervals. It is not possible to propagate this description of uncertainty through a mathematical model. Instead, the Bayesian methodology allows obtaining probability distributions for the model parameters, which can be easily used in uncertainty propagation. Therefore, the Bayesian methodology provides a framework in which epistemic uncertainty can be also addressed using probability theory, in contrast with the frequentist approach.

1.5 Probability fundamentals

The vast majority of the probability contents of this book are based on the concepts of conditional probability, total probability, and Bayes' theorem. These concepts are briefly explained in this section.

1.5.1 Bayes' Theorem

Though named after the 18th century mathematician and theologian Thomas Bayes [21], it was the French mathematician Pierre-Simon Laplace who pioneered and popularised what is now called Bayesian probability [176, 175]. For a brief history of Bayesian methods, refer to [66]. The law of conditional probability is fundamental to the development of Bayesian philosophy:

$$P(AB) = P(A|B)P(B) = P(B|A)P(A) \tag{1.5}$$

Consider a list of mutually exclusive and exhaustive events A_i ($i = 1$ to N) that form the sample space all together. Let B denote any other event from the sample space such that $P(B) > 0$. Based on Equation (1.5), it follows that:

$$P(A_i|B) = \frac{P(B|A_i)P(A_i)}{\sum_{j=1}^{N} P(B|A_j)P(A_j)} \tag{1.6}$$

What does Equation (1.6) mean? Suppose that the probabilities of all events A_i ($i = 1$ to N) are known before conducting any experiments. These probabilities are referred to as *prior probabilities* in the Bayesian context. Then the experiment is conducted and event B, which is *conditionally dependent* on A_i has been observed; therefore, it can be probabilistically expressed as $P(B|A_i)$. In the light of this information, the reciprocal event $P(A_i|B)$ ($i = 1$ to N), known as the *posterior probability* in the Bayesian approach, can be calculated using Bayes' theorem given by Equation (1.6). The quantity $P(B|A_i)$ is the probability of observing the event B conditioned on A_i. It can be argued that event B has "actually been observed" and there is no uncertainty regarding its occurrence, which renders the probability $P(B|A_i)$ meaningless. Hence, researchers "invented" new terminology in order to denote this quantity. In earlier days, this quantity was referred to as "inverse probability", and since the advent of Fisher [103, 5] and Edwards [62], this terminology has become obsolete and has been replaced by the term "likelihood". In fact, it is also common to write $P(B|A_i)$ as $L(A_i)$.

In general terms, the conditional probability $P(B|A_i)$ is interpreted as the *degree of belief* of proposition B given that proposition A_i holds. Observe that for the definition of conditional probability, it is not required for the conditional proposition to be *true* or to happen; for example, it is not essential that the event A_i (e.g. an earthquake) has occurred in order to define the probability $P(B|A_i)$ (e.g. building collapse given an earthquake); instead, this probability is simply conditionally asserted: "if A_i occurs, then there is a corresponding probability for B, and this probability is denoted as $P(B|A_i)$." In addition, this notion of conditional probability does not necessarily imply a *cause-consequence* relationship between the two propositions. For example, the occurrence of A_i does not lead to the occurrence of B. It is obviously meaningless from a causal dependence (physical) point of view. Instead, $P(B|A_i)$ refers to the degree of plausibility of proposition B given the information in proposition A_i, whose truth we need not know. In the extreme situation, that is, if A_i implies B, then proposition A_i gives complete information about B, and thus $P(B|A_i) = 1$; otherwise, when A_i implies not B, then $P(B|A_i) = 0$. This *information dependence* instead of *causal dependence* between conditional propositions brings us to the Cox-Jaynes interpretation of subjective probability as a multi-valued logic for plausible inference [52, 99], which is adopted in this book.

1.5.2 Total probability theorem

Let us assume that one of the conditional propositions, for example, A, can be partitioned into N mutually exclusive propositions, $A_1, A_2, \cdots, A_N$. Then, the total probability theorem allows us to obtain the *total* probability of proposition B, as follows:

$$P(B) = \sum_{i=1}^{N} P(B|A_i)P(A_i) \tag{1.7}$$

Example 1 *Suppose that we can classify proposition A "it rains" into three different intensity levels; for example, A_1: "low rainfall intensity", A_2 : "medium rainfall intensity", A_3 : "extreme rainfall intensity", whose plausibilities $P(A_i)$, $i = 1, 2, 3$, are known. Then, the plausibility of a new proposition B: "traffic jam" can be obtained as*

$$P(B) = P(B|A_1)P(A_1) + P(B|A_2)P(A_2) + P(B|A_3)P(A_3) \tag{1.8}$$

where $P(B|A_i)$ is the plausibility of a traffic jam given (conditional to) a particular rainfall intensity A_i. If the conditional probabilities $P(B|A_i)$ are known, then the total probability $P(B)$ can be obtained using Equation (1.8).

In the cases where the conditional proposition is represented by a continuous real-valued variable $x \in \mathcal{X}$, (e.g. the rainfall intensity), the total probability theorem is rewritten as following

$$P(B) = \int_{\mathcal{X}} P(B|x)p(x)dx \tag{1.9}$$

where $p(x)$ is the probability density function previously presented of continuous variable x. In what follows, $P(\cdot)$ is used to denote probability, whereas a PDF is expressed as $p(\cdot)$.

When both propositions can be expressed by continuous-valued variables $x \in \mathcal{X}$ and $y \in \mathcal{Y}$, (e.g. traffic intensity and rainfall intensity, respectively), then

$$p(y) = \int_{\mathcal{X}} p(y|x)p(x)dx \tag{1.10}$$

where $p(y|x)$ is the conditional probability density function between the two propositions.

In the context of mutually exclusive propositions, Bayes' theorem and total probability theorem can be combined to obtain the conditional probability of a particular proposition A_i, as follows

$$P(A_i|B) = \frac{P(B|A_i)P(A_i)}{\underbrace{\sum_{j=1}^{N} P(B|A_j)P(A_j)}_{P(B)}} \tag{1.11}$$

The same applies for conditional propositions described by discrete and continuous-valued variables, as

$$p(x|B) = \frac{P(B|x)p(x)}{\int_{\mathcal{X}} P(B|x)p(x)dx} \tag{1.12}$$

or reciprocally

$$P(B_i|x) = \frac{p(x|B_i)P(B_i)}{\sum_{j=1}^{N} p(x|B_j)P(B_j)} \tag{1.13}$$

For continuous-valued propositions $x \in \mathcal{X}$, $y \in \mathcal{Y}$, Bayes' theorem can be written as

$$p(x|y) = \frac{p(y|x)p(x)}{\int_{\mathcal{X}} p(y|x)p(x)dx} \tag{1.14}$$

or reciprocally

$$p(y|x) = \frac{p(x|y)p(y)}{\int_{\mathcal{Y}} p(x|y)p(y)dy} \tag{1.15}$$

The integrals in Equations (1.12), (1.14) and (1.15) are nontrivial except for some particular cases [22]. Section 1.7.1 below provides sampling methods to numerically approximate these equations.

1.6 The Bayesian approach to inverse problems

The fundamentals of Bayesian philosophy are well-established in several textbooks [117, 118, 30, 172], and the Bayesian approach is being increasingly applied to engineering problems in recent times, especially to solve *inverse problems*. This leads to two important questions: What is an inverse problem? Why is it important? Prior to understanding the concepts related to an "inverse problem" and their practical implications, it is first necessary to understand the idea of a "forward problem".

1.6.1 The forward problem

In general, a forward problem can be defined as predicting how an engineering system will behave based on some knowledge about the system. This includes knowledge regarding the system's characteristics/properties, inputs experienced or to be experienced by the system, a model that represents an input-output relationship, where the input includes the system's loading conditions, operating conditions, usage conditions, etc., and the output corresponding to the system behaviour, model parameters (if any), etc. Sometimes, the input-output relationship is explicitly available in the form of a mathematical model, as follows:

$$\mathbf{x} = \mathbf{g}(\mathbf{u}, \boldsymbol{\theta}) \tag{1.16}$$

In the expression (1.16), $\mathbf{g}$ represents the mathematical expression of the model, the vector $\mathbf{u}$ represents all inputs (including loading and usage conditions), the vector $\boldsymbol{\theta}$ represents the model parameters (such parameters are specific to the model being developed; sometimes they have an explicit physical meaning whereas some other times, they are simply "fitting" or "calibration" parameters without physical meaning), and $\mathbf{x}$ represents the system output, also referred to as the *system response.*

The goal of the forward problem is to predict $\mathbf{x}$ based on existing knowledge about $\mathbf{g}$, $\mathbf{u}$, and $\boldsymbol{\theta}$. Note that Equation (1.16) is deliberately chosen to be very simplistic for the purpose of explanation; however it may be much more complicated in practise. For example, contrary to an explicit model such as Equation (1.16), several practical engineering systems are studied using implicit models, that is, $\mathbf{x}$ is not an explicit function of $\mathbf{u}$, and $\boldsymbol{\theta}$. Sometimes they are described using a complicated partial differential equation. This renders the evaluation of a forward model challenging. A simple but practically challenging example of such a forward problem is the prediction of the deflection of an aircraft wing in flight. In this problem, $\mathbf{g}$ is essentially composed of a coupled, multi-disciplinary partial differential equation system where some equations are based on fluid mechanics, while others refer to structural and solid mechanics.

Apart from these complexities, the forward problem is affected by several sources of uncertainty, as explained earlier in Section 1.2. Rarely do we know all about $\mathbf{g}$, $\mathbf{u}$, or $\boldsymbol{\theta}$ precisely. In fact, the famous statistician George Box says: "All models are wrong; some are useful." Therefore, $\mathbf{g}$ may be prone to *prediction errors.* Moreover, there may even be not just one single model $\mathbf{g}$ but multiple competing $\mathbf{g}$'s, and it may be necessary to employ a hybrid combination of these in order to make a rigorous and robust prediction. The inputs $\mathbf{u}$ and parameters $\boldsymbol{\theta}$ could be imprecisely known, and such imprecision could be represented using probability distributions. This is how forward problem becomes stochastic in nature, and probabilistic methods need to be employed in order to deal with such sources of uncertainty. Therefore, the goal of a forward problem is to determine the probability distribution of $\mathbf{x}$ given the best knowledge about $\mathbf{g}$, $\mathbf{u}$, and $\boldsymbol{\theta}$. Typically, forward problems are solved with quantified uncertainty using Monte Carlo sampling [85] and related-advanced methods [14]. Model-based reliability analysis methods like first-order reliability analysis (FORM) and second-order reliability analysis (SORM) can be also useful to solve the forward problem [86, 57, 56, 185].

Notwithstanding, a straightforward yet rigorous way to obtain a probabilistic forward model from any deterministic I/O model is by *stochastic embedding* [22]. This can be achieved by introducing an uncertain model error term $\mathbf{e}$ as a discrepancy between the actual system output $\mathbf{y}$ and the modelled output $\mathbf{x} = \mathbf{g}(\mathbf{u}, \boldsymbol{\theta})$, as follows:

$$\underbrace{\mathbf{y}}_{\text{system output}} = \underbrace{\mathbf{g}(\mathbf{u}, \boldsymbol{\theta})}_{\text{model output}} + \underbrace{\mathbf{e}}_{\text{error}} \tag{1.17}$$

Since $\mathbf{e}$ is uncertain, it can be modelled by a probability distribution (e.g. a Gaussian distribution), and this in turns determines the probability model for the system output $\mathbf{y}$. For example, if the error term is assumed to be modelled as a zero-mean Gaussian distribution, that is, $\mathbf{e} \sim \mathcal{N}(0, \boldsymbol{\Sigma}_{\mathbf{e}})$, then the system output $\mathbf{y}$ will be distributed as a Gaussian distribution, as follows:

$$\mathbf{e} = (\mathbf{y} - \mathbf{g}(\mathbf{u}, \boldsymbol{\theta})) \sim \mathcal{N}(0, \boldsymbol{\Sigma}_{\mathbf{e}}) \implies \mathbf{y} \sim \mathcal{N}(\mathbf{g}(\mathbf{u}, \boldsymbol{\theta}), \boldsymbol{\Sigma}_{\mathbf{e}}) \tag{1.18}$$

where $\boldsymbol{\Sigma}_{\mathbf{e}}$ is the covariance matrix of the error term. See Figure 1.1 for a graphical description of the stochastic embedding process.

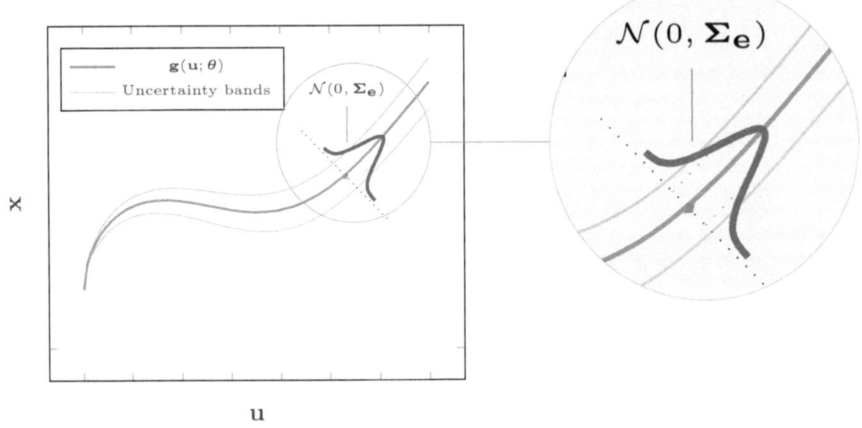

Figure 1.1: Illustration of the stochastic embedding process represented in Eq. (1.17).

A rational way to establish a probability model for the error term is given by the Principle of Maximum Information Entropy (PMIE) [102], which states that a probability model should be conservatively selected to produce the most prediction uncertainty (largest Shannon entropy), subjected to parameterised constraints. Thus, if the error variable $\mathbf{e}$ is constrained to a particular mean value $\boldsymbol{\mu}_{\mathbf{e}}$ (e.g. $\boldsymbol{\mu}_{\mathbf{e}} = 0$) and a variance or covariance matrix $\boldsymbol{\Sigma}_{\mathbf{e}}$, then it can be shown by PMIE that the maximum-entropy probability model for $\mathbf{e}$ is the Gaussian PDF with mean $\boldsymbol{\mu}_{\mathbf{e}}$ and covariance matrix $\boldsymbol{\Sigma}_{\mathbf{e}}$, i.e. $\mathbf{e} \sim \mathcal{N}(\boldsymbol{\mu}_{\mathbf{e}}, \boldsymbol{\Sigma}_{\mathbf{e}})$. In this context, it follows from expression (1.18) that a probabilistic forward model can be obtained from the deterministic model $\mathbf{g}(\mathbf{u}, \boldsymbol{\theta})$ as

$$p(\mathbf{y}|\mathbf{u}, \boldsymbol{\theta}) = \left((2\pi)^{N_o} |\boldsymbol{\Sigma}_{\mathbf{e}}|\right)^{-\frac{1}{2}} \exp\left(-\frac{1}{2} (\mathbf{y} - \mathbf{x})^T \boldsymbol{\Sigma}_{\mathbf{e}}^{-1} (\mathbf{y} - \mathbf{x})\right) \tag{1.19}$$

where N_o is the size of the $\mathbf{y}$ vector and $\mathbf{x} = \mathbf{g}(\mathbf{u}, \boldsymbol{\theta})$ is the output of the deterministic forward model.

It should be noted that Equation (1.17) implicitly assumes that both the model error and the measurement error are subsumed into $\mathbf{e}$. Such an assumption can be adopted when the measurement error is negligible as compared to the model error, or when an independent study about the measurement error is not available. Otherwise, $\mathbf{e}$ in Equation (1.17) can be expressed as the sum of the two independent errors $\mathbf{e} = \mathbf{e}_m + \mathbf{e}_d$, with $\mathbf{e}_m$ being the model error, and $\mathbf{e}_d$ the measurement error. Under the maximum-entropy assumption of zero-mean Gaussian errors, the consideration of the two independent errors, namely, measurement and modelling errors, would lead to Equation (1.19) with $\mathbf{\Sigma_e} = \mathbf{\Sigma_{e_m}} + \mathbf{\Sigma_{e_d}}$ as covariance matrix.

1.6.2 The inverse problem

In contrast to the forward problem, an inverse problem can be defined as the process of inferring unknown parts of the system (which may include the inputs, the parameters, or even the model-form itself) based on a limited and possibly noisy set of observations/measurements of the system output. The processes of "model parameter estimation", "statistical model calibration", "model selection", "system identification" (commonly encountered in system/structural health monitoring), etc. are all examples of inverse problems. In the context of Equation (1.16), an inverse problem is the process of inferring $\mathbf{g}$ and/or $\mathbf{u}$ and/or $\boldsymbol{\theta}$ based on observed measurements of $\mathbf{y}$.

A natural question is: Why is solving the inverse problem important? It turns out that, in order to accurately solve the forward problem and predict the system response, it is first essential to solve the inverse problem. For instance, in the context of Equation (1.16), how would an analyst know the probability distributions of $\mathbf{u}$ and $\boldsymbol{\theta}$, to be used to simulate the forward problem? Sometimes, the inputs $\mathbf{u}$ can be known based on system-knowledge; on the contrary, the parameters $\boldsymbol{\theta}$ are almost always estimated through an inverse problem. Sometimes, even inputs are estimated using inverse problems; for example, based on the damage experienced by a building, some characteristics of the earthquake can be inferred. Inverse problems are not only useful for estimating $\mathbf{u}$ and $\boldsymbol{\theta}$, but also for estimating the goodness of the underlying model form "$\mathbf{g}$"; when there are competing models, inverse problems provide a systematic approach to probabilistic *model selection*. Therefore, in practical engineering applications, it is perhaps necessary to iteratively solve the forward problem as well as the inverse problem in order to achieve desirable results with acceptable confidence.

Another related issue is: Even when there is no stochasticity/uncertainty involved, it is preferred to solve an inverse problem using a probabilistic methodology. This is because, while the forward problem may commonly have a unique solution, the inverse problem may have many possible solutions [181], that is, there are several models, model parameters, and even different *model classes* that may be consistent with the same set of observed measurements/data. Figure 1.2 depicts this concept. In this context, providing deterministic single-valued solutions for inverse problems has a limited meaning if one considers that the model itself is just an idealisation of reality, and furthermore, the existence of measurements errors. Instead, probability-based solutions should be provided, which carry information about the *degree of plausibility* of the models and hypotheses which are more consistent with the observations.

Several methods based on classical statistics [29] and Bayesian statistics [27, 22] have been proposed to deal with probabilistic inverse problems. It has been widely recognised that the Bayesian approach to inverse problems provides the most adequate and rigorous framework for "well-developed" inverse problems [102]. The Bayesian approach to inverse problems aims at estimating the *posterior probability* of a particular hypothesis or model (e.g. unknown model form and/or parameters), conditional on all the evidence at hand (typically, experimental data). This is accomplished by Bayes' theorem, which converts the available *prior information* about the hypotheses or missing parts of the model, into *posterior information*, based on the observed system response. This model inference process is explained in Section 1.7 for model parameter estimation, and in Section 1.8 for model class selection.

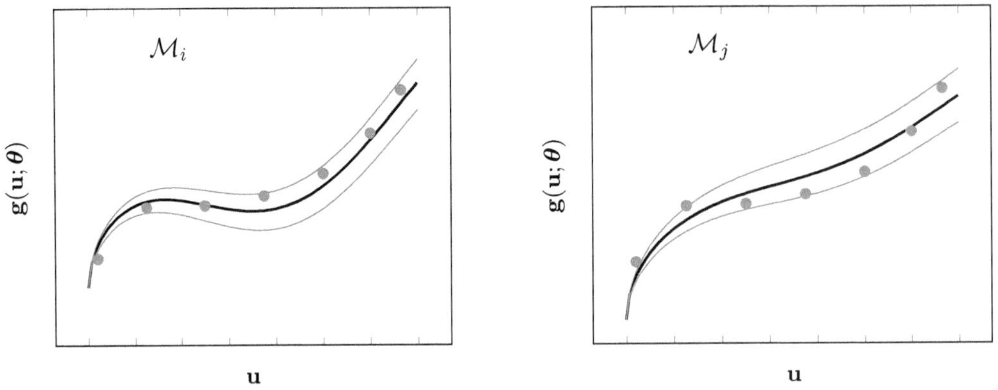

(a) Complex model (b) Simpler model

Figure 1.2: Illustrative example of different model classes consistent with the data.

1.7 Bayesian inference of model parameters

A typical goal in physical sciences and engineering is to estimate the values of the uncertain parameters $\boldsymbol{\theta}$, such that a model $\mathbf{g}(\mathbf{u}, \boldsymbol{\theta})$ parameterised by $\boldsymbol{\theta}$ belonging to a particular model class $\mathcal{M}$, is more likely to reproduce the observed system response data $\mathcal{D} = \{\hat{\mathbf{y}}_1, \hat{\mathbf{y}}_2, \ldots, \hat{\mathbf{y}}_n\}$. An engineering example for this problem would be the estimation of the plausible values for the effective mechanical properties of a degraded structure based on ambient vibration data. Assuming continuous-valued parameters $\boldsymbol{\theta}$, this plausibility is represented by the conditional PDF $p(\boldsymbol{\theta}|\mathcal{D}, \mathcal{M})$, which reads: *if model class $\mathcal{M}$ is adopted and information $\mathcal{D}$ from the system response is available* (conditional proposition), *then the model specified by $\boldsymbol{\theta}$ will predict the system output with probability $p(\boldsymbol{\theta}|\mathcal{D}, \mathcal{M})$*.

From the above description, we can highlight three important pieces of information in the Bayesian inverse problem, which are described here:

$\mathcal{D}$: Data set containing the system output or input-output data, depending on the experimental setup.

$\mathcal{M}$: Candidate model class among a set of possible model classes hypothesised to represent the system (e.g. analytical model, semi-empirical model, finite-element model, etc.)

$\boldsymbol{\theta}$: Set of uncertain model parameters belonging to a specific model class $\mathcal{M}$, that calibrate the idealised relationships between the input and output of the system.

Our interest here is to obtain the *posterior* PDF $p(\boldsymbol{\theta}|\mathcal{D}, \mathcal{M})$. This is given by Bayes' theorem, as follows

$$p(\boldsymbol{\theta}|\mathcal{D}, \mathcal{M}) = \frac{p(\mathcal{D}|\boldsymbol{\theta}, \mathcal{M})p(\boldsymbol{\theta}|\mathcal{M})}{p(\mathcal{D}|\mathcal{M})} \tag{1.20}$$

Note that Bayes' theorem takes the initial quantification of the plausibility of each model specified by $\boldsymbol{\theta}$ in $\mathcal{M}$, which is expressed by the *prior* PDF $p(\boldsymbol{\theta}|\mathcal{M})$, and *updates* this plausibility to obtain the *posterior* PDF $p(\boldsymbol{\theta}|\mathcal{D},\mathcal{M})$ by using information from the system response expressed through the PDF $p(\mathcal{D}|\boldsymbol{\theta},\mathcal{M})$, known as the *likelihood function*. The likelihood function provides a measure about how likely the data $\mathcal{D}$ are reproduced by model class $\mathcal{M}$ specified by θ[1]. It is obtained by evaluating the data $\mathcal{D}$ as the outcome of the stochastic forward model given by Equation (1.19). Figure 1.3 illustrates the concepts for prior and posterior information of model parameters by means of their associated probability density functions.

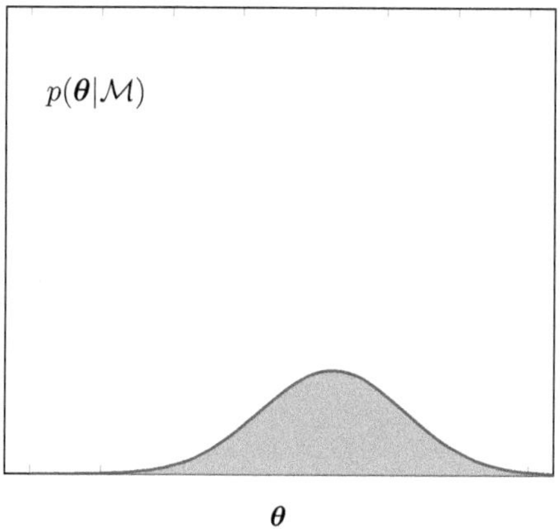

(a) Prior PDF

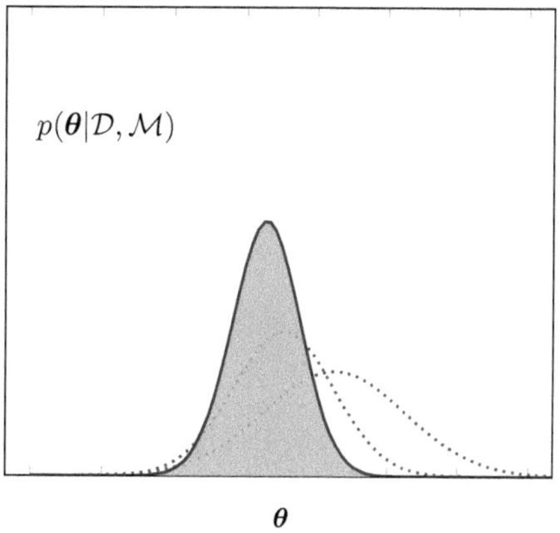

(b) Posterior PDF

Figure 1.3: Illustration of the prior and posterior information of model parameters. Observe that after assimilating the data $\mathcal{D}$, the *posterior* probabilistic information about $\boldsymbol{\theta}$ is concentrated over a narrower space.

Example 2 *Suppose that we are asked to estimate the gravitational acceleration g of a comet based on a sequence of measurements $\mathcal{D} = \{\hat{y}_1, \hat{y}_2, \cdots, \hat{y}_T\}$ about the angular displacement of a pendulum mounted on a spacecraft which has landed on the comet. These measurements were obtained using an on-board sensor whose precision is known to be given by a zero-mean Gaussian noise $e_t \sim \mathcal{N}(0,\sigma)$, with σ being independent of time.*

The (deterministic) physical model $\mathcal{M}$ describing the angular displacement of the pendulum as a function of time is assumed to be given by

$$x_t = \sin(\theta t) \tag{1.21}$$

[1]The concept of likelihood is used both in the context of physical probabilities (frequentist) and subjective probabilities, especially in the context of parameter estimation. From a *frequentist* point of view (the underlying parameters are deterministic), the likelihood function can be maximised in order to obtain the maximum likelihood estimate of the parameters. According to Fisher [67], the popular least squares approach is an indirect approach to parameter estimation and one can "solve the real problem directly" by maximising the "probability of observing the given data" conditioned on the parameter θ [67, 4]. On the other hand, the likelihood function can also be interpreted using *subjective* probabilities. Singpurwalla [167, 168] explains that the likelihood function can be viewed as a collection of weights or masses and therefore, it is meaningful only up to a proportionality constant [62]. In other words, if $p(\mathcal{D}|\theta^{(1)}) = 10$, and $p(\mathcal{D}|\theta^{(2)}) = 100$, then $\mathcal{D}$ is ten times more likely to be reproduced by $\theta^{(2)}$ than by $\theta^{(1)}$.

where $\theta = \sqrt{\frac{g}{L}}$ is the uncertain model parameter, with g being the actually unknown variable, and L is the length of the pendulum (known). Based on some theoretical considerations, the gravitational acceleration g is known to be bounded within the interval $[g_1, g_2]$; thus, we can use this information to define a uniform prior PDF for this parameter, as

$$p(\theta|\mathcal{M}) = \frac{1}{\theta_2 - \theta_1} \tag{1.22}$$

where $\theta_j = \sqrt{\frac{g_j}{L}}$, $j = 1, 2$. Given the existence of measurement errors, the observed system response would actually be represented by the equation

$$y_t = \sin(\theta t) + e_t \tag{1.23}$$

where $e_t \sim \mathcal{N}(0, \sigma)$. Therefore, the stochastic forward model of the system response will be given by

$$p(y_t|\theta, \mathcal{M}) = \frac{1}{\sqrt{2\pi\sigma^2}} \exp\left(-\frac{1}{2}\frac{(y_t - \sin\theta t)^2}{\sigma^2}\right) \tag{1.24}$$

As explained above, the likelihood function $p(\mathcal{D}|\theta, \mathcal{M})$ can be obtained by evaluating the data $\mathcal{D}$ as the outcome of the stochastic forward model, therefore

$$p(\mathcal{D}|\theta, \mathcal{M}) = p(y_1 = \hat{y}_1, y_2 = \hat{y}_2, \ldots, y_T = \hat{y}_T|\theta, \mathcal{M}) \tag{1.25a}$$

$$= \prod_{t=1}^{T} \frac{1}{\sqrt{2\pi\sigma^2}} \exp\left(-\frac{1}{2}\frac{(\hat{y}_t - \sin\theta t)^2}{\sigma^2}\right) \tag{1.25b}$$

Then, based on Bayes' theorem, the updated information about the gravitational acceleration of the comet can be obtained as

$$p(\theta|\mathcal{D}, \mathcal{M}) \propto \underbrace{p(\mathcal{D}|\theta, \mathcal{M})}_{\text{Eq. (1.25b)}}\ \underbrace{p(\theta|\mathcal{M})}_{\text{Eq. (1.22)}} \tag{1.26}$$

which can be readily simulated using the stochastic simulated methods explained below.

Note from Equation (1.25b) that we implicitly assume there is no dependence between the observations, that is, $p(\hat{y}_1, \ldots, \hat{y}_T|\boldsymbol{\theta}, \mathcal{M}) = \prod_{t=1}^{T} p(\hat{y}_t|\boldsymbol{\theta}, \mathcal{M})$. It is emphasised that this stochastic independence refers to information independence and should not be confused with causal independence. It is equivalent to asserting that if the modelling or measurement errors at certain discrete times are given, this does not influence the error values at other times.

Apart from the likelihood function, another important factor in Equation (1.20) is $p(\mathcal{D}|\mathcal{M})$, which is known as the *evidence* (also called the *marginal likelihood*) for the model class $\mathcal{M}$. This factor expresses how likely the observed data will be reproduced if model class $\mathcal{M}$ is adopted. The evidence can be obtained by total probability theorem as

$$p(\mathcal{D}|\mathcal{M}) = \int_{\boldsymbol{\Theta}} p(\mathcal{D}|\boldsymbol{\theta}, \mathcal{M}_j) p(\boldsymbol{\theta}|\mathcal{M}_j) \mathrm{d}\boldsymbol{\theta} \tag{1.27}$$

In most practical situations, the evaluation of the multi-dimensional integral in Equation (1.27) is analytically intractable. Stochastic simulation methods such as the family of Markov Chain Monte Carlo (MCMC) methods [134, 76] can be used to draw samples from the posterior PDF in Equation (1.20) while circumventing the evaluation of $p(\mathcal{D}|\mathcal{M})$. By means of this, the posterior PDF can be straightforwardly approximated as a probability mass function, mathematically described as follows

$$p(\boldsymbol{\theta}|\mathcal{D},\mathcal{M}) \approx \frac{1}{K}\sum_{k=1}^{K}\delta(\boldsymbol{\theta}-\widetilde{\boldsymbol{\theta}}^{(k)}) \tag{1.28}$$

where $\delta(\boldsymbol{\theta}-\widetilde{\boldsymbol{\theta}}^{(k)})$ is the Dirac function, which makes 1 when $\boldsymbol{\theta}=\widetilde{\boldsymbol{\theta}}^{(k)}$ and 0 otherwise, with $\widetilde{\boldsymbol{\theta}}^{k}$, $k=1\dots,K$ being samples drawn from $p(\boldsymbol{\theta}|\mathcal{D},\mathcal{M})$ using an appropriate stochastic simulation method. See Figure 1.4 for a graphical illustration of this method. Further insight about MCMC simulation methods is given in Section 1.7.1 below.

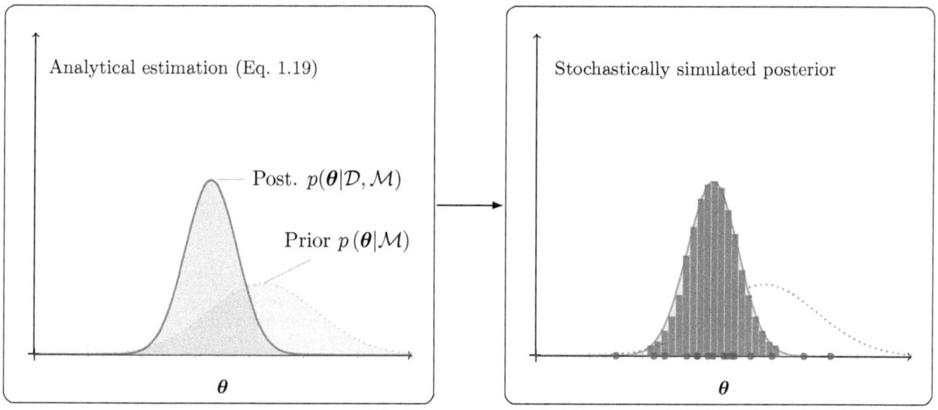

Figure 1.4: Illustration of stochastic simulation for Bayesian inference.

Finally, it should be noted that, although the posterior PDF of parameters provides full information about the plausibility of model parameters among the full range of possibilities, most of the times engineering decisions are made based on single-valued engineering parameters. This fact does not restrict the practical applicability of the Bayesian inverse problem explained here. On the contrary, not one but several single-valued "representative values" can be extracted from the full posterior PDF of parameters (e.g. mean, median, percentiles, etc.), which would enrich the decision-making process with further information. Among them, a value of particular interest is the *maximum a posteriori* (MAP), which can be obtained as the value $\boldsymbol{\theta}_{MAP}\in\boldsymbol{\Theta}$ which maximises the posterior PDF, that is, $\boldsymbol{\theta}_{MAP}=\arg\max_{\boldsymbol{\theta}}p(\boldsymbol{\theta}|\mathcal{D},\mathcal{M})$. Note from Equation (1.20) that the MAP is equivalent to the widely known *maximum likelihood estimation* method (MLE), namely, $\boldsymbol{\theta}$ that maximises $p(\mathcal{D}|\boldsymbol{\theta},\mathcal{M})$, when the prior PDF is an uniform distribution.

1.7.1 Markov Chain Monte Carlo methods

While the power of the Bayesian inverse problem for model updating is well recognised, there are several computational issues during its implementation that need dedicated solutions. As explained above, the main difficulty when applying Bayes' theorem is that the factor $p(\mathcal{D}|\mathcal{M})$ in Equation (1.20) cannot be evaluated analytically nor is it readily calculated by numerical integration methods. To tackle this problem, Markov Chain Monte Carlo (MCMC) methods [134, 76] have been widely used for their ability to estimate the posterior PDF while avoiding the evaluation of $p(\mathcal{D}|\mathcal{M})$. In general, the goal of these stochastic simulation methods is to generate samples that are distributed according to an arbitrary probability distribution, commonly known as the *target* PDF, which is known up to a scaling constant [132, 121]. In the context of the Bayesian inverse problem, such target corresponds to the posterior PDF given by Equation (1.20), and the scaling constant is the evidence $p(\mathcal{D}|\mathcal{M})$.

There are two required properties for the MCMC algorithms to obtain correct statistical estimations about the target: (1) *ergodicity*, which is concerned with whether the generated samples can populate the regions of significant probability of the target PDF; and (2) *stationarity*, which ensures that the forward samples of the Markov Chain are more equally distributed than the previous samples, provided that the later are distributed as the target PDF. Under the assumption of ergodicity, the samples generated from MCMC will converge to the target distribution (provided that a large amount of samples are used) even if the initial set of samples are simulated from a PDF different from the target. The theoretical demonstration of ergodicity and stationarity for MCMC is out of the scope of this book, but the interested reader is referred to [178] for a comprehensive theoretical treatment of MCMC methods.

1.7.1.1 Metropolis-Hasting algorithm

Several MCMC algorithms have been proposed in the literature such as the Metropolis-Hastings (M-H), Gibbs Sampler or Slice Sampling, among others [121]. In this chapter, we pay special attention to the Metropolis-Hastings algorithm [131, 90] for its versatility and implementation simplicity. Notwithstanding, the selection of the "best" MCMC algorithm is case specific, and several other MCMC algorithms can provide an excellent performance in the context of some particular Bayesian inverse problems.

The M-H algorithm generates samples from a specially constructed Markov Chain whose stationary distribution is the posterior PDF $p(\boldsymbol{\theta}|\mathcal{D},\mathcal{M})$. In M-H, a candidate model parameter $\boldsymbol{\theta}'$ is sampled from a *proposal distribution* $q(\boldsymbol{\theta}'|\boldsymbol{\theta}^{(k-1)})$, given the state of the Markov Chain at step $k-1$. At the next state of the chain k, the candidate parameter $\boldsymbol{\theta}'$ is accepted (i.e. $\boldsymbol{\theta}^{(k)} = \boldsymbol{\theta}'$) with probability $\min\{1, r\}$, and rejected ($\boldsymbol{\theta}^{(k)} = \boldsymbol{\theta}^{(k-1)}$) with probability $1 - \min\{1, r\}$, where r is calculated as:

$$r = \frac{p(\mathcal{D}|\boldsymbol{\theta}',\mathcal{M})p(\boldsymbol{\theta}'|\mathcal{M})q(\boldsymbol{\theta}^{(k-1)}|\boldsymbol{\theta}')}{p(\mathcal{D}|\boldsymbol{\theta}^{(k-1)},\mathcal{M})p(\boldsymbol{\theta}^{(k-1)}|\mathcal{M})q(\boldsymbol{\theta}'|\boldsymbol{\theta}^{(k-1)})} \tag{1.29}$$

The process is repeated until N_s samples have been generated. An algorithmic description of this method is provided in Algorithm 1 below.

Algorithm 1: M-H algorithm

1. Initialize $\boldsymbol{\theta}^{(0)}$ by sampling from the prior: $\boldsymbol{\theta}^{(0)} \sim p(\boldsymbol{\theta}|\mathcal{M})$
for $k = 1$ to N_s **do**
  2. Sample from the proposal: $\boldsymbol{\theta}' \sim q(\boldsymbol{\theta}'|\boldsymbol{\theta}^{(k-1)})$
  3. Compute r from Equation (1.29)
  4. Generate a uniform random number: $\alpha \sim \mathcal{U}[0, 1]$
  if $r \geqslant \alpha$ **then**
    5. Set $\boldsymbol{\theta}^{(k)} = \boldsymbol{\theta}'$
  else
    6. Set $\boldsymbol{\theta}^{(k)} = \boldsymbol{\theta}^{(k-1)}$
  end if
end for

An important consideration for the proper implementation of the M-H algorithm is the specification of the variance σ_q^2 of the proposal distribution $q(\boldsymbol{\theta}'|\boldsymbol{\theta}^{(k-1)})$, typically Gaussian, which has a significant impact on the speed of convergence of the algorithm [90]. Small values tend to produce candidate samples that are accepted with high probability, but may result in highly dependent chains that explore the space very slowly. In contrast, large values of the variance tend to produce large steps and therefore a fast space exploration, but result in small acceptance rates and thus, a larger time of convergence. Therefore, it is often worthwhile to select appropriate proposal variances by controlling the *acceptance rate* (e.g. number of accepted samples over total amount of samples) in a certain range, depending on the dimension d of the proposal PDF, via some pilot runs [73, 152]. The interval $[20\% - 40\%]$ is suggested for the acceptance rate in low dimensional spaces, say $d \leqslant 10$ [152]. Note also that if a Gaussian distribution is chosen as the proposal, then q has the symmetry property, that is, $q(\boldsymbol{\theta}'|\boldsymbol{\theta}^{(k-1)}) = q(\boldsymbol{\theta}^{(k-1)}|\boldsymbol{\theta}')$, then Equation (1.29) simplifies as

$$r = \frac{p(\mathcal{D}|\boldsymbol{\theta}', \mathcal{M})p(\boldsymbol{\theta}'|\mathcal{M})}{p(\mathcal{D}|\boldsymbol{\theta}^{(k-1)}, \mathcal{M})p(\boldsymbol{\theta}^{(k-1)}|\mathcal{M})} \tag{1.30}$$

Furthermore, in case of adopting a uniform probability model for the prior PDFs of the model parameters, then the M-H test in Equation (1.29) reduces as

$$r = \frac{p(\mathcal{D}|\boldsymbol{\theta}', \mathcal{M})}{p(\mathcal{D}|\boldsymbol{\theta}^{(k-1)}, \mathcal{M})} \tag{1.31}$$

which corresponds to the ratio between likelihoods.

1.8 Bayesian model class selection

As explained in Section 1.6.2, the Bayesian inverse problem provides a rigorous and systematic approach to the problem of model class selection. This approach is motivated by the fact that the model itself may not necessarily reproduce the observed system, but it is just an approximation [102]. Therefore, there may exist not only different values for model parameters but physically different models classes that may be consistent with the observations of such a system. As with the inverse problem of model parameter estimation, the goal here is to use the available information from the system response $\mathcal{D}$ to compare and rank the relative performance of a set of candidate model classes $\mathfrak{M} = \{\mathcal{M}_1, \ldots, \mathcal{M}_j, \ldots, \mathcal{M}_{N_M}\}$ in reproducing the data. This performance can be compared and ranked using the conditional posterior probability $P(\mathcal{M}_j|\mathcal{D}, \mathfrak{M})$, which provides information about the relative extent of support of model class $\mathcal{M}_j$ for representing the data $\mathcal{D}$ among the set of candidates $\mathfrak{M} = \{\mathcal{M}_1, \ldots, \mathcal{M}_j, \ldots, \mathcal{M}_{N_M}\}$.

The requested posterior probability of the overall model can be obtained extending Bayes' theorem at the model class level [22], as

$$P(\mathcal{M}_j|\mathcal{D}, \mathfrak{M}) = \frac{p(\mathcal{D}|\mathcal{M}_j)P(\mathcal{M}_j|\mathfrak{M})}{\sum_{i=1}^{N_M} p(\mathcal{D}|\mathcal{M}_i)P(\mathcal{M}_i|\mathfrak{M})} \tag{1.32}$$

where $P(\mathcal{M}_j|\mathfrak{M})$ is the prior probability of the jth model class in the set $\mathfrak{M}$, satisfying $\sum_{j=1}^{N_M} P(\mathcal{M}_j|\mathfrak{M}) = 1$. This prior probability expresses the initial modeller's judgement on the relative degree of belief on $\mathcal{M}_j$ within the set $\mathfrak{M}$. An important point to remark here is that either the prior and posterior probabilities, $P(\mathcal{M}_j|\mathfrak{M})$ and $P(\mathcal{M}_j|\mathcal{D}, \mathfrak{M})$, respectively, refer to the *relative probability* in relation to the set of models $\mathfrak{M}$, thus satisfying $\sum_{j=1}^{N_M} P(\mathcal{M}_j|\mathfrak{M}) = 1$ and $\sum_{j=1}^{N_M} P(\mathcal{M}_j|\mathcal{D}, \mathfrak{M}) = 1$, respectively. In this context, computing the posterior plausibility of a model class would automatically require to compute it for *all* models classes in $\mathfrak{M}$. Notwithstanding, if the interest is to compare the performance of two competing model classes $\mathcal{M}_i$ and $\mathcal{M}_j$, this can be straightforwardly done using the concept of *Bayes' factor*, as follows

$$\frac{P(\mathcal{M}_i|\mathcal{D}, \mathfrak{M})}{P(\mathcal{M}_j|\mathcal{D}, \mathfrak{M})} = \frac{p(\mathcal{D}|\mathcal{M}_i, \mathfrak{M})}{p(\mathcal{D}|\mathcal{M}_j, \mathfrak{M})} \frac{P(\mathcal{M}_i)}{P(\mathcal{M}_j)} \tag{1.33}$$

which does not require computing the posterior over all possible model classes. When the prior plausibilities of the two candidate model classes are identical, that is, $P(\mathcal{M}_i) = P(\mathcal{M}_j)$, then Bayes' factor reduces to the ratio of evidences of the model classes.

Example 3 *Figure 1.5 shows an illustration of a typical problem of model class assessment using four generic model classes, that is,* $\mathfrak{M} = \{\mathcal{M}_1, \mathcal{M}_2, \mathcal{M}_3, \mathcal{M}_4\}$. *Observe that* $\sum_{j=1}^{4} P(\mathcal{M}_j|\mathfrak{M}) = P(\mathcal{M}_j|\mathcal{D}, \mathfrak{M}) = 1$. *Initially, the prior plausibility of the four model classes is identical, that is,* $P(\mathcal{M}_j|\mathfrak{M}) = 0.25$, $j = 1, \ldots, 4$. *After the updating process, model class* $\mathcal{M}_2$ *results to be the most plausible. Should another model class, let say* $\mathcal{M}_5$, *be added to the set of candidates, then the values for both the prior and the posterior plausibilities would be different to satisfy* $\sum_{j=1}^{5} P(\mathcal{M}_j|\mathfrak{M}) = P(\mathcal{M}_j|\mathcal{D}, \mathfrak{M}) = 1$.

An important element in any Bayesian model class selection problem is the evidence $p(\mathcal{D}|\mathcal{M}_j)$, previously explained in Section 1.7. This factor expresses how likely the observed system response ($\mathcal{D}$) is reproduced if the overall model class $\mathcal{M}_j$ is adopted instead of an alternative model class. The evidence is obtained by total probability theorem given by Equation (1.27). It can be observed that the evidence is equal to the normalising constant in Equation (1.20) for model parameter estimation.

Once the evidences for each model class are computed, their values allow us to rank the model classes according to the posterior probabilities given in Equation (1.32). However, as explained before, the evaluation of the multi-dimensional integral in Equation (1.27) is analytically intractable in most of the cases except for some cases where the Laplace's method of asymptotic approximation

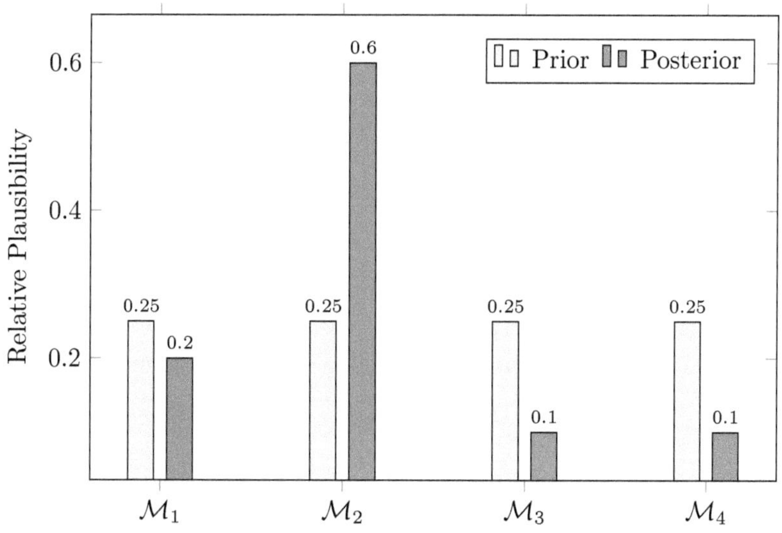

Figure 1.5: Example of relative prior and posterior probabilities for model classes.

can be used [27]. Details about stochastic simulation methods to compute the evidence are given in Section 1.8.1 below.

1.8.1 Computation of the evidence of a model class

The calculation of the evidence given in Equation (1.27) is the most difficult task when assessing the relative plausibility of a model class. In the case of globally identifiable model classes based on the data [26, 110], the posterior PDF in Equation (1.20) may be accurately approximated by a Gaussian distribution, and the evidence term can be obtained by Laplace's approximation. The reader is referred to [27, 26, 140] for details about this technique. However, in the more general case, stochastic simulation methods are required.

One straightforward way to approximate the evidence is by considering the probability integral in Equation (1.27) as a mathematical expectation of the likelihood $p(\mathcal{D}|\boldsymbol{\theta}, \mathcal{M}_j)$ with respect to the prior PDF $p(\boldsymbol{\theta}|\mathcal{M}_j)$. This approach leads to the Monte Carlo method of numerical integration, as follows

$$p(\mathcal{D}|\mathcal{M}_j) \approx \frac{1}{N_1} \sum_{k=1}^{N_1} p(\mathcal{D}|\boldsymbol{\theta}^{(k)}, \mathcal{M}_j) \tag{1.34}$$

where the $\boldsymbol{\theta}^{(k)}$ are N_1 samples drawn from the prior PDF. However, although this calculation can be easily implemented and can provide satisfactory results, it may result in a computationally inefficient method, since the region of high probability content of $p(\boldsymbol{\theta}|\mathcal{M}_j)$ is usually very different from the region where the likelihood $p(\mathcal{D}|\boldsymbol{\theta}, \mathcal{M}_j)$ has its largest values. To overcome this problem, some techniques for calculating the evidence based on samples from the posterior $p(\boldsymbol{\theta}|\mathcal{D}, \mathcal{M}_j)$ are available [135, 72, 40]. Among them, we reproduce in this chapter the method proposed by Cheung and Beck [40] based on an analytical approximation of the posterior [40], which is presented here with uniform notation in the context of the Metropolis-Hastings algorithm.

Let $K(\boldsymbol{\theta}|\boldsymbol{\theta}^*)$ be the transition PDF of any MCMC algorithm with stationary PDF $\pi(\boldsymbol{\theta}) = p(\boldsymbol{\theta}|\mathcal{D}, \mathcal{M}_j)$. The stationarity condition for the MCMC algorithm satisfies the following relation

$$\pi(\boldsymbol{\theta}) = \int K(\boldsymbol{\theta}|\boldsymbol{\theta}^*)\pi(\boldsymbol{\theta}^*)d\boldsymbol{\theta}^* \tag{1.35}$$

A general choice of $K(\boldsymbol{\theta}|\boldsymbol{\theta}^*)$ that applies to many MCMC algorithms, can be defined as

$$K(\boldsymbol{\theta}|\boldsymbol{\theta}^*) = T(\boldsymbol{\theta}|\boldsymbol{\theta}^*) + (1 - a(\boldsymbol{\theta}^*))\delta(\boldsymbol{\theta} - \boldsymbol{\theta}^*) \tag{1.36}$$

where $T(\boldsymbol{\theta}|\boldsymbol{\theta}^*)$ is a smooth function that does not contain delta functions and $a(\boldsymbol{\theta}^*)$ is the acceptance probability which must satisfy $a(\boldsymbol{\theta}^*) = \int T(\boldsymbol{\theta}|\boldsymbol{\theta}^*)d\boldsymbol{\theta} \leqslant 1$. By substituting Equation (1.36) into (1.35), an analytical approximation of the target posterior results as follows

$$\pi(\boldsymbol{\theta}) = p(\boldsymbol{\theta}|\mathcal{D}, \mathcal{M}_j) = \frac{\int T(\boldsymbol{\theta}|\boldsymbol{\theta}^*)\pi(\boldsymbol{\theta}^*)d\boldsymbol{\theta}^*}{a(\boldsymbol{\theta})} \approx \frac{1}{a(\boldsymbol{\theta})N_1}\sum_{k=1}^{N_1} T(\boldsymbol{\theta}|\boldsymbol{\theta}^{(k)}) \tag{1.37}$$

where the $\boldsymbol{\theta}^{(k)}$ are N_1 samples distributed according to the posterior. For the special case of the Metropolis-Hastings algorithm, the function $T(\boldsymbol{\theta}|\boldsymbol{\theta}^*)$ can be defined as $T(\boldsymbol{\theta}|\boldsymbol{\theta}^*) = r(\boldsymbol{\theta}|\boldsymbol{\theta}^*)q(\boldsymbol{\theta}|\boldsymbol{\theta}^*)$, where $q(\boldsymbol{\theta}|\boldsymbol{\theta}^*)$ is the proposal PDF, and $r(\boldsymbol{\theta}|\boldsymbol{\theta}^*)$ is given by

$$r(\boldsymbol{\theta}|\boldsymbol{\theta}^*) = \min\left\{1, \frac{p(\mathcal{D}|\boldsymbol{\theta}, \mathcal{M}_j)p(\boldsymbol{\theta}|\mathcal{M}_j)q(\boldsymbol{\theta}^*|\boldsymbol{\theta})}{p(\mathcal{D}|\boldsymbol{\theta}^*, \mathcal{M}_j)p(\boldsymbol{\theta}^*|\mathcal{M}_j)q(\boldsymbol{\theta}|\boldsymbol{\theta}^*)}\right\} \tag{1.38}$$

Additionally, for this algorithm, the denominator $a(\boldsymbol{\theta})$ in Equation (1.37) can be approximated by an estimator that uses samples from the proposal distribution as follows

$$a(\boldsymbol{\theta}) = \int r(\tilde{\boldsymbol{\theta}}|\boldsymbol{\theta})q(\tilde{\boldsymbol{\theta}}|\boldsymbol{\theta})d\tilde{\boldsymbol{\theta}} \approx \frac{1}{N_2}\sum_{k=1}^{N_2} r(\tilde{\boldsymbol{\theta}}^{(k)}|\boldsymbol{\theta}) \tag{1.39}$$

where the $\tilde{\boldsymbol{\theta}}^{(k)}$ are N_2 samples from $q(\tilde{\boldsymbol{\theta}}|\boldsymbol{\theta})$, when $\boldsymbol{\theta}$ is fixed. Once the analytical approximation to the posterior in Equation (1.37) is set, then Equation (1.20) can be used to evaluate the evidence, as follows

$$\log p(\mathcal{D}|\mathcal{M}_j) \approx \log p(\mathcal{D}|\boldsymbol{\theta}, \mathcal{M}_j) + \log p(\boldsymbol{\theta}|\mathcal{M}_j) - \underbrace{\log p(\boldsymbol{\theta}|\mathcal{D}, \mathcal{M}_j)}_{\text{Analytical approx.}} \tag{1.40}$$

The last expression is obtained by taking logarithms of Bayes' theorem, explained earlier in Equation (1.20). Observe that, except for the posterior PDF $p(\boldsymbol{\theta}|\mathcal{D}, \mathcal{M}_j)$ whose information is based on samples, the rest of terms can be evaluated analytically for any $\boldsymbol{\theta} \in \boldsymbol{\Theta}$. Bayes' theorem ensures that the last equation is valid for all $\boldsymbol{\theta} \in \boldsymbol{\Theta}$, so it is possible to use only one value for this parameter. However a more accurate estimate for the log-evidence can be obtained by averaging the results from Equation (1.40) using different values for $\boldsymbol{\theta}$ [40, 43]. The method is briefly summarised by the pseudo-code given in Algorithm 2, which specifically focuses on the proposed implementation for the inverse problem based on the M-H algorithm.

Algorithm 2: Evidence computation by [40]

1.- Take $\left\{\boldsymbol{\theta}^{(k)}\right\}_{k=1}^{N_1}$ samples from $p(\boldsymbol{\theta}|\mathcal{D},\mathcal{M}_j)$
2.- Choose a model parameter vector $\boldsymbol{\theta} \in \boldsymbol{\Theta}$
for $k = 1$ to N_1 **do**
  3.- Evaluate $q(\boldsymbol{\theta}|\boldsymbol{\theta}^{(k)})$
  4.- Evaluate $r(\boldsymbol{\theta}|\boldsymbol{\theta}^{(k)})$ (Eq. (1.29))
end for
5.- Take $\left\{\boldsymbol{\theta}^{(\ell)}\right\}_{\ell=1}^{N_2}$ samples from $q(\cdot|\boldsymbol{\theta})$
for $\ell = 1$ to N_2 **do**
  6.-Evaluate $r(\boldsymbol{\theta}^{(\ell)}|\boldsymbol{\theta})$ (Eq. (1.29))
end for
7.- Obtain $p(\boldsymbol{\theta}|\mathcal{D},\mathcal{M}_j) \approx \frac{\frac{1}{N_1}\sum_{k=1}^{N_1} q(\boldsymbol{\theta}|\boldsymbol{\theta}^{(k)})r(\boldsymbol{\theta}|\boldsymbol{\theta}^{(k)})}{\frac{1}{N_2}\sum_{\ell=1}^{N_2} r(\boldsymbol{\theta}^{(\ell)}|\boldsymbol{\theta})}$
8.- Evaluate $\log p(\mathcal{D}|\mathcal{M}_j)$ (Eq. (1.40))

1.8.2 Information-theory approach to model-class selection

From the perspective of forward modelling problems, more complex models tend to be preferred over simpler models because they are considered more realistic. For inverse problems, however, this may not be the case. A "complex" model may lead to over-fitting of the data where the model is unnecessarily complex in relation to the data. It means that the model does not generalise well when making predictions since it depends too much on the details of the data. Figure 1.2 illustrates this model complexity concept.

A common principle in science and engineering is that, if data can be explained by several models, then the "simpler" one should be preferred over more complex models that lead to only slightly better agreement with the data. This is often referred to as the Principle of Model Parsimony or *Ockham's razor*. The Bayesian approach to model class selection explained here shows that the evidence of a model class automatically enforces a quantitative expression of Ockham's razor [82, 102]. This was formally shown by Muto and Beck [132], which led to an expression for the evidence that allows an information-theoretic interpretation of Ockham's razor, as follows

$$\log p(\mathcal{D}|\mathcal{M}_j) = \int_{\boldsymbol{\Theta}} \left[\log p(\mathcal{D}|\boldsymbol{\theta},\mathcal{M}_j)\right] p(\boldsymbol{\theta}|\mathcal{D},\mathcal{M}_j)d\boldsymbol{\theta} - \int_{\boldsymbol{\Theta}} \left[\log \frac{p(\boldsymbol{\theta}|\mathcal{D},\mathcal{M}_j)}{p(\boldsymbol{\theta}|\mathcal{M}_j)}\right] p(\boldsymbol{\theta}|\mathcal{D},\mathcal{M}_j)d\boldsymbol{\theta} \quad (1.41)$$

The first term of the right side of Equation (1.41) is the log-likelihood function averaged by the posterior PDF, which can be interpreted as a measure of the *average goodness of fit* (AGF) of the model $\mathcal{M}_j$. The second term is the *relative entropy* between the posterior and the prior PDFs, which measures the "difference" between those PDFs. This difference will be larger for models that extract more information from the data to update their prior information, and determines the *expected information gained* (EIG) about the model class $\mathcal{M}_j$ from the data. This term is always non-negative, and, since it appears subtracting the data-fit (AGF) term, it provides a penalty against more complex model classes, which extract more information from the data to update their prior information. Therefore, the log-evidence of a model class is comprised of a data-fit term (AGF) and a term (EIG) that provides a penalty against more complex model classes. This interpretation of the evidence allows us to find a correct trade-off between fitting accuracy and model complexity for a particular model class, and gives an intuitive understanding of why the computation of the evidence automatically enforces a quantitative expression of the Principle of Model Parsimony or Ockham's razor [102].

1.9 Concluding remarks

This chapter presented an overview of uncertainty quantification and its application to engineering systems, with a special emphasis on Bayesian methods. It was explained that probability theory provides a systematic foundation to deal with uncertainty in engineering systems, and the fundamentals of probability theory were explained in detail. Further, the interpretation of probability was discussed, drawing the distinction between frequentist probabilities, commonly encountered in the context of aleatory uncertainty, and subjective probabilities, commonly encountered in Bayesian analysis. It was shown that subjective probabilities are versatile enough to deal with both aleatory uncertainty and epistemic uncertainty, by means of expressing the degree of belief regarding certain variables or hypotheses. The topics of forward problem and inverse problem were discussed in detail, and it was also explained that the Bayesian approach provides a systematic, rigorous framework to solve inverse problems including the problem of model class selection.

The following chapters in this book will delve into the topic of Bayesian inverse problems in greater detail, and will illustrate these methods using practical engineering applications.

2

Solving Inverse Problems by Approximate Bayesian Computation

Manuel Chiachío-Ruano, Juan Chiachío-Ruano* and *María L. Jalón*

This chapter aims at supplying information about the theoretical basis of Approximate Bayesian Computation (ABC), which is an efficient computational tool to solve inverse problems without the need to formulate, nor evaluate the likelihood function. By ABC, the posterior PDF can be computed in those cases where the likelihood function is intractable, impossible to formulate, or computationally demanding. Several ABC pseudo-codes are included in this chapter and an example of application is provided. Finally, the ABC-SubSim algorithm, which was initially proposed by Chiachío et al. [*SIAM Journal of Scientific Computing, Vol. 36, No. 3, pp. A1339–A1358*], is explained within the context of an example of application.

2.1 Introduction to the ABC method

As explained in Chapter 1, the Bayesian inverse problem aims at updating the *a priori* information about a set of parameters $\boldsymbol{\theta} \in \boldsymbol{\Theta} \subset \mathbb{R}^d$ for a parameterised model class $\mathcal{M}_j$, based on the available information from the system response contained in the data $\mathcal{D} \in \mathfrak{D} \subset \mathbb{R}^\ell$. This is achieved by Bayes' theorem which yields the posterior PDF $p(\boldsymbol{\theta}|\mathcal{D}, \mathcal{M}_j)$ of the model specified by $\boldsymbol{\theta}$ in the model class $\mathcal{M}_j$. However, in engineering practise, there are situations where the Bayesian inverse problem involves a likelihood function that is not completely known or it is computationally unaffordable, perhaps because it requires the evaluation of an intractable multi-dimensional integral [127]. Approximate Bayesian Computation (ABC) was conceived with the aim of evaluating the posterior PDF in those cases where the likelihood function is intractable [182]. In the Bayesian literature, such a method is also called as a *likelihood-free computation algorithm*, since it circumvents the explicit evaluation of the likelihood function by using a stochastic simulation approach. In this section, the method ABC is briefly described.

Let $\mathbf{x} \in \mathbb{R}^\ell$ denote a simulated outcome from $p(\mathbf{x}|\boldsymbol{\theta}, \mathcal{M}_j)$, the stochastic forward model for model class $\mathcal{M}_j$ parameterised by $\boldsymbol{\theta}$, formerly explained in Chapter 1, Equation (1.19). ABC aims at evaluating the posterior $p(\boldsymbol{\theta}|\mathcal{D}, \mathcal{M}_j) \propto p(\mathcal{D}|\boldsymbol{\theta}, \mathcal{M}_j)p(\boldsymbol{\theta}|\mathcal{M}_j)$ by applying Bayes' theorem to the

University of Granada, Spain.
* Corresponding author: mchiachio@ugr.es

pair $(\boldsymbol{\theta}, \mathbf{x}) \in \boldsymbol{\Theta} \times \mathfrak{D} \subset \mathbb{R}^{d+\ell}$:

$$p(\boldsymbol{\theta}, \mathbf{x}|\mathcal{D}) \propto p(\mathcal{D}|\mathbf{x}, \boldsymbol{\theta}) p(\mathbf{x}|\boldsymbol{\theta}) p(\boldsymbol{\theta}) \tag{2.1}$$

In the last equation, the conditioning on model class $\mathcal{M}_j$ has been omitted for simplicity, given that the ABC theory is valid irrespective of a chosen model class. The basic form of the ABC algorithm to generate samples from the posterior given by Equation (2.1), is a rejection algorithm that consists in jointly generating $\boldsymbol{\theta}' \sim p(\boldsymbol{\theta})$ and $\mathbf{x}' \sim p(\mathbf{x}|\boldsymbol{\theta}')$, and accepting them conditional on the equality $\mathbf{x}' = \mathcal{D}$ being fulfilled. This is due to the fact that the PDF $p(\mathcal{D}|\mathbf{x}, \boldsymbol{\theta})$ in Equation (2.1) gives higher density values for the posterior in those regions where $\mathbf{x}$ is close to $\mathcal{D}$. Of course, obtaining the sample $\mathbf{x}' = \mathcal{D}$ is unlikely in most of the cases, and it is only feasible if $\mathfrak{D}$ consists of a finite set of values rather than a continuous region in $\mathbb{R}^{\ell}$. To address the above difficulty, two main approximations have been conceived in ABC theory [128]: a) replace the equality $\mathbf{x} = \mathcal{D}$ by the approximation $\mathbf{x} \approx \mathcal{D}$ and introduce a tolerance parameter ϵ that accounts for the quality of such approximation through a suitable metric ρ; and b) introduce a low-dimensional vector of summary statistics $\eta(\cdot)$ that allows a weak comparison of the closeness between $\mathbf{x}$ and $\mathcal{D}$. Through this approach, the posterior $p(\boldsymbol{\theta}, \mathbf{x}|\mathcal{D})$ in Equation (2.1) is approximated by $p_\epsilon(\boldsymbol{\theta}, \mathbf{x}|\mathcal{D})$, which assigns higher probability density to those values of $(\boldsymbol{\theta}, \mathbf{x}) \in \boldsymbol{\Theta} \times \mathfrak{D}$ that satisfy the condition $\rho\big(\eta(\mathbf{x}), \eta(\mathcal{D})\big) \leqslant \epsilon$.

The standard version of the ABC algorithm defines an approximate likelihood function given by $P_\epsilon(\mathcal{D}|\boldsymbol{\theta}, \mathbf{x}) \triangleq P(\mathbf{x} \in \mathcal{B}_\epsilon(\mathcal{D})|\boldsymbol{\theta}, \mathbf{x})$ [46], where $\mathcal{B}_\epsilon(\mathcal{D})$ is a region of the data space $\mathfrak{D}$ defined as

$$\mathcal{B}_\epsilon(\mathcal{D}) = \big\{\mathbf{x} \in \mathfrak{D} : \rho\big(\eta(\mathbf{x}), \eta(\mathcal{D})\big) \leqslant \epsilon\big\} \tag{2.2}$$

In the expression of the approximate likelihood function and also in what follows, $P(\cdot)$ is adopted to denote probability whereas $p(\cdot)$ denotes a PDF. Thus, from Bayes' theorem, the approximate posterior $p_\epsilon(\boldsymbol{\theta}, \mathbf{x}|\mathcal{D})$ can be obtained as

$$p_\epsilon(\boldsymbol{\theta}, \mathbf{x}|\mathcal{D}) \propto P(\mathbf{x} \in \mathcal{B}_\epsilon(\mathcal{D})|\mathbf{x}, \boldsymbol{\theta}) p(\mathbf{x}|\boldsymbol{\theta}) p(\boldsymbol{\theta}) \tag{2.3}$$

The approximate likelihood can be formulated as $P(\mathbf{x} \in \mathcal{B}_\epsilon(\mathcal{D})|\mathbf{x}, \boldsymbol{\theta}) = \mathbb{I}_{\mathcal{B}_\epsilon(\mathcal{D})}(\mathbf{x})$, with $\mathbb{I}_{\mathcal{B}_\epsilon(\mathcal{D})}(\mathbf{x})$ being an indicator function for the set $\mathcal{B}_\epsilon(\mathcal{D})$ that assigns the unity when $\rho\big(\eta(\mathbf{x}), \eta(\mathcal{D})\big) \leqslant \epsilon$, and 0 otherwise. It follows that the approximate posterior $p_\epsilon(\boldsymbol{\theta}, \mathbf{x}|\mathcal{D})$ can be readily computed as

$$p_\epsilon(\boldsymbol{\theta}, \mathbf{x}|\mathcal{D}) \propto p(\mathbf{x}|\boldsymbol{\theta}) p(\boldsymbol{\theta}) \mathbb{I}_{\mathcal{B}_\epsilon(\mathcal{D})}(\mathbf{x}) \tag{2.4}$$

Since the ultimate interest of the Bayesian inverse problem is typically the posterior of model parameters $p_\epsilon(\boldsymbol{\theta}|\mathcal{D})$, it can be obtained by marginalising the approximate posterior PDF in Equation (2.4)

$$p_\epsilon(\boldsymbol{\theta}|\mathcal{D}) \propto p(\boldsymbol{\theta}) \int_{\mathfrak{D}} p(\mathbf{x}|\boldsymbol{\theta}) \mathbb{I}_{\mathcal{B}_\epsilon(\mathcal{D})}(\mathbf{x}) d\mathbf{x} = P(\mathbf{x} \in \mathcal{B}_\epsilon(\mathcal{D})|\boldsymbol{\theta}) p(\boldsymbol{\theta}) \tag{2.5}$$

Note that this integration need not be done explicitly since samples from this marginal PDF are obtained by taking the $\boldsymbol{\theta}$-component of the samples from the joint PDF in Equation (2.4) [151]. A pseudo-code implementation of ABC algorithm is given below as Algorithm 3.

Algorithm 3: Standard ABC

Inputs ϵ {Tolerance value}, $\eta(\cdot)$ {Summary statistic}, $\rho(\cdot)$ {metric}, K {number of simulations}
Begin
for $k = 1$ to K **do**
  repeat
    1.- Simulate $\boldsymbol{\theta}'$ from the prior $p(\boldsymbol{\theta}|\mathcal{M}_j)$
    2.- Generate $\mathbf{x}'$ from the stochastic forward model $p(\mathbf{x}|\boldsymbol{\theta}', \mathcal{M}_j)$
  until $\rho\big(\eta(\mathbf{x}'), \eta(\mathcal{D})\big) \leqslant \epsilon$
  Accept $(\boldsymbol{\theta}', \mathbf{x}')$ as $(\boldsymbol{\theta}^{(k)}, \mathbf{x}^{(k)})$
end for

Output $\{(\boldsymbol{\theta}, \mathbf{x})\}_{k=1}^{K} \sim p_\epsilon(\boldsymbol{\theta}, \mathbf{x}|\mathcal{D})$

Example 4 *Let us consider a* 2 *[m] length column with* 0.4 *[m] square cross section, which is loaded with* $F = 1$ *[kN] at the top, as illustrated in Figure 2.1. Let also consider that, for some reason, the column is made of a degrading material so that its Young's modulus decreases at an unknown constant rate* ξ *from an initial value* $E_0 = 40$ *[MPa], such that*

$$E_n = e^{-\xi} E_{n-1} + v_n \tag{2.6}$$

where E_n *is the Young's modulus at time or instant* $n \in \mathbb{N}$ *expressed in weeks, and* v_n *is an unknown model error term, which is assumed to be distributed as a zero-mean Gaussian with uncertain standard deviation, that is,* $v_n \sim \mathcal{N}(0, \sigma)$*. Next, a sensor is assumed to be placed at the top of the column to register deflections, and the following measurement equation can be considered:*

$$\delta_n = f(E_n) + w_n \tag{2.7}$$

where $f : \mathbb{R}_{\geqslant 0} \to \mathbb{R}_{\geqslant 0}$ *is a mathematical function that provides the deflection of the column as a function of* E_n*. Assuming the linear elasticity theory, this function can be expressed as* $f = \frac{FL^3}{3E_n I}$*, where* I *is the inertia momentum of the cross section. In Equation* (2.7)*, the term* w_n *is the measurement error, which is assumed to be negligible as compared to the model error term, so that it*

Figure 2.1: Scheme of structure used for Example 4.

is subsumed into the error term v_n. *In this example, the degradation rate and the standard deviation of the error term are selected as the uncertain model parameters, so that* $\boldsymbol{\theta} = \{\theta_1, \theta_2\} = \{\xi, \sigma\}$, *whose prior information can be represented by the uniform piece-wise PDFs* $p(\theta_1) = \mathcal{U}[0.0001, 0.02]$, *and* $p(\theta_2) = \mathcal{U}[0.01, 2]$, *respectively. The data in this example are given as a recorded time-history of deflections over a period of time* $T = 200$ *weeks, that is,* $\mathcal{D} = \left\{\delta_{n,meas}\right\}_{n=0}^{200}$. *These data are synthetically generated from Equations (2.6) and (2.7) considering* $\boldsymbol{\theta}_{true} = (0.005, 0.1)$, *shown in Figure (2.2), panels (a) and (b). The ABC-rejection algorithm is adopted with* $K = 20,000$ *samples to obtain the approximate posterior of* $\boldsymbol{\theta}$ *based on the referred data. The results are shown in Figure 2.2c.*

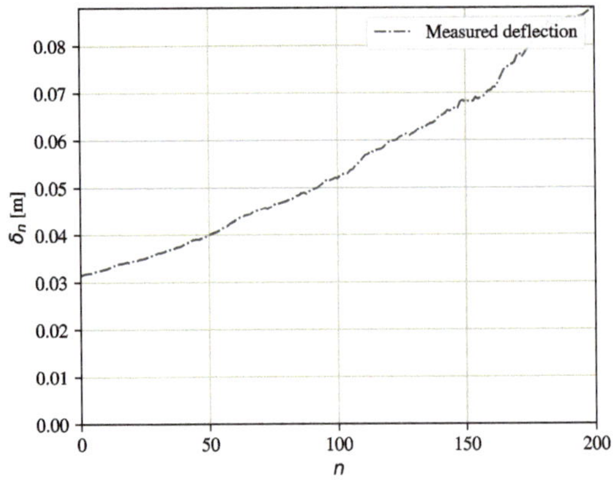

(a) Measured deflections.

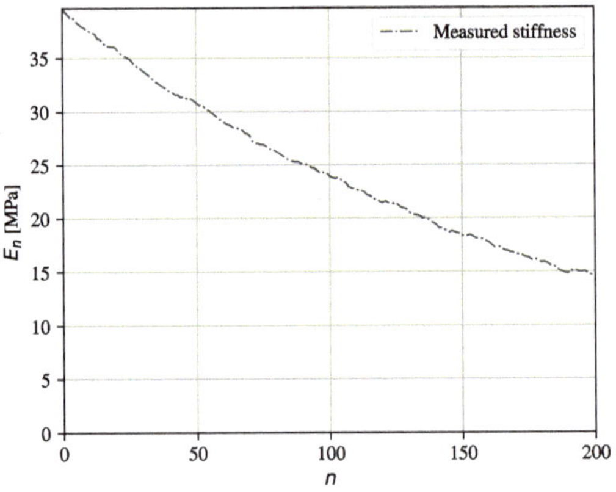

(b) True stiffness values.

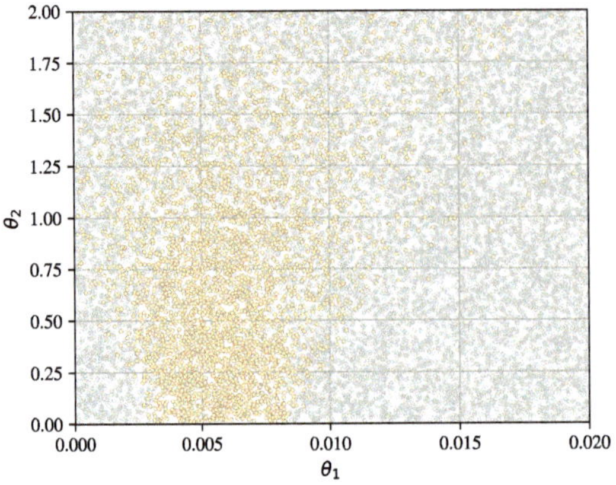

(c) ABC Output.

Figure 2.2: Output of the ABC rejection algorithm in application to Example 4. In panels (a) and (b), the history plot of measured deflections and stiffness values based on $\boldsymbol{\theta}_{\text{true}}$, are shown. Circles represents values in the $\boldsymbol{\theta}$-space.

In Figure 2.2, the grey circles represent samples from the prior, whilst the darker-colored circles represent samples from $\boldsymbol{\theta}$ distributed according to the approximate posterior PDF $p_\epsilon(\boldsymbol{\theta}|\mathcal{D})$ (recall Eq. (2.5)), where $\epsilon = 1100$, using a ℓ_1 distance as metric, that is $\rho = \sum_{n=0}^{200} \|E_{n,true} - E_n\|$. A total amount of 5,000 samples lie within the approximate posterior samples, which populate a region close to the $\boldsymbol{\theta}_{true}$ values. Observe that, in average, the approximate posterior gives us a region of plausible values of $\boldsymbol{\theta} \approx \boldsymbol{\theta}_{true}$, which in view of Figure 2.2 is relatively large. Hence, better results would be expected if smaller ϵ values are considered, as will be shown further below.

Note that the quality of the posterior approximation in ABC depends on a suitable selection of the metric ρ, the tolerance parameter ϵ and, of special importance, the summary statistic $\eta(\cdot)$ [65]. The choice of the tolerance parameter ϵ is basically a matter of the amount of computational effort that the modeller is willing to expend. Indeed, for ϵ sufficiently small $\eta(\mathbf{x}') \to \eta(\mathcal{D})$, and thus all accepted samples simulated according to Equation (2.5) come from the closest approximation to the posterior density $p(\boldsymbol{\theta}|\mathcal{D})$, where the exactness is achieved when $\eta(\cdot)$ is a sufficient statistic. This desirable fact is at the expense of a high computational effort to get $\eta(\mathbf{x}') = \eta(\mathcal{D})$ under the stochastic forward model $p(\mathbf{x}|\boldsymbol{\theta}, \mathcal{M}_j)$, usually prohibitive. On the contrary, as $\epsilon \to \infty$, all accepted simulations $\mathbf{x}'$ come from the prior density. Hence, the choice of the tolerance parameter ϵ reflects a trade-off between computational cost and accuracy of the ABC posterior approximation.

For that reason, several computational improvements have been proposed in the literature addressing this trade-off. Among them, combining ABC with MCMC methods (commonly called the ABC-MCMC method) has demonstrated to be efficient [128] in those cases where the probability content of the posterior is concentrated over a small region in relation to a diffuse prior. In fact, in ABC-MCMC, a proposal PDF $q(\boldsymbol{\theta}'|\boldsymbol{\theta}^{(\zeta)})$ is used to sample a new parameter $\boldsymbol{\theta}'$ based on a previous accepted one $\boldsymbol{\theta}^{(\zeta)}$, targeting the stationary distribution $p_\epsilon(\boldsymbol{\theta}|\mathcal{D})$. A pseudo-code is provided below as Algorithm 4.

Algorithm 4: ABC-MCMC

Inputs ϵ {Tolerance value}, $\eta(\cdot)$ {Summary statistic}, $\rho(\cdot)$ {metric}, $q(\cdot)$ {proposal PDF}, K {number of simulations}

Begin

1.- Initialise $(\theta^{(0)}, \mathbf{x}^{(0)})$ from $p_\epsilon(\boldsymbol{\theta}, \mathbf{x}|\mathcal{D})$; e.g. use Algorithm 3.

for $k = 1$ to K **do**

  2.- Generate $\boldsymbol{\theta}' \sim q(\boldsymbol{\theta}|\boldsymbol{\theta}^{(k-1)})$ and $\mathbf{x}' \sim p(\mathbf{x}|\boldsymbol{\theta}')$.

  3.- Accept $(\boldsymbol{\theta}', \mathbf{x}')$ as $(\boldsymbol{\theta}^{(k)}, \mathbf{x}^{(k)})$ with probability:

$$\alpha = \min\left\{1, \frac{P_\epsilon(\mathcal{D}|\mathbf{x}', \boldsymbol{\theta}')p(\boldsymbol{\theta}')q(\boldsymbol{\theta}^{(k-1)}|\boldsymbol{\theta}')}{P_\epsilon(\mathcal{D}|\mathbf{x}^{(k-1)}, \boldsymbol{\theta}^{(k-1)})p(\boldsymbol{\theta}^{(k-1)})q(\boldsymbol{\theta}'|\boldsymbol{\theta}^{(k-1)})}\right\}$$

  else set $(\boldsymbol{\theta}^{(k)}, \mathbf{x}^{(k)}) = (\boldsymbol{\theta}^{(k-1)}, \mathbf{x}^{(k-1)})$

end for

Output: $\{(\boldsymbol{\theta}, \mathbf{x})\}_{k=1}^{K} \sim p_\epsilon(\boldsymbol{\theta}, \mathbf{x}|\mathcal{D})$

It should be noted that when the likelihood function is approximated as $P_\epsilon(\mathcal{D}|\mathbf{x},\boldsymbol{\theta}) = \mathbb{I}_{\mathcal{B}_{\epsilon}(\mathcal{D})}(\mathbf{x})$, as in our case, the acceptance probability α (Step 3 in Algorithm 4) can be rewritten as the product of a MCMC acceptance ratio and the indicator function as follows

$$\alpha = \min\left\{1, \frac{p(\boldsymbol{\theta}')q(\boldsymbol{\theta}^{(n-1)}|\boldsymbol{\theta}')}{p(\boldsymbol{\theta}^{(n-1)})q(\boldsymbol{\theta}'|\boldsymbol{\theta}^{(n-1)})}\right\}\mathbb{I}_{\mathcal{B}_{\epsilon}(\mathcal{D})}(\mathbf{x}') \tag{2.8}$$

where Equation (2.8) clearly shows that the dependence upon ϵ in the indicator function may lead to an inefficient algorithm for a good approximation of the true posterior. In fact, given that α can only be non-zero if $\mathbb{I}_{\mathcal{B}_{\epsilon}(\mathcal{D})}(\mathbf{x}') = 1$ (i.e. $\rho\big(\eta(\mathbf{x}'), \eta(\mathcal{D})\big) \leqslant \epsilon$), then the Markov chain may persist in distributional tails for long periods of time if ϵ is sufficiently small, due to the acceptance probability being zero in Step 3 of Algorithm 4. Again, despite some attempts to effectively overcome this limitation [28, 169], the desirable fact $\epsilon \to 0$ in ABC again implies heavy computation.

For that reason, a branch of computational techniques have emerged in the literature to obtain high accuracy ($\epsilon \to 0$) with a feasible computational cost by combining sequential sampling algorithms adapted for ABC [55]. These techniques share a common principle of achieving computational efficiency by learning about intermediate target distributions determined by a decreasing sequence of tolerance levels $\epsilon_1 > \epsilon_2 > \ldots > \epsilon_m = \epsilon$, where the last is the desired tolerance ϵ. The reader is referred to [46] for an overview of the main contributions to the literature on this topic. Within the available methods, the so-called ABC-SubSim algorithm published by Chiachío et al. in [46] has demonstrated efficiency as compared with the available ABC algorithms in the literature [188]. The methodological bases of ABC-SubSim algorithm are provided below.

2.2 Basis of ABC using Subset Simulation

2.2.1 Introduction to Subset Simulation

Subset Simulation (SS) [14] is an efficient method for simulation of rare-events in high-dimensional spaces, which makes this method applicable to a broad range of areas of science and engineering [17, 47]. By Subset Simulation, conditional samples can be efficiently generated corresponding to specified levels of a performance function $g : \mathcal{Z} \subset \mathbb{R}^d \to \mathbb{R}$ in a progressive manner, converting a problem involving rare-event simulation into a sequence of problems involving higher probability events [14]. This method was originally proposed to compute small failure probabilities encountered in reliability analysis of engineering systems (e.g. [16, 47]). For illustration purposes, the Subset Simulation method is presented in this section using its primary aim of small failure probabilities estimation. In the next section, it will be specialised for ABC simulation.

Let us consider that $\mathcal{F}$ is a failure region in a **z**-space, corresponding to the exceedance of the performance function g above some specified threshold value b

$$\mathcal{F} = \{\mathbf{z} \in \mathcal{Z} : g(\mathbf{z}) > b\} \tag{2.9}$$

Let us now assume that $\mathcal{F}$ can be defined as the intersection of m regions $\mathcal{F} = \bigcap_{j=1}^{m} \mathcal{F}_j$, such that they are arranged as a nested sequence $\mathcal{F}_1 \supset \mathcal{F}_2 \ldots \supset \mathcal{F}_{m-1} \supset \mathcal{F}_m = \mathcal{F}$, where $\mathcal{F}_j = \{\mathbf{z} \in \mathcal{Z} : g(\mathbf{z}) > b_j\}$, with $b_{j+1} > b_j$, such that $p(\mathbf{z}_j|\mathcal{F}_j) \propto p(\mathbf{z})\mathbb{I}_{\mathcal{F}_j}(\mathbf{z})$, $j = 1, \ldots, m$. The term $p(\mathbf{z})$ is to denote the probability model for $\mathbf{z}$. When the event $\mathcal{F}_j$ holds, then $\{\mathcal{F}_{j-1}, \ldots, \mathcal{F}_1\}$ also hold, henceforth $P(\mathcal{F}_j|\mathcal{F}_{j-1}, \ldots, \mathcal{F}_1) = P(\mathcal{F}_j|\mathcal{F}_{j-1})$, so that

$$P(\mathcal{F}) = P\Big(\bigcap_{j=1}^{m} \mathcal{F}_j\Big) = P(\mathcal{F}_1) \prod_{j=2}^{m} P(\mathcal{F}_j|\mathcal{F}_{j-1}) \tag{2.10}$$

where, for simpler notation, we use $P(\mathcal{F}) \triangleq P(\mathbf{z} \in \mathcal{F})$ and $P(\mathcal{F}_j|\mathcal{F}_{j-1}) \triangleq P(\mathbf{z} \in \mathcal{F}_j|\mathbf{z} \in \mathcal{F}_{j-1})$, with $P(\mathbf{z} \in \mathcal{F}_j|\mathbf{z} \in \mathcal{F}_{j-1})$ being the conditional failure probability at the $(j-1)^{th}$ conditional level. Notice that although the probability $P(\mathcal{F})$ can be relatively small, by choosing the intermediate regions appropriately, the conditional probabilities involved in Equation (2.10) can be made large, thus avoiding the simulation of rare events.

In the last equation, apart from $P(\mathcal{F}_1)$ (which can be readily estimated by the standard Monte Carlo method (MC)), the remaining factors cannot be efficiently estimated because of the conditional sampling involved. However, MCMC methods can be used for sampling from the PDF $p(\mathbf{z}_{j-1}|\mathcal{F}_{j-1})$ when $j \geqslant 2$, although it is at the expense of generating K-dependent samples, giving

$$P(\mathcal{F}_j|\mathcal{F}_{j-1}) \approx \bar{P}_j = \frac{1}{K} \sum_{k=1}^{K} \mathbb{I}_{\mathcal{F}_j}(\mathbf{z}_{j-1}^{(k)}), \qquad \mathbf{z}_{j-1}^{(k)} \sim p(\mathbf{z}_{j-1}|\mathcal{F}_{j-1}) \tag{2.11}$$

where $\mathbb{I}_{\mathcal{F}_j}(\mathbf{z}_{j-1}^{(k)})$ is the indicator function for the region $\mathcal{F}_j, j = 1, \ldots, m$, that assigns a value 1 when $g(\mathbf{z}_{j-1}^{(k)}) > b_j$, and 0 otherwise. Observe that it is possible to obtain Markov chain samples that are generated at the $(j-1)$th level which lie in $\mathcal{F}_j$, so that they are distributed as $p(\cdot|\mathcal{F}_j)$. Hence these samples provide "seeds" for generating more samples according to $p(\mathbf{z}_j|\mathcal{F}_j)$ by using MCMC sampling. Because of the way the seeds are chosen, Subset Simulation exhibits the benefits of *perfect sampling* [207, 151] which is an important feature to avoid wasting samples during a burn-in period, in contrast to other MCMC algorithms such as the MH algorithm.

As explained below, $\mathcal{F}_j$ is actually chosen adaptively based on the samples $\{\mathbf{z}_{j-1}^{(k)}, k = 1, \ldots, K\}$ from $p(\mathbf{z}|\mathcal{F}_{j-1})$ in such a way that there are exactly NP_0 of these samples in $\mathcal{F}_j$ (so $\bar{P}_j = P_0$ in Equation (2.11)). Next, MCMC is applied to obtain $({}^1\!/_{P_0} - 1)$ samples from $p(\mathbf{z}_j|\mathcal{F}_j)$ starting at each seed, so that a total of N samples are obtained in $\mathcal{F}_j$. Repeating this process, one can compute the conditional probabilities of the higher-conditional levels until the final region of interest $\mathcal{F}_m = \mathcal{F}$ has been reached. Note that the intermediate threshold value b_j defining $\mathcal{F}_j$ is obtained in an adaptive manner as the $[(1-P_0)K]_{th}$ largest value of the sample values $g(\mathbf{z}_{j-1}^{(k)})$, $k = 1, \ldots, K$, in such a way that the sample estimate of $P(\mathcal{F}_j|\mathcal{F}_{j-1})$ in Equation (2.11) is equal to P_0. See Figure 2.3 for an illustrative example of the Subset Simulation method.

In Figure 2.3, panels (a) & (b) represent the initial set of samples of dimension two simulated according to any specified model, and the selection of seeds whereby $\mathcal{F}_1$ is defined, respectively. In panel (c), new samples simulated according to $p(\mathbf{z}|\mathcal{F}_1)$ are represented along with the seeds for defining the subsequent intermediate failure region, which is represented using samples in panel (d). Panel (e) represents the final step where the whole set of samples arranged through subsets is showed along with the sequence of intermediate threshold levels until the final one (solid line) is reached.

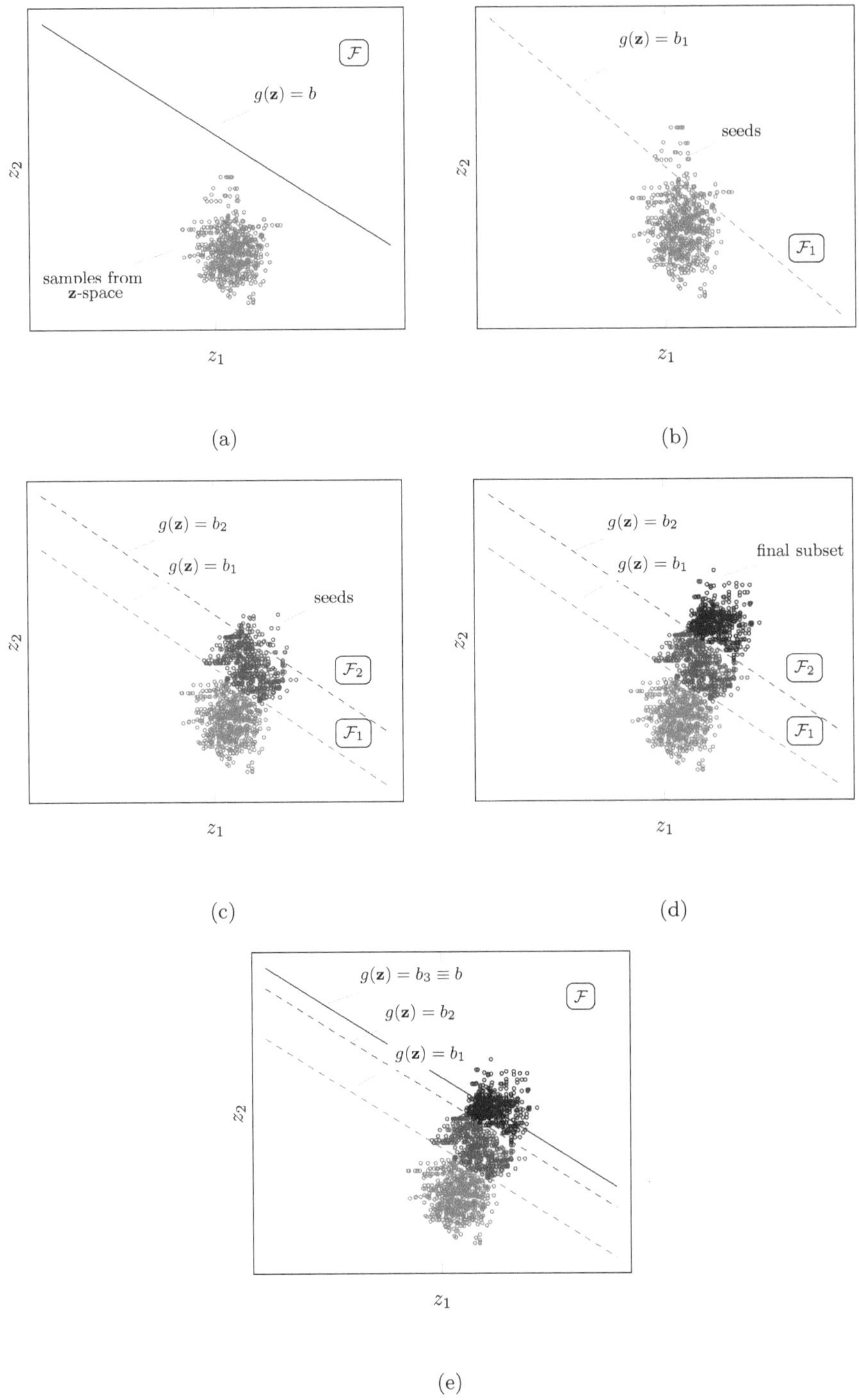

Figure 2.3: Conceptual example of Subset Simulation.

It should be noted that the choice of an adequate P_0 has a significant impact on the efficiency of Subset Simulation. Indeed, a small value for the conditional probability ($P_0 \to 0$) ensures that the distance between consecutive intermediate levels $b_j - b_{j-1}$ becomes too large, which leads to a rare-event simulation problem. If, on the contrary, the intermediate threshold values were chosen too close ($P_0 \to 1$), the algorithm would require more simulation levels m (and hence, large computational effort) to progress towards $\mathcal{F}$. In the literature about Subset Simulation, several works have reported about the scaling strategies for P_0. The value $P_0 = 0.1$ was recommended in the original presentation of Subset Simulation [14], and more recently in [207], the range $0.1 \leqslant P_0 \leqslant 0.3$ was proposed as near optimal range of values after a rigorous sensitivity study of Subset Simulation.

As stated before, to draw samples from the conditional PDF $p(\mathbf{z}_j|\mathcal{F}_j)$, MCMC methods like Metropolis-Hastings [131] are adequate. In the original version of Subset Simulation [14], the Modified Metropolis Algorithm (MMA) was proposed and it was shown to work well for very high dimensions (e.g. 10^3- 10^4). In MMA, a univariate proposal PDF is chosen for each component of the parameter vector and each candidate component is accepted or rejected separately, instead of drawing a full parameter vector from a multi-dimensional PDF as in the original algorithm. Later in [16], grouping of the parameters was considered when constructing a proposal PDF to allow for the case where small groups of components in the parameter vector are highly correlated. An appropriate choice for the proposal PDF for ABC-SubSim is introduced in the next section. Further insights for Subset Simulation using MMA as a conditional sampler are provided by Zuev et al. [207] where an exhaustive sensitivity study of MMA is presented along with enhancements for the Subset Simulation method.

2.2.2 Subset Simulation for ABC

In this section, the Subset Simulation method is described as a specialised efficient sampler for ABC. To this end, let us define the joint state-parameter vector $\mathbf{z} = (\mathbf{x}, \boldsymbol{\theta}) \in \mathcal{Z} \subset \mathbb{R}^{\ell+d}$, where $\mathbf{x}$ are simulated outcomes from the model class parameterised by $\boldsymbol{\theta}$, so that $p(\mathbf{z}) = p(\mathbf{x}|\boldsymbol{\theta})p(\boldsymbol{\theta})$. Let also $\mathcal{F}_j$ in Section 2.2.1 be replaced by a nested sequence of regions $\mathcal{Z}_j$, $j = 1 \ldots, m$, in $\mathcal{Z}$ defined by

$$\mathcal{Z}_j = \Big\{(\boldsymbol{\theta}, \mathbf{x}) : \rho\big(\eta(\mathbf{x}), \eta(\mathcal{D})\big) \leqslant \epsilon_j\Big\} \tag{2.12}$$

arranged so that $\mathcal{Z}_1, \mathcal{Z}_2, \ldots, \supset \mathcal{Z}_j \supset \mathcal{Z}_{j+1}, \ldots, \supset \mathcal{Z}_m = \mathcal{Z}$, and ρ a metric on the set $\big\{\eta(\mathbf{x}) : \mathbf{x} \in \mathcal{D}\big\}$. In ABC based on Subet Simulation, the sequence of tolerances $\epsilon_1, \ldots, \epsilon_j, \ldots, \epsilon_m$, with $\epsilon_{j+1} < \epsilon_j$, is chosen adaptively where the number of levels m is chosen so that $\epsilon_m \leqslant \epsilon$, a specified tolerance.

As stated by Equation (2.4), the sequence of intermediate posteriors $p(\boldsymbol{\theta}, \mathbf{x}|\mathcal{Z}_j)$, $j = 1, \ldots, m$ can be obtained by Bayes' theorem as

$$p(\boldsymbol{\theta}, \mathbf{x}|\mathcal{Z}_j) = \frac{P(\mathcal{Z}_j|\boldsymbol{\theta}, \mathbf{x})p(\mathbf{x}|\boldsymbol{\theta})p(\boldsymbol{\theta})}{P(\mathcal{Z}_j)} \propto \mathbb{I}_{\mathcal{Z}_j}(\boldsymbol{\theta}, \mathbf{x})p(\mathbf{x}|\boldsymbol{\theta})p(\boldsymbol{\theta}) \tag{2.13}$$

where $\mathbb{I}_{\mathcal{Z}_j}(\boldsymbol{\theta}, \mathbf{x})$ is the indicator function for the set $\mathcal{Z}_j$. Observe that when $\epsilon \to 0$, $\mathcal{Z}_m$ represents a small closed region in $\mathcal{Z}$ and hence the probability $P(\mathcal{Z}_m)$ is very small under the model $p(\boldsymbol{\theta}, \mathbf{x}) = p(\mathbf{x}|\boldsymbol{\theta})p(\boldsymbol{\theta})$. In this situation, using MCMC sampling directly is not efficient for the reason described above. Thus, this is the point at which the efficiency of Subset Simulation is exploited for ABC, given that such a small probability $P(\mathcal{Z}_m)$ is converted into a sequence of larger conditional probabilities, as stated in Equation (2.10).

2.3 The ABC-SubSim algorithm

The ABC-SubSim algorithm was conceived by combining the ABC principles with the technique of Subset Simulation for efficient rare-event simulation, first developed by Au and Beck [14]. Algorithm 5 provides a pseudo-code implementation of ABC-SubSim that is intended to be sufficient for most applications. The algorithm is implemented such that a maximum allowable number of simulation levels (m) is considered in case the specified ϵ is too small. Recent improvements for adaptive selecting ϵ have appeared in the literature [186, 187]

Algorithm 5: ABC-SubSim

Inputs:
$P_0 \in [0,1]$ // {gives percentile selection, chosen so that $KP_0, 1/P_0 \in \mathbb{Z}^+$;
$P_0 = 0.2$ // recommended value [46]}
m // {maximum number of simulation levels}
K // {number of samples per intermediate level}
Begin:
Sample $\left[(\boldsymbol{\theta}_0^{(1)}, \mathbf{x}_0^{(1)}), \ldots, (\boldsymbol{\theta}_0^{(k)}, \mathbf{x}_0^{(k)}), \ldots, (\boldsymbol{\theta}_0^{(K)}, \mathbf{x}_0^{(K)})\right]$, where $(\boldsymbol{\theta}, \mathbf{x}) \sim p(\boldsymbol{\theta})p(\mathbf{x}|\boldsymbol{\theta})$
for $j : 1, \ldots, m$ **do**
  for $k : 1, \ldots, K$ **do**
    Evaluate $\rho_j^{(k)} = \rho\big(\eta(\mathbf{x}_{j-1}^{(k)}), \eta(\mathcal{D})\big)$
  end for
  Renumber $\left[(\boldsymbol{\theta}_{j-1}^{(k)}, \mathbf{x}_{j-1}^{(k)}), k : 1, \ldots, K\right]$ so that $\rho_j^{(1)} \leqslant \rho_j^{(2)} \leqslant \ldots \rho_j^{(K)}$
  Fix $\epsilon_j = \frac{1}{2}\left(\rho_j^{(KP_0)} + \rho_j^{(KP_0+1)}\right)$
  for $k = 1, \ldots, KP_0$ **do**
    Select as a seed $(\boldsymbol{\theta}_j^{(k),1}, \mathbf{x}_j^{(k),1}) = (\boldsymbol{\theta}_{j-1}^{(k)}, \mathbf{x}_{j-1}^{(k)}) \sim p\big(\boldsymbol{\theta}, \mathbf{x}|(\boldsymbol{\theta}, \mathbf{x}) \in \mathcal{Z}_j\big)$
    Run Modified Metropolis Algorithm [14] to generate $1/P_0$ states of a Markov chain lying in $\mathcal{Z}_j$ (Eq. (2.12)): $\left[(\boldsymbol{\theta}_j^{(k),1}, \mathbf{x}_j^{(k),1}), \ldots, (\boldsymbol{\theta}_j^{(k),1/P_0}, \mathbf{x}_j^{(k),1/P_0})\right]$
  end for
  Renumber $\left[(\boldsymbol{\theta}_j^{(k),i}, \mathbf{x}_j^{(k),i}) : k = 1, \ldots, KP_0; i = 1, \ldots, 1/P_0\right]$ as $\left[(\boldsymbol{\theta}_j^{(1)}, \mathbf{x}_j^{(1)}), \ldots, (\boldsymbol{\theta}_j^{(K)}, \mathbf{x}_j^{(K)})\right]$
  if $\epsilon_j \leqslant \epsilon$ **then**
    End algorithm
  end if
end for

Output: $\{(\boldsymbol{\theta}, \mathbf{x})\}_{k=1}^{K} \sim p_\epsilon(\boldsymbol{\theta}, \mathbf{x}|\mathcal{D})$

One of the key aspects of ABC-SubSim is the selection of the ϵ_j values. Indeed, these values are adaptively chosen as in Subset Simulation [14], so that the sample estimate $\bar{P}_j$ of $P(\mathcal{Z}_j|\mathcal{Z}_{j-1})$ satisfies $\bar{P}_j = P_0$ [46]. By this way, the intermediate tolerance value ϵ_j can be simply obtained as the $100P_0$ percentile of the set of distances given by $\rho\big(\eta(\mathbf{x}_{j-1}^{(k)}), \eta(\mathcal{D})\big), k = 1, \ldots, K$, arranged in increasing order. Moreover, observe in Algorithm 5 that P_0 is chosen such that KP_0 and $1/P_0$ are integers, and so the size of the subset of samples generated in $\mathcal{Z}_{j-1}$ that lie in $\mathcal{Z}_j$ is known in advance and equal to KP_0, which provides a significant advantage for the implementation. These KP_0 samples in $\mathcal{Z}_j$ are used as seeds for KP_0 Markov chains of length $\frac{1}{P_0}$, where the new $(\frac{1}{P_0} - 1)$ samples in $\mathcal{Z}_j$ in each chain can be readily generated by MMA [14]. Hence the total number of samples of $(\boldsymbol{\theta}, \mathbf{x})$ lying in $\mathcal{Z}_j$ is K, but KP_0 of them were generated at the $(j-1)^{th}$ level. Because of the way the seeds are chosen, ABC-SubSim exhibits the benefits of *perfect sampling* [207, 151], which is an important feature to avoid wasting samples during a burn-in period, in contrast to ABC-MCMC.

Finally, it is important to remark that the important control parameters to be chosen in ABC-SubSim are P_0 and $\sigma_q^{(j)}$, the standard deviation in the Gaussian proposal PDF in MMA at the jth level. For P_0, the recommendation is to choose it within the interval $[0.1, 0.3]$, and more specifically $P_0 = 0.2$ [46]. In regards to the optimal variance of the local proposal PDF for the MMA sampler, the rule is to adaptively choose the standard deviation $\sigma_q^{(j)}$ of the jth intermediate level so that the monitored acceptance rate lies within the interval $[0.2, 0.4]$ based on an initial chain sample of small length (e.g. 10 states). The reader is referred to [186] for a detailed study about adaptive scaling of the ABC-SubSim algorithm, and also to [207], where the optimal scaling is addressed for the general Subset Simulation method.

Example 5 *Let us consider again the example of the column given in Example 4. In this case, we want to obtain samples from the posterior* $\boldsymbol{\theta} \sim p_\epsilon(\theta|\mathcal{D})$, *where* ϵ *is much smaller than that obtained in Example 4. The ABC-SubSim algorithm is used with* $K = 5,000$, $P_0 = 0.2$, *and* $m = 4$ *simulation levels, so that the final subset produces* $K = 5,000$ *samples distributed according to* $p_{\epsilon_m}(\theta|\mathcal{D})$. *The results are shown in Figure 2.4. In Figure 2.4c, the intermediate posterior samples are superimposed in increasing gray tones to reveal the uncertainty reduction. The final subset is depicted using darker circles. ABC-SubSim uses 17,000 simulations to generate 5,000 samples representing a final posterior that is better aligned with the values of* $\boldsymbol{\theta}_{true}$. *Moreover, to demonstrate the efficacy of the algorithm, the model is evaluated by picking a sample* $\hat{\boldsymbol{\theta}}^{ABC\text{-}SS}$ *from* $p_\epsilon(\theta|\mathcal{D})$. *Recall that in both cases, the final posterior has the same amount of samples, namely, 5,000 samples. The results for stiffness and deflections are presented superimposed in Figures 2.4a and 2.4b for better interpretation. Note that the results are very satisfactory in the sense that ABC-SubSim can reconstruct the true structural behaviour with high precision in comparison with the ABC-rejection algorithm with the same computational cost. In fact, in both cases, the computational burden of the algorithms are almost equivalent (indeed, ABC-SubSim uses 17,000 samples to produce a final posterior of 5,000 samples, whilst ABC-rejection needs 20,000), while the accuracy obtained by ABC-SubSim is, by far, superior to ABC-rejection, which corroborates the findings originally observed in [46].*

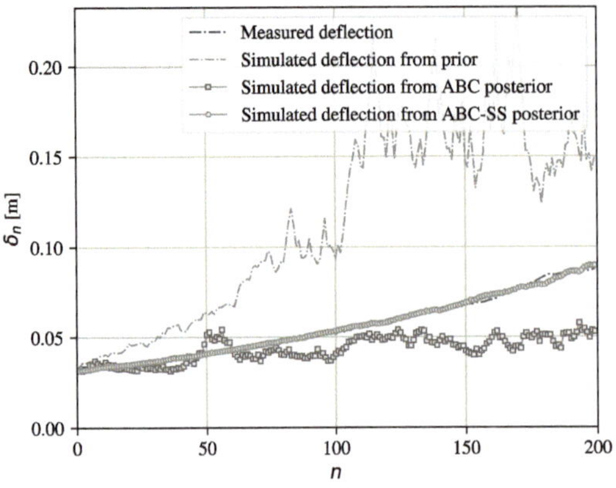

(a) Complex model

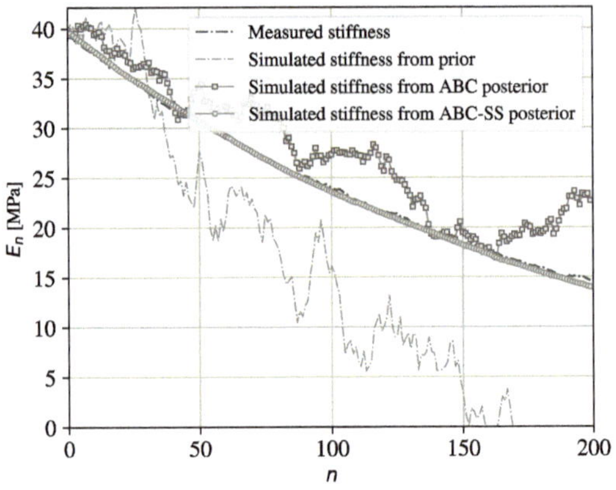

(b) Deflections

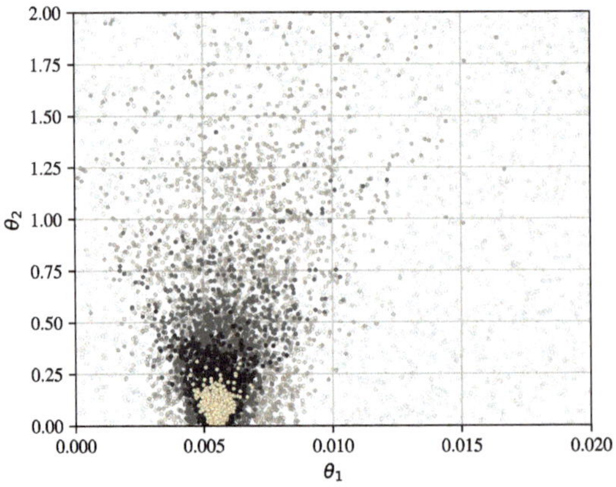

(c) Stiffness

Figure 2.4: Output of the ABC-SubSim algorithm in application to Example 5. In panels (a) and (b), the history plot of synthetic deflections and stiffness values based on $\boldsymbol{\theta}_{\text{true}}$, are shown together with the simulated plots of stiffness and deflections using the model. Grey circles represents values in the $\boldsymbol{\theta}$-space.

2.4 Summary

This chapter has shown that ABC methods are efficient for giving a solution to inverse problems where the likelihood function is unknown or directly computationally intractable. The resulting posterior PDF is obtained for a given tolerance value which provides a measure of how close the simulations are obtained using the posterior samples with respect to the data. The original ABC (rejection) algorithm is shown to be simple to implement but computationally inefficient overall when small tolerance values are adopted. Since the publication of the first ABC methodology, new ABC variants have been proposed to tackle the computational costs when adopting small tolerances, with the ABC-SubSim being a highly efficient one, based on the study published in [46], and further corroborated in works like [186, 188]. The ABC-SubSim algorithm has been explained in this chapter within the context of a structural engineering example based on a cantilever column under a constant lateral load, whose material is subject to a degradation process. The algorithm has been able to infer the damage model parameters with high fidelity with a reasonable computational cost irrespective of the big uncertainty considered. Moreover, the ABC-SubSim algorithm has been compared with the former ABC-rejection algorithm under similar computational cost, showing that ABC-SubSim can obtain a much finer posterior inference. Further research should be focused on investigating ways to adaptively scale the ABC-SubSim meta parameters, which have a strong influence on the posterior quality and the computational cost. Also, another fruitful research would be to explore the use of ABC-SubSim as a surrogate model for the evaluation of complex degradation engineering processes where multidimensional model parameters are involved.

3

Fundamentals of Sequential System Monitoring and Prognostics Methods

David E. Acuña-Ureta,[1,*] *Juan Chiachío-Ruano*,[2] *Manuel Chiachío-Ruano*[2] and *Marcos E. Orchard*[3]

This chapter provides a detailed description of the most significant constituent elements in the problem of sequential system monitoring and event prognostics, including aspects related to multi-step ahead predictions and the computation of the remaining useful life. The authors' main interest here is to provide a holistic view of the problem of uncertainty characterisation in prognostics, as well as the theory and algorithms that allow us to make use of real-time data to sequentially predict the current state (or *health*) of the monitored system and the characterisation of its future evolution in time. A numerical example is finally provided to exemplify the end of life calculation using a prognostics algorithm.

3.1 Fundamentals

Prognostics aims at determining the *end of life* (EOL) and *remaining useful life* (RUL) of components or systems given the information about the current degree of damage, the component's usage, and the anticipated future load and environmental conditions. In prognostics, the EOL is defined as the *limiting time when an asset is expected to depart from the serviceability conditions*. RUL is the *period of remaining time from the current time (or time of prediction) until estimated EOL*.

Prognostics can be seen as a natural extension of structural health monitoring (SHM) technology in the sense that the predictions of RUL and EOL are updated as long as data are available from a sensing system. It is rather a sequential process of *update-predict-reassess* where the user is not only concerned with detecting, isolating and sizing a fault mode, but also with (1) predicting the remaining time before the failure occurs, and (2) quantifying the uncertainty in the prediction, that can be further used for risk assessment and rational decision-making [44]. Henceforth, prognostics requires periodic health monitoring measurements to reassess and improve the quality of the predictions of EOL and RUL as time goes by.

[1] Pontificia Universidad Católica de Chile, Chile.
[2] University of Granada, Spain.
[3] University of Chile, Chile.
[*] Corresponding author: david.acuna@ing.puc.cl

It is important to remark that techniques and methodologies for prognostics are application-specific [166] and hence, global solutions for particular problems are rarely available. In fact, designing a prognostics framework requires (in a broad sense): (a) the choice of an adequate sensing method capable of providing health information given the practical constraints of the specific application; (b) development of an ad-hoc probabilistic modelling framework for prediction; and (c) a quantifiable criterion for what constitutes failure along with some prognostics metrics that can be used for decision-making. The following sections overview several of the aforementioned aspects about prognostics.

3.1.1 Prognostics and SHM

Using suitable SHM sensors that can interrogate the system health state and assess in real time any change in fault severity are of paramount importance. Since damage predictions are sequentially updated from periodical measurements, the higher the accuracy expected from prognostics, the better the quality required for the information obtained from the sensing system. However, this information comes at the expense of more targeted sensing and significant computational requirements. Complex systems subjected to a variety of fault modes (cracks, voids, delamination, corrosion, etc.) often require dedicated sensors and sensor networks for detection as no sensor can provide sufficient information to cover all fault modes. The choice of the sensing method is typically guided by the feature or set of features to be monitored. For example, weight loss or power demand sensors on-board airspace systems lead to a different sensor choice than the one made for monitoring vibrations in buildings or corrosion in bridge structures [166].

Sensor locations are chosen such that the expected type of damage produces observable and statistically significant effects in features derived from the measurements at these locations, which is often determined through numerical simulations or physical tests. Low-level local response caused by damage (e.g. cracks onset and closing) must be separated from a large-amplitude global response, such as that caused by heavy traffic loads on pavements, by determining the required sensitivity and dynamic range through analysis or experimentation. There are well-known methods for optimal placement of sensors that consider the uncertainty in both measurements and model response (see for example [139, 141, 33]).

3.1.2 Damage response modelling

As mentioned above, underlying any prediction is a model that describes how the component or system behaves under nominal conditions and how it will evolve as it experiences wear or a fault condition. To that end, one needs to represent that process in a mathematical fashion. The model might be derived from laws of physics, encapsulated by empirical relations, or learned from data. Models can also be formed using a combination of these approaches. For example, parts of a component that are well-understood may be constructed using physics-based models, with unknown physical parameters learned from data using an appropriate inference technique, as pointed out in Chapter 1.

Modelling of damage or degradation can be accomplished at different levels of granularity; for example, micro and macro levels. At the micro level, models are embodied by a set of constitutive equations that define relationships about primary physical variables and the observed component damage under specific environmental and operational conditions. Since measurements of critical damage properties (such as micro-cracks, voids, dislocations, etc.) are rarely available, sensed system parameters have to be used to infer the damage properties. Micro-level models need to account for the assumptions and simplifications in the uncertainty management. In contrast, macro-level models are characterised by a somewhat abridged representation that makes simplifying assumptions, thus reducing the complexity of the model (typically at the expense of accuracy). Irrespectively, as pointed out in Chapter 1, a model is just an idealisation to the reality that suffices for an approximation, henceforth the disparity with respect to the reality (manifested through the data) should be accounted for via explicit uncertainty management [160].

3.1.3 Interpreting uncertainty for prognostics

Developing methods for uncertainty quantification in the context of prognostics is challenging because uncertainty methods are generally computationally expensive, whereas prognostics requires real time computation for anticipated decision-making. An important aspect of prognostics is the prediction of RUL and EOL, and several publications [63, 156] have discussed the importance of quantifying the uncertainty of these measures.

Typically, probabilistic methods have widely been used for uncertainty quantification and management in various engineering applications, although the interpretation of probability is, at times, not straightforward. As mentioned in Chapter 1, there are two main interpretations of probability: frequentist versus subjective. In the context of prognostics and health management, the frequentist interpretation of uncertainty is applicable only for testing-based methods, mostly when several nominally identical specimens are tested to understand the inherent (random) variability across them. On the other hand, in *condition-based prognostics*, the focus is typically on one particular unit (e.g. a bridge). At any specific time-instant, this particular unit is in a given state with no variability, basically because there is not any other identical individual to compare it with. Nonetheless, there exists an associated uncertainty regarding this state, which is simply reflective of our degree of belief about it. Bayesian tracking methods (Kalman filtering, particle methods, etc.) [9] used for sequential state estimation, which constitutes the previous step before EOL/RUL prediction, are called *Bayesian* not only because they use Bayes' theorem for state estimation but also because they are based on probabilities that express the degree of plausibility of a proposition (damage state) conditioned on the given information (SHM data), which provides a rigorous foundation in a subjective uncertainty context [52]. For example, the future degradation of a component is subjected to epistemic (lack of knowledge) uncertainty; hence, the resulting RUL estimate that accounts for such future degradation needs to be interpreted probabilistically considering such uncertainty. Thus, it can be clearly seen that the subjective interpretation of uncertainty adopted across the entire book is also consistent in the domain of condition-based prognostics presented in this chapter [160].

3.1.4 Prognostic performance metrics

Quantifying the prediction performance is a natural step for an efficient prognostics framework after a component or subsystem is being monitored using an appropriate sensor system. Decisions based on poor and/or late predictions may increase the risk of system failure, whereas wrong predictions of failure (false positives) trigger unnecessary maintenance actions with unavoidable cost increase. Saxena et al. [163] provided a detailed discussion about deriving prognostics requirements from top level system goals, so the reader is referred to this work for further insight. These requirements are generally specified in terms of a prediction performance that prognostics must satisfy for providing a desired level of safety or cost benefit. A variety of prognostics performance evaluation metrics have been defined in the literature, like the prediction horizon (PH), the α-λ accuracy measure, and other relative accuracy measures [162, 161]. As described by [164], prognostics performance can be evaluated using the following three main attributes, namely:

correctness, which is related to the prediction accuracy using the data as a benchmark;

timeliness, which accounts for how fast a prognostics method (or algorithm) produces the output as compared to the rate of upcoming outcomes from the system being monitored;

confidence, which deals with the uncertainty in a prognostics output, typically from a prognostics algorithm.

Among the metrics proposed by Saxena et al. [162, 161], the PH and the α-λ accuracy measures are widely used in the prognostics literature and are also adopted in the majority of works about prognostics. The PH serves to determine the maximum early warning capability that a prediction algorithm can provide with a user-defined confidence level denoted by α. Typically, a graphical representation using a straight line with a negative slope serves to illustrate the "true RUL", that decreases linearly as time progresses. The predicted PDFs of RUL are plotted against time using error bars (e.g. by 5%–95% error bars), as depicted in Figure 3.1a. The median of the RUL predictions should stay within the accuracy regions specified by α, and ideally on the dotted line (RUL^*) that represents the true RUL. By means of this representation, the PH can be directly determined as shown in Figure 3.1a. The PH metric can be further parameterised by a parameter β (thus denoted by $\text{PH}_{\alpha,\beta}$) that specifies the minimum acceptable probability of overlap between the predicted RUL and the α accuracy bands delimited by the dashed lines in Figure 3.1a. Both α and β are scaling parameters for the prognostics which should be fixed considering the application scenario.

For the α-λ accuracy metric, a straight line with negative slope is also used to represent the true RUL. Predicted PDFs of RUL are plotted against time of prediction (which is termed as λ in the original paper by Saxena et al. [162]) using error bars. As in Figure 3.1a, accurate predictions should lie on this line as long as they are sequentially updated using SHM data. In this case, the accuracy region is determined by parameter α representing a percentage of the true RUL, in the

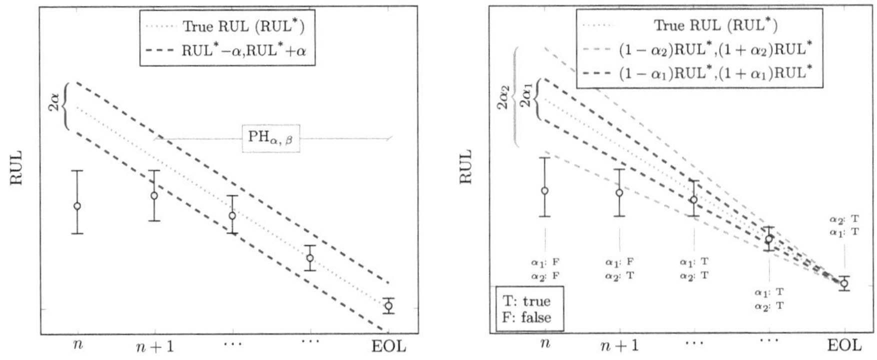

(a) (b)

Figure 3.1: Illustrations of (a) PH and (b) $\alpha - \lambda$ prognostics metrics.

sense that the accuracy of prediction becomes more critical as EOL approaches. See Figure 3.1b for illustration. In this case, two confidence regions are employed by adopting $0 < \alpha_1 < \alpha_2 < 1$, so that each predicted RUL can be validated depending on whether or not it belongs to any of the α_1 or α_2 regions. The interested reader is referred to [45, 44, 42] for further insight about using prognostics metrics in the context of engineering case studies.

3.2 Bayesian tracking methods

Condition monitoring aims at assessing the capability of machinery to safely perform required tasks while in operation, to identify any significant change which could be indicative of a developing fault. Condition monitoring techniques assume that the system health degrades in time due to cumulative damage and usage, and that it is possible to build a measure (or health indicator) using information from manufacturers, equipment operators, historical records, and real-time measurements. On this subject, it must be noted that Bayesian tracking methods [58, 87, 149, 88] offer a rigorous and remarkably intuitive framework for the implementation of condition monitoring algorithms, as they allow researchers to merge prior knowledge on the equipment operation (i.e. a system model) and information that can be captured from noisy (and sometimes corrupt) data acquisition systems.

Bayesian tracking methods incorporate modern system theory to build a paradigm for the characterisation of uncertainty sources in physical processes. In this paradigm, processes are conceived as dynamical systems and described, typically, using partial differential equations. The system state vector (usually denoted by x) corresponds to a mathematical representation of a condition indicator of the system, whereas the interaction with the immediate physical medium is summarised by the definition of input and output variables (usually denoted by u and y, respectively). Uncertainty sources, particularly those related to measurement noises and model inaccuracies are also explicitly incorporated in the model. Bayesian tracking methods allow researchers to compute *posterior* estimates of the state vector, after merging *prior* information (provided by the system model) and evidence from measurements (acquired sequentially in real-time). Whenever Bayesian tracking methods are used in fault diagnosis, the states are defined as a function of critical variables whose future evolution in time might significantly affect the system health, thus yielding to a failure condition.

Although Bayesian tracking methods have also been developed for continuous-time systems, most applications in failure diagnosis and prognostics consider discrete-time models. Discrete-time models allow researchers to easily incorporate sequential (sometimes unevenly sampled) measurements, enabling recursive updates of estimates and speeding up the computational time. For this reason, this section provides a brief review on popular discrete-time Bayesian tracking methods utilised in model-based fault diagnostic schemes.

Bayesian tracking methods constitute practical solutions to the *optimal filtering problem*, which focuses on the computation of the "best estimate" of the system state, given a set of noisy measurements. In other words, it aims at calculating the probability distribution of the state vector conditional on acquired observations, constituting an "optimal" estimate of the state. In mathematical terms, let $X = \{X_k\}_{k \in \mathbb{N} \cup \{0\}}$ be a Markov process denoting a n_x-dimensional system state vector with initial distribution $p(x_0)$ and transition probability $p(x_k|x_{k-1})$. Also, let $Y = \{Y_k\}_{k \in \mathbb{N}}$

denote n_y-dimensional conditionally independent noisy observations. It is assumed a state-space representation of the dynamic nonlinear system as the following

$$x_k = f(x_{k-1}, u_{k-1}, \omega_{k-1}) \tag{3.1}$$

$$y_k = g(x_k, u_k, \nu_k) \tag{3.2}$$

where ω_k and ν_k denote independent not necessarily Gaussian random vectors representing the uncertainty in both the modelling and the observations respectively. The objective of the optimal filtering problem is to obtain the posterior distribution $p(x_k|y_{1:k})$ sequentially as long as new observations are available. As this is a difficult task to achieve, estimators require the implementation of structures based on Bayes' rule where, under Markovian assumptions, the filtering posterior distribution can be written as

$$p(x_k|y_{1:k}) = \frac{p(y_k|x_k)p(x_k|y_{1:k-1})}{p(y_k|y_{1:k-1})} \tag{3.3}$$

where $p(y_k|x_k)$ is the *likelihood*, $p(x_k|y_{1:k-1})$ is the prior distribution and $p(y_k|y_{1:k-1})$ is a constant known as the *evidence*. Since the evidence is a constant, the posterior distribution is usually written as

$$p(x_k|y_{1:k}) \propto p(y_k|x_k)p(x_k|y_{1:k-1}) \tag{3.4}$$

then $p(x_k|y_{1:k})$ is obtained by normalising $p(y_k|x_k)p(x_k|y_{1:k-1})$.

The feasibility of a solution for the optimal filtering problem depends on the properties associated with both the state transition and the measurement models, as well as on the nature of the uncertainty sources (i.e. functions f and g, and random vectors ω_k and ν_k in Equations (3.1) and (3.2)). There are a few cases in which the solution for the optimal filtering problem has analytic expressions. For example, if the system is linear and has uncorrelated white Gaussian sources of uncertainty, the solution is provided by the widespread and well known *Kalman Filter* (KF). On the other hand, for cases where it is impossible to obtain an analytic expression for the state posterior distribution (e.g. nonlinear, non-Gaussian systems), the problem is referred to as *nonlinear filtering*, and has been properly addressed by some variants of the KF – such as the *Extended Kalman Filter* (EFK) or the *Unscented Kalman Filter* (UKF); though Sequential Monte Carlo, also known as *Particle Filters* (PFs), has lately caught the special attention of the members of the Prognostics and Health Management (PHM) community.

We now proceed to review some of the most frequently used Bayesian tracking methods in fault diagnostic and condition monitoring algorithms: KF, UKF, and PFs.

3.2.1 Linear Bayesian Processor: The Kalman Filter

The *Kalman Filter* (KF), proposed in 1960 by Rudolf E. Kálmán [107], triggered a huge revolution by that time in a wide range of disciplines such as signal processing, communication networks and modern control theory. It offers an analytic expression for the solution of the optimal filtering problem of linear Gaussian systems. Indeed, the structure of Equations (3.1) and (3.2) in this setting is

$$x_k = A_{k-1}x_{k-1} + B_{k-1}u_{k-1} + \omega_{k-1} \tag{3.5}$$

$$y_k = C_k x_k + \nu_k \tag{3.6}$$

with matrices $A_k \in \mathcal{M}_{n_x \times n_x}(\mathbb{R})$, $B_k \in \mathcal{M}_{n_x \times n_u}(\mathbb{R})$, $C_k \in \mathcal{M}_{n_y \times n_x}(\mathbb{R})$ and $D_k \in \mathcal{M}_{n_y \times n_u}(\mathbb{R})$. The random vectors $\omega_k \sim \mathcal{N}(0, Q_k)$ and $\nu_k \sim \mathcal{N}(0, R_k)$ are uncorrelated zero mean white process and measurement noises, respectively.

Provided the system is linear and Gaussian, if x_0 has a Gaussian distribution and considering that the sum of Gaussian distributions remains Gaussian, the linearity of the system guarantees x_k to be Gaussian for all $k \geq 0$. In fact, since the state posterior distribution $p(x_k|y_{1:k})$ is always Gaussian, then we also know that the conditional expectation

$$\mathbb{E}_{p(x_k|y_{1:k})}\{x_k\} = \int_{\mathbb{R}^{n_x}} x_k p(x_k|y_{1:k}) dx_k \tag{3.7}$$

and the conditional covariance matrix

$$\mathbb{C}ov_{p(x_k|y_{1:k})}\{x_k\} = \int_{\mathbb{R}^{n_x}} (x_k - \mathbb{E}_{p(x_k|y_{1:k})})(x_k - \mathbb{E}_{p(x_k|y_{1:k})})^T p(x_k|y_{1:k}) dx_k \tag{3.8}$$

are the only elements that fully determine the state posterior distribution.

The KF is a recursive algorithm that provides an optimal unbiased estimate of the state of a system – under the aforementioned setting – in the sense of minimising the mean squared error. It recursively updates the conditional expectation and covariance matrix as new measurements are acquired sequentially with each sample time. In this setting, the minimum mean squared error estimate, $\hat{x}_k^{MMSE}$, coincides with the conditional expectation, so it is directly obtained from the KF's recursion.

Let us denote the conditional expectation and covariance matrix at time k given measurements until time $\bar{k}$ by $\hat{x}_{k|\bar{k}}$ and $P_{k|\bar{k}}$, respectively. The KF algorithm is summarised below.

Algorithm 6: Kalman Filter

Initial Conditions:

$$\hat{x}_{0|0}, \; P_{0|0}$$

Prediction:

$$\begin{aligned} \hat{x}_{k|k-1} &= A_{k-1}\hat{x}_{k-1|k-1} + B_{k-1}u_{k-1} \\ P_{k|k-1} &= A_{k-1}P_{k-1|k-1}A_{k-1}^T + Q_{k-1} \end{aligned}$$

Innovation:

$$\begin{aligned} e_k &= y_k - C_{k-1}\hat{x}_{k|k-1} \\ S_k &= CP_{k|k-1}C_{k-1}^T + R_k \end{aligned}$$

Kalman Gain:

$$K_k = P_{k|k-1}C_{k-1}^T S_k^{-1}$$

Update:

$$\begin{aligned} \hat{x}_{k|k} &= \hat{x}_{k|k-1} + K_k e_k \\ \tilde{P}_{k|k} &= (1 - K_k C_{k-1})P_{k|k-1} \end{aligned}$$

3.2.2 Unscented Transformation and Sigma Points: The Unscented Kalman Filter

When linearity and Gaussianity conditions are not fulfilled, the KF cannot ensure optimal state estimates in the mean square error sense. Two other alternative Bayesian tracking methods inspired on KF have been developed to offer sub-optimal solutions: the *Extended Kalman Filter* (EKF) and the *Unscented Kalman Filter* (UKF). On the one hand, the EKF proposes a first order approximation of the system model, to allow a setup where the standard KF could still be used. On the other hand, the UKF focuses on approximating the statistics of the random state vectors after undergoing non-linear transformations.

Indeed, the *Unscented Transform* (UT) [105, 106] is a method designed for approximating nonlinear transformations of random variables. Although some nonlinear functions can be approximated by truncating Taylor expansions, UT focuses on preserving some of the statistics related to the nonlinear transformation, using sets of deterministic samples called *sigma-points*. For illustrative purposes, let us suppose $x \in \mathbb{R}^{n_x}$ is a random variable with mean $\bar{x}$ and covariance matrix P_x, and $g(\cdot)$ is a nonlinear real-valued function. The objective is to approximate $y = g(x)$. The UT method states the following: Firstly, compute a set of $2n_x + 1$ deterministically chosen weighted samples, namely the sigma-points $\mathcal{S}_i = \{\mathcal{X}^{(i)}, \mathcal{W}^{(i)}\}$, $i = 0, \ldots, 2n_x$, such that

$$\mathcal{X}^{(i)} = \bar{x} \qquad i = 0 \tag{3.9}$$

$$\mathcal{X}^{(i)} = \bar{x} + \left(\sqrt{(n_x + \kappa)P_x}\right)_i \qquad i = 1, \ldots, n_x \tag{3.10}$$

$$\mathcal{X}^{(i)} = \bar{x} - \left(\sqrt{(n_x + \kappa)P_x}\right)_i \qquad i = n_x + 1, \ldots, 2n_x \tag{3.11}$$

where $\left(\sqrt{(n_x + \kappa)P_x}\right)_i$ is the ith column of the matrix square root of $(n_x + \kappa)P_x$. On the other hand, there are weights defined as

$$\mathcal{W}^{(i)} = \frac{\kappa}{n_x + \kappa} \qquad i = 0 \tag{3.12}$$

$$\mathcal{W}^{(i)} = \frac{1}{2(n_x + \kappa)} \qquad i = 1, \ldots, 2n_x \tag{3.13}$$

The variable i indexes the deterministic weighted samples $\{\mathcal{X}^{(i)}, \mathcal{W}^{(i)}\}_{i=0}^{2n_x}$ that approximate the first two moments of x as

$$\bar{x} = \sum_{i=0}^{2n_x} \mathcal{W}^{(i)} \mathcal{X}^{(i)}$$

$$P_x = \sum_{i=0}^{2n_x} \mathcal{W}^{(i)} (\mathcal{X}^{(i)} - \bar{x})(\mathcal{X}^{(i)} - \bar{x})^T$$

Weights satisfy $\sum_{i=0}^{2n_x} \mathcal{W}^{(i)} = 1$. On the other hand, κ is a parameter (commonly used to adjust higher order moments) that determines how far the sigma-points set are from the mean.

Applying the nonlinear transformation on each sigma-point

$$\mathcal{Y}^{(i)} = g(\mathcal{X}^{(i)}) \qquad i = 0, \ldots, 2n_x, \tag{3.14}$$

then the statistics of y can be approximated as

$$\bar{y} \approx \sum_{i=0}^{2n_x} \mathcal{W}^{(i)}\mathcal{Y}^{(i)}, \tag{3.15}$$

$$P_y \approx \sum_{i=0}^{2n_x} \mathcal{W}^{(i)}(\mathcal{Y}^{(i)} - \bar{y})(\mathcal{Y}^{(i)} - \bar{y})^T \tag{3.16}$$

$$P_{xy} \approx \sum_{i=0}^{2n_x} \mathcal{W}^{(i)}(\mathcal{X}^{(i)} - \bar{x})(\mathcal{Y}^{(i)} - \bar{y})^T \tag{3.17}$$

As it can be seen in the expressions above, the variable i now indexes the transformed weighted samples that define the mean, covariance, and cross-covariance matrices of the random variable y, respectively.

To avoid non-positive definite covariance matrices, weights in Equations (3.12) and (3.13) can be optionally modified as follows:

$$\mathcal{W}_m^{(i)} = \frac{\lambda}{n_x + \lambda} \qquad i = 0 \tag{3.18}$$

$$\mathcal{W}_c^{(i)} = \frac{\lambda}{n_x + \lambda} + (1 - \alpha^2 + \beta) \qquad i = 0 \tag{3.19}$$

$$\mathcal{W}_m^{(i)} = \mathcal{W}_c^{(i)} = \frac{1}{2(n_x + \lambda)} \qquad i = 1, \ldots, 2n_x \tag{3.20}$$

so that the set of weights $\{\mathcal{W}_m^{(i)}\}_{i=0}^{2n_x}$ are used to compute mean vectors, whereas the set of weights $\{\mathcal{W}_c^{(i)}\}_{i=0}^{2n_x}$ are used to compute covariance (or cross-covariance) matrices instead. Note that, in comparison to Equations (3.12) and (3.13), we have replaced the tuning parameter $\kappa \to \lambda = \alpha^2(n_x + \kappa) - n_x$, leading to a scaled form of the UT. The parameter α regulates the speediness of sigma-points around the mean vector and according to [31], typical values are set to $0.01 \leq \alpha \leq 1$. Similarly, κ is commonly chosen as 0 or $3 - n_x$, whereas β provides and additional degree of freedom for further adjustments if there is prior knowledge about the distribution of the random vector to be non-linearly transformed.

During the filtering stage, uncertainty arises from i) initial state condition, ii) state transition model (process noise), and iii) observation model (measurement noise). Hence, in order to incorporate all these uncertainty sources, the implementation of the UKF for a general nonlinear system (see Equations (3.1) and (3.2)) includes the usage of an augmented vector:

$$x_k^a = \begin{bmatrix} {x_k}^T & {\omega_k}^T & {\nu_k}^T \end{bmatrix}^T \tag{3.21}$$

In this sense, uncertainty is jointly propagated trough nonlinear transformations. The computational cost may be lowered by assuming that both the process and measurement noises are purely additive, which is very convenient in the context of online monitoring. The resulting system would be described as

$$x_k = f(x_{k-1}, u_{k-1}) + \omega_{k-1} \tag{3.22}$$

$$y_k = g(x_k, u_k) + \nu_k \tag{3.23}$$

with ω_k and ν_k zero mean process and measurement noises; with second moments denoted as Q_k and R_k, respectively. The UKF version for systems with additive sources of uncertainty is summarised in Algorithm 7.

Algorithm 7: Unscented Kalman Filter

Initial Conditions:

$$\hat{x}_{0|0} = \mathbb{E}_{p(x_{0|0})}\{x_{0|0}\} \tag{3.24}$$

$$P_{0|0} = \mathbb{E}_{p(x_{0|0})}\{(x_{0|0} - \hat{x}_{0|0})(x_{0|0} - \hat{x}_{0|0})^T\} \tag{3.25}$$

Weights:

$$\mathcal{W}^{(i)} = \frac{\kappa}{n_x + \kappa} \qquad i = 0 \tag{3.26}$$

$$\mathcal{W}^{(i)} = \frac{1}{2(n_x + \kappa)} \qquad i = 1, \ldots, 2n_x \tag{3.27}$$

State Sigma-points and Prediction:

$$\mathcal{X}^{(i)}_{k-1|k-1} = \hat{x}_{k-1|k-1} \qquad i = 0 \tag{3.28}$$

$$\mathcal{X}^{(i)}_{k-1|k-1} = \hat{x}_{k-1|k-1} + \left(\sqrt{(n_x + \kappa)P_{k-1|k-1}}\right)_i \qquad i = 1, \ldots, n_x \tag{3.29}$$

$$\mathcal{X}^{(i)}_{k-1|k-1} = \hat{x}_{k-1|k-1} - \left(\sqrt{(n_x + \kappa)P_{k-1|k-1}}\right)_i \qquad i = n_x + 1, \ldots, 2n_x \tag{3.30}$$

$$\mathcal{X}^{(i)}_{k|k-1} = f(\mathcal{X}^{(i)}_{k-1|k-1}, u_{k-1}) \qquad i = 0, \ldots, 2n_x \tag{3.31}$$

$$\hat{x}_{k|k-1} = \sum_{i=0}^{2n_x} \mathcal{W}^{(i)} \mathcal{X}^{(i)}_{k|k-1} \tag{3.32}$$

Covariance Prediction:

$$\tilde{\mathcal{X}}^{(i)}_{k|k-1} = \mathcal{X}^{(i)}_{k|k-1} - \hat{x}_{k|k-1} \qquad i = 0, \ldots, 2n_x \tag{3.33}$$

$$P_{k|k-1} = \sum_{i=0}^{2n_x} \mathcal{W}^{(i)} \tilde{\mathcal{X}}^{(i)}_{k|k-1} \tilde{\mathcal{X}}^{(i)}_{k|k-1}{}^T + Q_{k-1} \tag{3.34}$$

Measurement Sigma-points and Prediction:

$$\hat{\mathcal{X}}^{(i,j)}_{k|k-1} = \mathcal{X}^{(i)}_{k|k-1} \qquad i, j = 0 \tag{3.35}$$

$$\hat{\mathcal{X}}^{(i,j)}_{k|k-1} = \mathcal{X}^{(i)}_{k|k-1} + \left(\kappa\sqrt{Q_{k-1}}\right)_j \qquad i, j = 1, \ldots, n_x \tag{3.36}$$

$$\hat{\mathcal{X}}^{(i,j)}_{k|k-1} = \mathcal{X}^{(i)}_{k|k-1} - \left(\kappa\sqrt{Q_{k-1}}\right)_j \qquad i, j = n_x + 1, \ldots, 2n_x \tag{3.37}$$

$$\mathcal{Y}_{k|k-1}^{(i,j)} = g(\hat{\mathcal{X}}_{k|k-1}^{(i,j)}, u_k) \qquad i, j = 0, \ldots, 2n_x \tag{3.38}$$

$$\hat{y}_{k|k-1} = \sum_{i=0}^{2n_x} \mathcal{W}^{(i)} \sum_{j=0}^{2n_x} \mathcal{W}^{(j)} \mathcal{X}_{k|k-1}^{(i,j)} \tag{3.39}$$

Residual Prediction

$$\xi_{k|k-1}^{(i)} = \sum_{j=0}^{2n_x} \mathcal{W}^{(j)} (\mathcal{Y}_{k|k-1}^{(i,j)} - \hat{y}_{k|k-1}) \qquad i = 0, \ldots, 2n_x \tag{3.40}$$

$$R_{k|k-1}^{\xi\xi} = \sum_{i=0}^{2n_x} \mathcal{W}^{(i)} \xi_{k|k-1}^{(i)} \xi_{k|k-1}^{(i)}{}^T + R_{k-1} \tag{3.41}$$

Gain:

$$R_{k|k-1}^{\tilde{x}\xi} = \sum_{i=0}^{2n_x} \mathcal{W}^{(i)} \tilde{\mathcal{X}}_{k|k-1}^{(i)} \xi_{k|k-1}^{(i)}{}^T \tag{3.42}$$

$$\mathcal{K}_k = R_{k|k-1}^{\tilde{x}\xi} R_{k|k-1}^{\xi\xi}{}^{-1} \tag{3.43}$$

Update:

$$e_k = y_k - \hat{y}_{k|k-1} \tag{3.44}$$

$$\hat{x}_{k|k} = \hat{x}_{k|k-1} + \mathcal{K}_k e_k \tag{3.45}$$

$$P_{k|k} = P_{k|k-1} - \mathcal{K}_k R_{k|k-1}^{\xi\xi} \mathcal{K}_k^T \tag{3.46}$$

3.2.3 Sequential Monte Carlo methods: Particle Filters

Except from a reduced number of cases, it is generally not possible to obtain an analytic solution for the filtering problem. Nonlinear systems may either be approximated by linearised versions within bounded regions of the state space (e.g. EKF) or even characterised in terms statistical approximations of state transitions (e.g. the UKF). However, KF-based implementations can only offer a sub-optimal solution and, moreover, underlying uncertainties may even distribute differently from Gaussian probability density functions (PDFs). In this regard, Sequential Monte Carlo methods, also known as *Particle Filters* (PFs) [9, 58], provide an interesting alternative for the solution of the filtering problem in the context of nonlinear non-Gaussian systems. PFs approximate the underlying PDF by a set of weighted samples, recursively updated, providing an empirical distribution from where statistical inference is performed.

PFs efficiently simulate complex stochastic systems to approximate the posterior PDF by a collection of N_p weighted samples or *particles* $\{x_{0:k}^{(i)}, w_k^{(i)}\}_{i=1}^{N_p}$, $\sum_{i=1}^{N_p} w_k^{(i)} = 1$, such that

$$p(x_{0:k}|y_{1:k}) \approx \hat{p}_{N_p}(x_{0:k}|y_{1:k}) = \sum_{i=1}^{N_p} w_k^{(i)} \delta_{x_{0:k}^{(i)}}(x_{0:k}) \tag{3.47}$$

where $\delta_{x_{0:k}^{(i)}}(x_{0:k})$ is a Dirac delta distribution located at the ith state trajectory $x_{0:k}^{(i)}$. The weighting process is performed by applying *sequential importance resampling* (SIR) algorithms in two stages: sequential importance sampling and resampling. The resulting algorithm is commonly known as "bootstrap filter", after the celebrated paper by Gordon et al. [79]. A pseudo-code implementation for the PF is given as Algorithm 8, which includes a systematic resampling step [9] to limit the well-known degeneracy problem[1].

3.2.3.1 Sequential importance sampling

In practice, obtaining samples directly from $p(x_{0:k}|y_{1:k})$ is not possible since this PDF is seldom known exactly, hence an *importance sampling distribution* $\pi(x_{0:k}|y_{1:k})$ is introduced to generate samples from, since it is easier to simulate. This leads to the *sequential importance sampling* (SIS) approach, and to compensate for the difference between the importance density and the true posterior density, the unnormalised weights are defined as follows

$$\tilde{w}_k^{(i)} = \frac{p(x_{0:k}^{(i)}|y_{1:k})}{\pi(x_{0:k}^{(i)}|y_{1:k})} \tag{3.48}$$

where $w_k^{(i)} = \frac{\tilde{w}_k^{(i)}}{\sum_{i=1}^{N_p} \tilde{w}_k^{(i)}}$, $i = 1, \ldots, N$, are then defined as normalised weights. In most of the practical applications, the importance sampling distribution is conveniently chosen so that it admits a sample procedure by adopting $\pi(x_{0:k}|y_{1:k}) = \pi(x_{0:k}|y_{1:k-1})$, hence it can be factorised in a form similar to that of the updating PDF, that is, $\pi(x_{0:k}|y_{1:k-1}) = \pi(x_k|x_{k-1}, y_{1:k-1})\pi(x_{0:k-1}|y_{1:k-1})$. Thus, the unnormalised importance weight for the ith particle at time k can be rewritten as

$$\tilde{w}_k^{(i)} \propto \underbrace{\frac{p(x_{0:k-1}^{(i)}|y_{1:k-1})}{\pi(x_{0:k-1}^{(i)}|y_{1:k-1})}}_{\tilde{w}_{k-1}^{(i)}} \frac{p(x_k^{(i)}|x_{k-1}^{(i)})p(y_k|x_k^{(i)})}{\pi(x_k^{(i)}|x_{k-1}^{(i)}, y_{1:k-1})} \tag{3.49}$$

[1]During resampling, particles are either dropped or reproduced that may result in a loss of diversity of the particle paths [9]. If necessary, a control step for limiting degeneracy by quantifying the effective sample size (ESS) [114] may be incorporated before the resampling.

Moreover, $\pi(x_k|x_{k-1}, y_{1:k-1})$ is typically chosen so as to coincide with $p(x_k|x_{k-1})$. Considering that $w_{k-1}^{(i)} \propto \tilde{w}_{k-1}^{(i)}$, the Equation (3.49) simplifies as

$$\tilde{w}_k^{(i)} \propto w_{k-1}^{(i)} p(y_k|x_k^{(i)}) \tag{3.50}$$

3.2.3.2 Resampling

In the SIS framework, the variance of particle weights tends to increase as time goes on. Moreover, most of the probability mass ends concentrating in only a few samples. This problem, known as *sample degeneracy,* is addressed by including a *resampling* step, leading to the *sequential importance resampling* (SIR) algorithm.

A measure of particle degeneracy is provided in terms of the *effective particle sample size*:

$$\hat{N}_{eff}(k) = \frac{1}{\sum_{i=1}^{N_p} (w_k^{(i)})^2} \tag{3.51}$$

The resampling step [59, 120] samples the particle population using probabilities that are proportional to the particle weights. Resampling is applied if $\hat{N}_{eff} \leq N_{thres}$, with N_{thres} a fixed threshold.

Algorithm 8: SIR Particle Filter

Importance Sampling

for $i = 1, \ldots, N_p$ **do**

• Sample $x_k^{(i)} \sim \pi(x_k|x_{0:k-1}^{(i)}, y_{1:k})$ and set $x_{0:k}^{(i)} \triangleq (x_{0:k}^{(i)}, x_k^{(i)})$

• Compute the importance weights $\tilde{w}_k^{(i)} = w_{k-1}^{(i)} \cdot \frac{p(y_k|x_k^{(i)}) p(x_k^{(i)}|x_{k-1}^{(i)})}{\pi(x_k^{(i)}|x_{0:k-1}^{(i)}, y_{1:k})}$

end for

for $i = 1, \ldots, N_p$ **do**

• Normalise $w_k^{(i)} = \frac{\tilde{w}_k^{(i)}}{\sum_{i=1}^{N_p} \tilde{w}_k^{(i)}}$

end for

Resampling

if $\hat{N}_{eff} \leq N_{thres}$ **then**

for $i = 1, \ldots, N_p$ **do**

• Store the values of $x_{0:k}^{(i)}$ $\tilde{x}_{0:k}^{(i)} = x_{0:k}^{(i)}$

end for

for $i = 1, \ldots, N_p$ **do**

• Sample an index $j(i)$ distributed according to the discrete distribution satisfying $\mathbb{P}(j(i) = l) = w_k^{(i)}$ for $l = 1, \ldots, N_p$
• Update the samples as
$x_{0:k}^{(i)} = \tilde{x}_{0:k}^{(j(i))}$
end for
for $i = 1, \ldots, N_p$ **do**
• Reset the weights
$w_k^{(i)} = \frac{1}{N_p}$
end for
end if

3.3 Calculation of EOL and RUL

The calculation of the EOL and RUL of a degrading system requires long-term predictions to describe its future evolution. These long-term predictions require, in turn, a thorough understanding of the underlying degradation processes and the anticipated future usage, as well as an effective characterisation of all associated uncertainty sources. This characterisation becomes a key piece of information to asses the risk associated with the future operation of the system and the optimal time when specific predictive maintenance actions need to take place. In this regard, we recognise at least four primary uncertainty sources that need to be correctly quantified in order to rigorously calculate the EOL and RUL:

- structure and parameters of the degradation model;
- true health status of the system at the time when the algorithm is executed (related to accuracy and precision of acquired measurements);
- future operating profiles (i.e. characterisation of future system inputs);
- hazard zone of the system (commonly associated with a failure threshold).

A conceptual scheme for RUL and EOL calculation is given in Figure 3.2 below. Note that the calculation of RUL and EOL becomes functionally dependent of the above features which are, in principle, uncertain. As mentioned in Section 3.1.3, the consideration of uncertainty in prognostics is not a trivial task even when the forward propagation models are linear and the uncertain variables follow Gaussian distributions [37, 53]. This is because the combination of the previously mentioned uncertain quantities within a fault propagation model will generally render a non-linear function of the predicted states [159]. That is the main reason explaining the limited applicability of analytical methods for prognostics in real life applications [156, 158]. In contrast, sampling-based algorithms such as the previously explained *particle filters* (PF) [9, 60] are best-suited to most of the situations [205, 133, 137]. Next, the failure prognosis problem is presented, where the probability distributions for the EOL and RUL are mathematically determined.

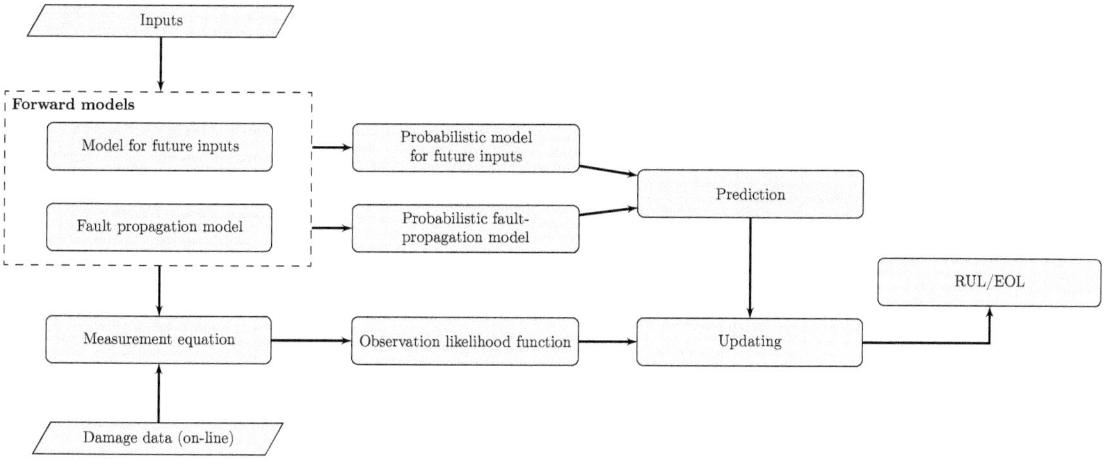

Figure 3.2: Conceptual scheme for RUL and EOL calculation.

3.3.1 The failure prognosis problem

Given that both the future conditions in which the systems operate as well as the dynamic models that seek to characterise them have various sources of uncertainty, the EOL of a system should therefore be expressed in terms of a random variable, which is denoted here as $\tau_{\mathcal{E}}$. If E_k is a binary random variable that indicates at each future time instant whether the event $\mathcal{E}$ = "System failure" occurs or not (which in the last case would be equivalent to saying that the complement event occurs, i.e. $\mathcal{E}^c$ = "System without failure"), then $\tau_{\mathcal{E}}$ is defined as [3]:

$$\tau_{\mathcal{E}}(k_p) := \inf\{k \in \mathbb{N} : \{k > k_p\} \wedge \{E_k = \mathcal{E}\}\} \tag{3.52}$$

where k_p is the current prediction time. In other words, $\tau_{\mathcal{E}}$ corresponds to the first future time instant k (since it is greater than k_p) in which there is a system failure. This definition is that of a random variable and, consequently, induces an associated probability distribution. According to [3], this probability distribution (applicable to any type of system) is equivalent to the probability that the system is in a faulty condition at time k and that before it has necessarily been operating under a healthy condition:

$$\mathbb{P}(\tau_{\mathcal{E}} = k) := \mathbb{P}\big(\underbrace{\{E_k = \mathcal{E}\}}_{\text{Faulty at } k}, \underbrace{\{E_j = \mathcal{E}^c\}_{j=k_p}^{k-1}}_{\substack{\text{Healthy from} \\ k_p \text{ up to } k-1}}\big) \tag{3.53}$$

If we further develop this expression by applying conditional probabilities, then we get

$$\mathbb{P}(\tau_{\mathcal{E}} = k) = \mathbb{P}\left(E_k = \mathcal{E} \middle| \{E_j = \mathcal{E}^c\}_{j=k_p}^{k-1}\right) \mathbb{P}\left(\{E_j = \mathcal{E}^c\}_{j=k_p}^{k-1}\right) \tag{3.54}$$

$$\vdots$$

$$= \mathbb{P}\left(E_k = \mathcal{E} \middle| \{E_j = \mathcal{E}^c\}_{j=k_p}^{k-1}\right) \prod_{j=k_p+1}^{k-1} \mathbb{P}\left(E_j = \mathcal{E}^c \middle| \{E_i = \mathcal{E}^c\}_{i=k_p}^{j-1}\right) \cancelto{1}{\mathbb{P}\left(E_{k_p} = \mathcal{E}^c\right)} \tag{3.55}$$

$$= \mathbb{P}\left(E_k = \mathcal{E} | \tau_{\mathcal{E}} \geq k\right) \prod_{j=k_p+1}^{k-1} \mathbb{P}\left(E_j = \mathcal{E}^c | \tau_{\mathcal{E}} \geq j\right) \tag{3.56}$$

$$= \mathbb{P}\left(E_k = \mathcal{E} | \tau_{\mathcal{E}} \geq k\right) \prod_{j=k_p+1}^{k-1} \left(1 - \mathbb{P}\left(E_j = \mathcal{E} | \tau_{\mathcal{E}} \geq j\right)\right) \tag{3.57}$$

However, since the binary random process $\{E_k\}_{k\in\mathbb{N}}$ depends on the system condition, the final expression is given by

$$\mathbb{P}(\tau_{\mathcal{E}} = k) = \int_{\mathbb{X}_{k_p+1:k}} \mathbb{P}\left(E_k = \mathcal{E} | x_k\right) \prod_{j=k_p+1}^{k-1} \left(1 - \mathbb{P}\left(E_j = \mathcal{E} | x_j\right)\right) p(x_{k_p+1:k} | y_{1:k_p}) dx_{k_p+1:k} \tag{3.58}$$

where $p(x_{k_p+1:k}|y_{1:k_p})$ corresponds to the joint probability density of the possible future trajectories of the system up to an instant k, with $k > k_p$, given the evidence in terms of measurements. The expression $\mathbb{P}(E_k = \mathcal{E}|x_k)$ refers to the probability, at the time instant k, with which the event $\mathcal{E}$ = "System failure" occurs given a system condition denoted by x_k.

There is a version of this probability distribution for continuous-time systems as well, for which the reader can refer to [3].

Since it is generally assumed that system failures are triggered when the condition indicators of such systems cross certain thresholds, then in those cases the probability $\mathbb{P}(E_k = \mathcal{E}|x_k)$ is defined as

$$\mathbb{P}(E_k = \mathcal{E}|x_k) = \mathbb{1}_{\{x\in\mathbb{R}^{n_x}: x > \bar{x}\}}(x_k) = \begin{cases} 1, & \text{if } x_k \geq \bar{x} \\ 0, & \sim \end{cases} \tag{3.59}$$

That is, if x_k exceeds an upper threshold $\bar{x}$ (it can also be set as a lower threshold depending on the context), the event $\mathcal{E}$ occurs with probability 1, which is the same as simply saying that it occurs with certainty.

Therefore, if failures are triggered by a threshold, as in most cases, then the statistics of the random variable $\tau_{\mathcal{E}}$ associated with the EOL time are determined by

$$\mathbb{P}(\tau_{\mathcal{E}} = k) = \int_{\mathbb{X}_{k_p+1:k}} \mathbb{1}_{\{x\in\mathbb{R}^{n_x}: x > \bar{x}\}}(x_k) \prod_{j=k_p+1}^{k-1} \left(1 - \mathbb{1}_{\{x\in\mathbb{R}^{n_x}: x > \bar{x}\}}(x_j)\right) p(x_{k_p+1:k}|y_{1:k_p}) dx_{k_p+1:k} \tag{3.60}$$

Once the probability distribution of $\tau_{\mathcal{E}}$ has been characterised, then decision-making becomes possible considering the risk of the future operation of the system. Each possible decision leads to consequences of varying severity, which must be analysed in proportion to the risk associated with incurring in each one of them.

There are even more general cases in which the failure condition is not determined by the crossing of thresholds. In this regard, the reader can refer to [3] for a detailed explanation.

3.3.2 Future state prediction

As we saw earlier, to calculate the probability that a system will fail in the future we must use Equation (3.60). However, this in turn requires:

- To define the time k_p from which the prediction is made onward.
- To define a failure threshold $\bar{x}$.
- To know the joint probability distribution $p(x_{k_p+1:k}|y_{1:k_p})$.

The first point is associated with the detection of an anomalous behaviour in the system that suggests the incursion of an imminent failure. The second point, on the other hand, will depend on the context and the specific system that is being monitored; the failure threshold can be defined either by an expert or through *post-mortem* analysis, which would require historical data. But what about the third point? Frequently, degradation processes can be represented through Markovian models as shown in Equations (3.1)–(3.2). In such cases we have

$$p(x_{k_p+1:k}|y_{1:k_p}) = \int_{\mathbb{X}_{0:k_p}} p(x_{0:k}|y_{1:k_p})dx_{0:k_p} \tag{3.61}$$

$$= \int_{\mathbb{X}_{0:k_p}} p(x_{k_p+1:k}|x_{0:k_p}, y_{1:k_p})p(x_{0:k_p}|y_{1:k_p})dx_{0:k_p} \tag{3.62}$$

$$= \int_{\mathbb{X}_{0:k_p}} \underbrace{\left(\prod_{j=k_p+1}^{k} p(x_j|x_{j-1})\right)}_{\text{Future}} \underbrace{p(x_{0:k_p}|y_{1:k_p})}_{\text{Present \& Past}} dx_{0:k_p} \tag{3.63}$$

Thus, the calculation of $p(x_{k_p+1:k}|y_{1:k_p})$ has a part related to the prediction itself, and another part that accounts for the current state of the system and how this state was reached. The latter is usually obtained using a Bayesian processor like those presented in Section 3.2. In this regard, the PF provides an expression for the joint probability density $p(x_{0:k_p}|y_{1:k_p})$, but the KF and its variants (such as the EKF and the UKF) provide the marginal probability density $p(x_{k_p}|y_{1:k_p})$. Therefore, $p(x_{k_p+1:k}|y_{1:k_p})$ can be calculated in either of the following two ways:

$$p(x_{k_p+1:k}|y_{1:k_p}) = \begin{cases} \int_{\mathbb{X}_{0:k_p}} \left(\prod_{j=k_p+1}^{k} p(x_j|x_{j-1})\right) p(x_{0:k_p}|y_{1:k_p})dx_{0:k_p} & \text{, if filtering with a PF.} \\ \int_{\mathbb{X}_{k_p}} \left(\prod_{j=k_p+1}^{k} p(x_j|x_{j-1})\right) p(x_{k_p}|y_{1:k_p})dx_{k_p} & \text{, if filtering with a KF variant.} \end{cases} \tag{3.64}$$

The two best-known methods for numerically solving the above integrals are *random walks* (sometimes referred to as "Monte Carlo simulations") and *particle-filtering-based prognostics*, which we present below.

1. **Random walks (Monte Carlo simulations)**

   This is the most popular method for predicting future events. In simple words, the method consists of simulating many (where "many" depends on the problem to be solved) possible future state trajectories of the system. Each trajectory crosses the failure threshold for the first time at a specific point in time. Then, with the information of all these instants of time, a histogram can be constructed, which corresponds to an approximation of the probability distribution for the failure time. An illustration of this process is provided in Figure 3.3. By the *Law of Large Numbers*, this approximation converges to the true probability distribution when the number of simulated state trajectories tends to infinity. The disadvantage of this method is that the computational cost it requires prevents its implementation in real-time monitoring systems.

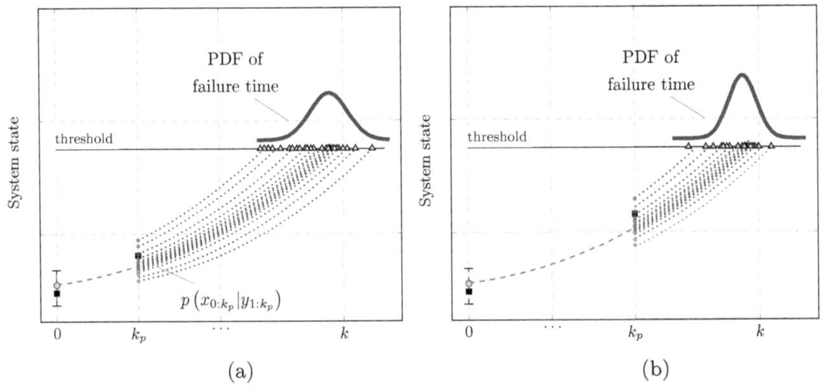

Figure 3.3: Conceptual Monte Carlo procedure for obtaining the PDF of the failure time. Panels (a) and (b) depict the process starting from different prediction times.

The method consists of the following. Suppose that the state of the system at time k_p is $x_{k_p}^{(i)}$ and is obtained by making one of the following approximations to $p(x_{0:k_p}|y_{1:k_p})$ or $p(x_{k_p}|y_{1:k_p})$, depending on the Bayesian tracking method used:

- $p(x_{0:k_p}|y_{1:k_p}) \approx \sum_{i=1}^{N} w_{k_p}^{(i)} \delta_{x_{0:k_p}^{(i)}}(x_{0:k_p})$, if filtering with a PF.
- $p(x_{k_p}|y_{1:k_p}) \approx \sum_{i=1}^{N} w_{k_p}^{(i)} \delta_{x_{k_p}^{(i)}}(x_{k_p})$, if filtering with a KF variant.

These approximations can be made by sampling either the probability distribution $p(x_{0:k_p}|y_{1:k_p})$ or $p(x_{k_p}|y_{1:k_p})$. If statistically independent samples are generated, using *inverse transform sampling*, for example, then in both cases $w_{k_p}^{(i)} = \frac{1}{N}$, $i \in \{1, \ldots, N\}$. Under other sampling approaches such as importance sampling, the weights $\{w_{k_p}^{(i)}\}_{i=1}^{N}$ might not be equal.

If we knew the future inputs of the system u_k, $k > k_p$, and, at each future time instant, we take a realisation of the process noise $\omega_k^{(i)} \sim \mathcal{N}(0, \sigma_\omega^2)$, then we can simulate random future state trajectories with the state transition Equation (3.1) as

$$x_{k+1}^{(i)} = f(x_k^{(i)}, u_k, \omega_k^{(i)}), \tag{3.65}$$

which is equivalent to sample $x_{k+1}^{(i)} \sim p(x_{k+1}|x_k^{(i)})$. If we start with $x_{k_p}^{(i)}$, we can generate the future state trajectory

$$x_{k_p:k}^{(i)} = \{x_{k_p}^{(i)}, x_{k_p+1}^{(i)}, \ldots x_k^{(i)}\} \tag{3.66}$$

Consequently, if this procedure is repeated N times, the probability distribution $p(x_{k_p+1:k}|y_{1:k_p})$ that contains information about the past, present, and future of the state of the system (see Equation (3.64)), is approximated as

$$p(x_{k_p+1:k}|y_{1:k_p}) \approx \begin{cases} \sum_{i=1}^{N} w_{k_p}^{(i)} \delta_{x_{0:k}^{(i)}}(x_{0:k}) & \text{, if filtering with a PF.} \\ \sum_{i=1}^{N} w_{k_p}^{(i)} \delta_{x_{k_p:k}^{(i)}}(x_{k_p:k}) & \text{, if filtering with a KF variant.} \end{cases} \tag{3.67}$$

Finally, the approximation of the probability distribution of $\tau_\mathcal{E}$ yields

$$\mathbb{P}(\tau_\mathcal{E} = k) = \int_{\mathbb{X}_{k_p+1:k}} \mathbb{1}_{\{x \in \mathbb{R}^{n_x} : x > \bar{x}\}}(x_k) \prod_{j=k_p+1}^{k-1} (1 - \mathbb{1}_{\{x \in \mathbb{R}^{n_x} : x > \bar{x}\}}(x_j)) p(x_{k_p+1:k}|y_{1:k_p}) dx_{k_p+1:k} \tag{3.68}$$

$$\approx \sum_{i=1}^{N} w_{k_p}^{(i)} \mathbb{1}_{\{x \in \mathbb{R}^{n_x} : x > \bar{x}\}} \left(x_k^{(i)}\right) \prod_{j=k_p+1}^{k-1} \left(1 - \mathbb{1}_{\{x \in \mathbb{R}^{n_x} : x > \bar{x}\}} \left(x_j^{(i)}\right)\right). \tag{3.69}$$

This last equation can be interpreted as: *the probability of system failure at time $\tau_\mathcal{E} = k$ can be approximated by the fraction of future state trajectories that crossed the threshold $\bar{x}$ for the first time at time k.*

Note that in the common case where the future inputs of the system $u_{k_p+1:k}$, $k > k_p$, were uncertain (i.e. u_k can be described as a stochastic process) then the generation of N future system state trajectories should consider one different realisation of the future system inputs per each future system state trajectory. In other words, in order to get the ith future system state trajectory ${x_{k_p+1:k}}^{(i)}$, with $i = 1, \ldots, N$, we require an ith simulation of the future system inputs ${u_{k_p+1:k}}^{(i)}$.

2. **Particle-filtering-based prognostics**

   The computational cost associated with the previous method is humongous and impractical for real-time decision-making processes (such as the ones found in condition-based maintenance schemes). This fact forces the implementation of simplified methods and algorithms for the characterisation of future evolution of the uncertainty related to system states.

A particle-filtering-based prognostic algorithm was originally presented in [136] and became the *de facto* failure prognostic algorithm in the literature. This algorithm is very similar to the one we presented above, but it differs in that, in this case, the transitions of future states are regularised with Epanechnikov kernels.

The pseudo-code for this prognostic algorithm is as follows:

0) Resample either $p(x_{0:k_p}|y_{1:k_p})$ or $p(x_{k_p}|y_{1:k_p})$ as appropriate, and get a set of N_θ equally weighted particles. This is,
   - $p(x_{0:k_p}|y_{1:k_p}) \approx \frac{1}{N_\theta}\sum_{i=1}^{N_p} \delta_{x_{0:k_p}^{(i)}}(x_{0:k_p})$, if filtering with a PF.
   - $p(x_{k_p}|y_{1:k_p}) \approx \frac{1}{N_\theta}\sum_{i=1}^{N_p} \delta_{x_{k_p}^{(i)}}(x_{k_p})$, if filtering with a KF variant.

Then, for each future time instant k, $k > k_p$, perform the following steps:

1) Compute the expected state transitions $x_k^{*(i)} = \mathbb{E}\{f(x_{k-1}{}^{(i)}, u_{k-1}, \omega_{k-1})\}$, $\forall i \in \{1, \ldots, N_\theta\}$, and calculate the empirical covariance matrix

$$\hat{S}_k = \frac{1}{N_\theta - 1}\sum_{i=1}^{N_\theta}[x_k^{*(i)} - \overline{x}_k^*][x_k^{*(i)} - \overline{x}_k^*]^T$$

   with $\overline{x}_k^* = \frac{1}{N_\theta}\sum_{i=1}^{N_\theta} x_k^{*(i)}$

2) Compute $\hat{D}_k$ such that $\hat{D}_k\hat{D}_k^T = \hat{S}_k$.
3) (Regularisation step) Update the samples as

$$x_k^{(i)} = x_k^{*(i)} + h_\theta \hat{D}_k \varepsilon_k^{(i)},\ \varepsilon_k^{(i)} \sim \mathcal{E}$$

   where $\mathcal{E}$ is the Epanechnikov kernel and h_θ corresponds to its bandwidth.

The hyper-parameters vector of the prognostic algorithm is defined as $\theta^T = \begin{bmatrix} N_\theta & h_\theta \end{bmatrix}$, and has to be adjusted. A formal methodology for algorithm design in this regard can be found in [2, 3].

Finally, the approximation of the probability distribution of $\tau_{\mathcal{E}}$ yields

$$\mathbb{P}(\tau_{\mathcal{E}} = k) = \int_{\mathbb{X}_{k_p+1:k}} \mathbb{1}_{\{x\in\mathbb{R}^{n_x}:x>\bar{x}\}}(x_k) \prod_{j=k_p+1}^{k-1} (1 - \mathbb{1}_{\{x\in\mathbb{R}^{n_x}:x>\bar{x}\}}(x_j))p(x_{k_p+1:k}|y_{1:k_p})dx_{k_p+1:k} \tag{3.70}$$

$$\approx \frac{1}{N_\theta}\sum_{i=1}^{N_p} \mathbb{1}_{\{x\in\mathbb{R}^{n_x}:x>\bar{x}\}}\left(x_k^{(i)}\right) \prod_{j=k_p+1}^{k-1}\left(1 - \mathbb{1}_{\{x\in\mathbb{R}^{n_x}:x>\bar{x}\}}\left(x_j^{(i)}\right)\right) \tag{3.71}$$

Similar to the previous algorithm, this last equation can be interpreted as: *the probability of system failure at time* $\tau_{\mathcal{E}} = k$ *can be approximated by the fraction of future state trajectories that crossed the threshold* $\bar{x}$ *for the first time at time* k.

Example 6 *Consider an exponential degradation process described by the following discrete state-transition equation:*

$$x_{k+1} = e^{-2\zeta k} x_k + \omega_k \tag{3.72}$$

where $x_k \in \mathbb{R}$ are discrete system states for $k \in \mathbb{N}$, $\zeta \in \mathbb{R}$ is the decay parameter, a scaling constant that controls the degradation velocity, and $\omega_k \in \mathbb{R}$ is the model error term which is assumed to be modelled as a zero-mean Gaussian distribution, that is $\omega_k \sim \mathcal{N}(0, \sigma_\omega^2)$. Let us now assume that the degradation process can be measured over time and that, at a certain time k, the measured degradation can be expressed as a function of the latent damage state x_k, as follows:

$$y_k = x_k + \nu_k \tag{3.73}$$

where $\nu_k \in \mathbb{R}$ is the measurement error, which is also assumed to be modelled as a zero mean Gaussian PDF with standard deviation σ_ν. The initial system condition is assumed to be uncertain and characterised as $x_0 \sim \mathcal{N}(0.9, 0.05^2)$, expressed in arbitrary units. We use synthetic data for y_k by generating it from Equations (3.72) and (3.73) considering $\sigma_\omega = 0.02$ and $\sigma_\nu = 0.01$. Additionally, the decay parameter is assumed to be $\zeta = 0.0004$.

The prognostics results at prediction time $k_p = 20$ are presented in Figure 3.4. A particle filter (PF) with $N_p = 1000$ particles is used for the filtering stage (i.e. when $k \leq k_p$). The probability distribution approximation $p(x_{0:k_p}|y_{1:k_p}) \approx \sum_{i=1}^{N_p} w_{k_p}^{(i)} \delta_{x_{0:k_p}^{(i)}}(x_{0:k_p})$ yielded from the PF at time k_p is used as initial condition for the prognostics stage. In this regard, prognostics are performed via

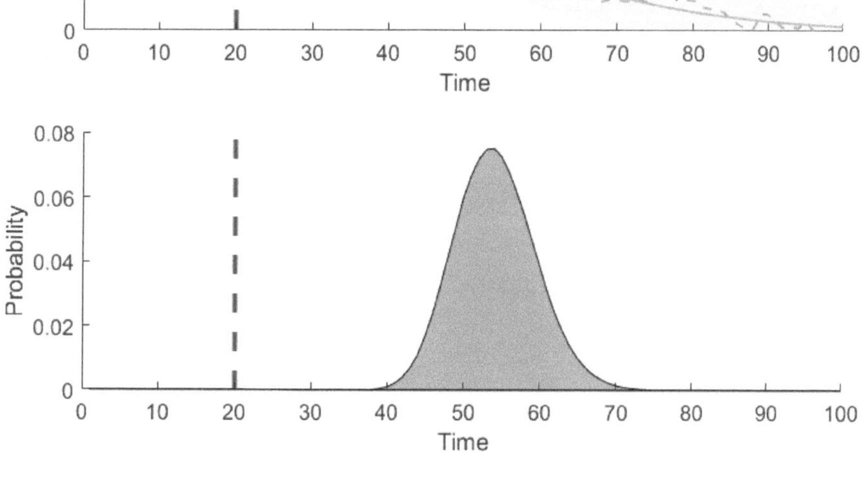

Figure 3.4: Failure time calculation using a random walk (Monte Carlo simulations) approach.

random walks (i.e. when $k > k_p$), for which $N = 10^6$ particles are resampled from the original PF approximation at k_p. Thus,

$$p(x_{0:k_p}|y_{1:k_p}) \approx \frac{1}{N}\sum_{i=1}^{N} \delta_{x_{0:k_p}^{(i)}}(x_{0:k_p})$$

The failure region is defined by fixing the threshold value $\bar{x} = 0.3$ in arbitrary units.

According to the results, we can expect a failure time at about 55 seconds (which can be visually obtained just by looking at the area of higher probability content in the PDF at Figure 3.4), which approximately coincides with the time where the real system gets into the failure domain. This simple example reveals the potential of the methodology presented in this chapter for important engineering applications such as failure time estimation.

3.4 Summary

This chapter presented a methodology for damage prognostics and time to failure estimation based on rigorous probabilistic Bayesian principles. First, an architecture for prognostics was developed; in this architecture, the prognostics problem was broken down into two important problems, namely, the sequential state estimation problem and the future state prediction. Since prognostics deals with the prediction of future events, a key feature of the above methodology was a systematic treatment of the various sources of uncertainty. In this sense, the Bayesian framework on which the presented methodology is grounded has shown to be extremely suitable to treat these uncertainties. It was illustrated that, in order to accurately calculate the probability distribution of the time of failure, the prediction needs to be posed as an uncertainty propagation problem that can be solved using a variety of statistical techniques. Some of these methods were briefly discussed, and algorithms were provided for both state estimation and time to failure prediction.

4

Parameter Identification Based on Conditional Expectation

Elmar Zander,[1,*] *Noémi Friedman*[2] and *Hermann G. Matthies*[1]

Parameter identification is an important issue in many scientific and technical disciplines. We present here a technique that updates our knowledge of the system parameters based on the so-called conditional expectation. It will be shown that for linear systems with normally distributed uncertainties this reproduces exactly the Bayes' posterior and is thus a non-linear extension of the Kalman filter.

4.1 Introduction

Estimation of model parameters is a very common problem in the natural and technical sciences. To eliminate problem specific details we can cast it into the following abstract form: Let q be a vector of parameters and u the complete state of the system; then there is a relation

$$A(u; q) = 0 \tag{4.1}$$

where A is a model of the system, given, for example, by a set of algebraic or differential equations. If the model is well-posed in the sense of Hadamard [84], then there is a unique solution for u given the parameters q and this dependence is continuous in q – that is, small changes in q invoke only small changes in u.

In many cases of interest, the full state of the system $\mathbf{u}$ is not directly observable. Rather, we can observe some measurements performed on the system that give us data

$$y = M(u; q) + \epsilon, \tag{4.2}$$

which depends on the state u and possibly also on the parameters q and is usually contaminated by measurement noise ϵ. Here, M signifies a mathematical model of the measurement process.

Example 7 *A simple example for demonstration is the flow of groundwater where parameters like permeability or boundary conditions shall be inferred from measurements inside of the domain. A standard linear model for groundwater flow is given by Darcy's equation*

$$-\nabla \cdot (\kappa \nabla u) = f$$

[1] Technische Universität Braunschweig, Germany.
[2] Institute for Computer Science and Control (SZTAKI), Budapest, Hungary.
[*] Corresponding author: elmar@zandere.de

on the domain $\mathcal{D}$ with Neumann conditions

$$\kappa \nabla u = g$$

on the boundary $\partial\mathcal{D}$. Here, u is the hydraulic head, f describes the sinks and sources inside the domain, and g denotes the inflow and outflow via the boundary.

Let the hydraulic head be measured at N locations $x_1, \ldots, x_N$ in $\mathcal{D}$ and the parameters to be inferred $q = (\kappa, f, g)$, which are assumed to be constant on the domain for simplicity. The system model is then given by

$$A(u; q) = \begin{pmatrix} -\nabla \cdot (\kappa \nabla u) - f \\ (\kappa \nabla u - g)|_{\partial\mathcal{D}} \end{pmatrix}$$

and the measurement operator by[1]

$$M(u; q) = (u(x_1), \ldots, u(x_n))$$

Using the system model A, the measurement model M, and values of the parameters q we can make predictions about the expected outcome of the measurements. We call this prediction the system response or response surface $G(q)$. However, given some actually observed measurements y_m, most likely they will not be identical to the predicted measurements $y = G(q)$. This disagreement between measurement and prediction can have multiple causes of which the most important ones are the aforementioned measurement error ϵ, wrong values for the parameters q not consistent with the true parameters q_{true}, numerical errors in computing u or y, and finally, an inadequate or inaccurate model for A or M, maybe omitting important physical effects.

In this chapter, we will only deal with errors of the first and second kind, that is, we assume the models are adequate for the process under investigation and the computational errors are negligible compared to the other types of errors. For the treatment and incorporation of modelling errors we refer the interested reader to the literature; see, for example [179, 170].

The disagreement between y and y_m can be seen as nuisance, but also as an opportunity to infer better knowledge about the true values of the parameters. In classical parameter estimation procedures we can use this information to update our belief about the value of the parameters q. A common approach is to choose a new set of parameters q' by minimising the distance between the predicted value y and y_m with respect to some loss or cost function. However, this is very often an ill-posed problem as the minimiser is usually not uniquely determined. A typical way to go is to use some regularisation scheme, for example, restricting q' in some norm or the difference between q and q' in another.

Though those schemes can turn the problem into a well-posed one, they suffer from being somewhat arbitrary and there is usually no good reason to choose one regularisation scheme over another – at least from a modelling point of view, maybe from a computational viewpoint there is.

In this chapter, we choose the Bayesian point of view to which this book is also devoted, acknowledging our lack of knowledge by modelling the parameter not as a single, deterministic value, but as a random variable (see Section 4.1.1). Our uncertainty about the true value is then determined by the variance of the random variable. Point estimates for q can then be given by the mean or median of the random variable.

In order to distinguish deterministic values from random variables we will use small letters for the former and capital letters for the latter. So, Q would denote the random variable corresponding

[1] For a more realistic and elaborate example see Chapter 8.

to the parameters and Y the one for the measurements. The aim of the present chapter is therefore to present a method to compute a new random variable Q', such that actual measurements y_m of Y are taken into account and the probability distribution of Q' is adjusted accordingly.

4.1.1 Preliminaries—basics of probability and information

Note, that a thorough treatment of the topics discussed here would need some familiarity with measure theory, which we do not assume here and only give an intuitive (mathematicians would call it sloppy) introduction to the definitions and ideas, that are needed in later sections. For more formal definitions and exact theorems see, for example, [80].

The mathematical definition of a probability space consists of three ingredients: a set of elementary events Ω, a set of events $\mathcal{F}$ – mathematically this is a structure called a σ-algebra – and a probability measure $\mathbb{P}$. The elementary events $\omega \in \Omega$ are the singular outcomes that will happen by chance or are the possible underlying events that characterise our insufficient knowledge, depending on whether we model with the probability space truly random events (aleatory uncertainty) or lack of knowledge (epistemic uncertainty).

For practical purposes it is often not necessary – and often also not possible – to know which elementary event exactly happened, but rather whether it lay in some specific subset of Ω or not. Such a set is commonly called an *event*. As the notion of an "event" is not really suitable for the case that the set describes uncertain knowledge, we call it here a "hypothesis", meaning the proposition whose true but unknown value lies inside the given set.

This also simplifies the definition of the probability measure, as it has to be defined only for those events in $\mathcal{F}$. Mathematically, the probability measure $\mathbb{P} : \mathcal{F} \to \mathbb{R}^+$ is thus a function from the set of events $\mathcal{F}$ to the positive real numbers. Since from all possible events in Ω, one must have happened (or equivalently, one value of ω must be the true value), we have the condition $\mathbb{P}(\Omega) = 1$, which is also called the *normalisation condition.*

The notion of hypotheses or events as subsets of the sample space Ω is also connected with that of *information.* Practically, we are usually not interested in which elementary event exactly will happen, but rather in the answer to questions like "Will it rain tomorrow?" or "Will the structure withstand the load?" If we put all outcomes ω for which the answer is "yes" into a set, say A, we can only answer questions about the probability of those hypotheses, when this set A is an element of $\mathcal{F}$. From this it becomes clear, that the more refined the subsets in $\mathcal{F}$ are, the more detailed questions we can answer within our probability model.

4.1.1.1 Random variables

In most elementary texts on probability, a random variable is a variable that "somehow" takes on different values according to some probability distribution. Though for basic results in probability, such a casual notion will suffice, we need a more formal approach here.

In the formal, axiomatic treatment of probability after Kolmogorov [112] a *random variable*, say X, is a so-called measurable function from the probability space $(\Omega, \mathcal{F}, \mathbb{P})$ into the real numbers. That means that every "sensible" subset of the real numbers[2] U has a preimage in $\mathcal{F}$, that is, there is a set $A \in \mathcal{F}$ such that $X(A) = U$, or equivalently $X^{-1}(U) \in \mathcal{F}$.

By collecting all preimages of a random variable, we can thus also define a sigma algebra. This is often denoted by $\sigma(X)$, the sigma algebra generated by the random variable X. The sets in

[2]Mathematically, those sets are called the Borel sets, but we don't need that kind of mathematical rigour here.

$\sigma(X)$ define thus what kind of information can be differentiated by observing values of the random variable X.

In some contexts, random variables that take values in a finite dimensional vector space are called *random vectors* or if the target space is a more general topological vector space (like a Banach or Hilbert space) they are called *random elements.* We will generally not make this distinction here and call all of them random variables.

4.1.2 Bayes' theorem

Although a detailed introduction to Bayes' theorem and Bayesian inverse problems was provided in Chapter 1, a summary is provided here for the sake of consistency and unified notation.

As discussed in Chapter 1, epistemic uncertainties due to a lack of knowledge are not intrinsic to the natural or technical system under investigation itself, but rather characterise our state of knowledge about that system. We encode that knowledge into specifying probabilities for different hypotheses over that state or over the parameters describing the system. Gaining knowledge by observing or performing measurements using the system will consequently increase our knowledge and thus modify the probability we will assign to these hypotheses $H \in \mathcal{F}$. We will call the probability measure before we have the new information of the *prior* probability and after incorporation of the information of the *posterior* probability.

Let us say that the information that we have – maybe from gathering data by performing measurements on the system – is codified in some set $E \in \mathcal{F}$, also called the *evidence.* We are interested now in the probability of some hypothesis H, given that we have already learned the evidence E – this probability will be denoted by $\mathbb{P}(H|E)$. Now, all events for which H is true, given that we have the additional information E, have to be in the intersection $H \cap E$, and thus the probability of the hypothesis H given knowledge of the evidence E must be proportional to that of $H \cap E$. Because the probability of E given that E is known must trivially be 1, it becomes obvious that the constant of proportionality must be $1/\mathbb{P}(E)$. Thus, we can define the conditional probability of H given E by

$$\mathbb{P}(H|E) = \frac{\mathbb{P}(H \cap E)}{\mathbb{P}(E)}. \tag{4.3}$$

As this relation must also be true when the roles of H and E are interchanged, that is,

$$\mathbb{P}(E|H) = \frac{\mathbb{P}(E \cap H)}{\mathbb{P}(H)}, \tag{4.4}$$

it follows easily that

$$\mathbb{P}(H|E) = \frac{\mathbb{P}(E|H)\mathbb{P}(H)}{\mathbb{P}(E)}, \tag{4.5}$$

which is the well-known *Bayes' theorem.* In the Bayesian framework, the terms appearing in Equation (4.5) have special meanings: $\mathbb{P}(H)$ is called the *prior probability* or just the *prior*, which is the state of knowledge before any measurements are available, while $\mathbb{P}(H|E)$, the *posterior probability* represents the state of knowledge after having learned the evidence. $\mathbb{P}(E|H)$ is called the *likelihood* and denotes the probability of measuring the evidence given that the hypothesis H is true.

The significance of the Bayes' theorem stems from the fact that it relates quantities that are easily calculable with quantities that are of interest. For example, it is generally easy to calculate the likelihood of getting the evidence E given that the hypothesis is true, since this is a direct consequence of the modelling assumptions. But what is really of interest (and this is in contrast to classical statistical methods such as maximum likelihood) is the probability of the hypothesis being true after acquiring the evidence, and this link can be established by Bayes' theorem. One possible drawback is that for the calculation of $\mathbb{P}(H|E)$, we need to make assumptions about the prior probability of the hypothesis $\mathbb{P}(H)$. The choice of the prior is often a subjective issue, which caused some controversy about the objectivity of Bayesian methods. However, in classical statistical methods there are also hidden assumptions which have to be made explicit in the Bayesian framework. Furthermore, there have been many advances to make the prior assumptions less subjective by using, for example, uninformative priors or empirical Bayes methods. For detailed accounts see [36, 74].

Example 8 *Referring back to the context given in the introduction where we had parameters $q \in \mathcal{Q}$ and measurements $y \in \mathcal{Y}$, the hypothesis H could correspond to the event that the parameters q are in some specific subset of the parameters space $\mathcal{Q}_H \subset \mathcal{Q}$, that is, $H = \{\omega | Q(\omega) \in \mathcal{Q}_H\} = Q^{-1}(\mathcal{Q}_H)$. The evidence E could correspond to the event that the measurements y lie in a specific subset of the space $\mathcal{Y}_E \subset \mathcal{Y}$, that is, $E = \{\omega | \mathcal{Y}_E(\omega) \in \mathcal{Y}_E\} = Y^{-1}(\mathcal{Y}_E)$.*

In many practical problems we need to condition on single measurements $Y_E = \{y_m\}$. Unfortunately, if the range of the measurements is continuous, the probability of such an event is zero in general. Since then the evidence $\mathbb{P}(E)$ and the likelihood $\mathbb{P}(E|H)$ is zero, the right hand side of Equation (4.5) is undefined. We can avoid this by conditioning on a finite set $\mathcal{Y}_E = \{y | d(y, y_m) \leq \Delta y\}$, where d is some distance function, for example, the Euclidean distance. Letting Δy go to zero and doing the equivalent thing for the hypothesis leads to

$$\pi(q|y) = \frac{\pi(y|q)\pi(q)}{\pi(y)}, \tag{4.6}$$

where π is the probability density of the given random variable. This is called the Bayes' theorem for densities.

Note that this limiting process can be somewhat problematic, as the density is not invariant under non-linear transformations of the measurement variable. For an example, the so-called Borel-Kolmogorov "paradox", see, for example, [102]. However, if the limit process is consistent with the measurement process, that is, the measured values are taken as is or only linearly transformed, the problems can be circumvented.

4.1.3 Conditional expectation

The paradoxes involved in using conditional probabilities can also be resolved by switching to conditional expectations instead. The classical way to define the conditional expectation is with respect to an event.

In the case that Q is a continuous random variable and $E \subset \mathcal{F}$ is an event, the conditional expectation of Q given E is given by

$$\mathbb{E}[Q|E] = \frac{1}{\mathbb{P}(E)} \int_E Q(\omega)\mathbb{P}(\mathrm{d}\omega). \tag{4.7}$$

Here, $\mathbb{P}(\mathrm{d}\omega)$ means the probability of the infinitesimally small set of size $\mathrm{d}\omega$ located at ω, which can also be written as $\pi(\omega)\mathrm{d}\omega$ if the $\mathbb{P}$ has continuous density π on Ω.

In case the event has zero probability, which again happens for events E of the form $Y^{-1}(\{y_m\})$, Equation (4.7) also becomes indefinite. The key in generalising this is to first rearrange the formula into

$$\int_Q \mathbb{E}[Q|E]\,\mathbb{P}(\mathrm{d}\omega) = \int_E Q(\omega)\mathbb{P}(\mathrm{d}\omega) \tag{4.8}$$

which is trivially true, if $\mathbb{P}(E) = 0$. Now, we define the conditional expectation as a random variable Q_Y by requiring that

$$\int_E Q_Y(\omega)\mathbb{P}(\mathrm{d}\omega) = \int_E Q(\omega)\mathbb{P}(\mathrm{d}\omega) \tag{4.9}$$

holds for all $E \in \sigma(Y)$. The conditional expectation Q_Y is very often denoted by $\mathbb{E}[Q|Y]$, but we prefer the notation Q_Y, because it makes it more evident, that this is a random variable and not a deterministic value.

It is apparent from the last equation that the conditional expectation $\mathbb{E}[Q|Y]$ depends only on the permitted sets $Q \in \sigma(Y)$ and is thus independent of transformations of the measurements Y. Counter-intuitive results such as the Borel-Kolmogorov paradox are thus impossible in this context and we will thus base our updating strategy rather on the conditional expectation than on conditional probabilities.

Note, that the definition of the conditional expectation as given above, only implicitly defines it, but does not say how to compute it. However, there is the closely related notion of the *minimum mean square error* estimator which allows efficient numerical approximations as will be shown in the next section.

4.2 The Mean Square Error Estimator

We call an *estimator* any possible function from the space of measurements $\mathcal{Y}$ to the space of unknowns $\mathcal{Q}$. From all of these functions – from which most would not even deserve the name estimator in the usual sense – we can select one by defining a measure on how close the estimates $\varphi(Y)$ come to the true value Q.

A common measure of closeness, which also has nice analytical properties, is the *mean squared error*, defined in the present setting by

$$e^2_{\mathrm{MSE}} = \mathbb{E}[\|Q - \varphi(Y)\|^2] \tag{4.10}$$

$$= \int_\Omega \|Q(\omega) - \varphi(Y(\omega))\|^2\, \mathbb{P}(\mathrm{d}\omega) \tag{4.11}$$

This assigns to each estimator φ the mean value of the squared error that is made, when trying to predict the parameter Q from any possible realisation of the measurement random variable Y. The minimum mean square error estimator $\hat{\varphi}$ is the one that minimises the mean square error e_{MSE}, and can be written as

$$\hat{\varphi} = \arg\min_{\varphi \in \mathcal{L}(\mathcal{Q},\mathcal{Y})} \mathbb{E}[\|Q - \varphi(Y)\|^2]. \tag{4.12}$$

Here, $\mathcal{L}(\mathcal{Q},\mathcal{Y})$ denotes the space of measurable functions, which includes a large amount of functions, for example, all the continuous functions from $\mathcal{Q}$ to $\mathcal{Y}$.

It can be shown that the MMSE estimator $\hat{\varphi}$ is the conditional expectation of Q given the measurements Y, that is,

$$\hat{\varphi}(Y) = Q_Y. \tag{4.13}$$

REMARK 1.1: A heuristic argument that helps to see the equality in (4.13) is, that for some random variable X, the real number u that minimises $\mathbb{E}[(X-u)^2]$ is given by $u = \mathbb{E}[X]$. For simple random variables, which take on only finitely many different values, this leads directly to the proof taking $X = Q \cdot \mathbb{1}_{Y=y}$ for some $y \in \mathcal{Y}$ and $\mathbb{1}_A$ being the indicator function for the set A. For a more rigorous treatment see, for example, [147] or [191]

4.2.1 Numerical approximation of the MMSE

In Equation (4.12) the minimisation is done over the whole space of measurable functions $\mathcal{L}(\mathcal{Q}, \mathcal{Y})$. Because this is in general an infinite dimensional space, we have to restrict it to a finite dimensional subspace $\mathcal{V}_\varphi \subset \mathcal{L}(\mathcal{Q}, \mathcal{Y})$ to make the problem computationally feasible. Let this space be defined by basis functions Ψ_γ, where $\gamma \subset \mathcal{J}$ is an index and $\mathcal{J}$ the set of indices. The functions Ψ_γ can be, for example, some sort of multivariate polynomials and the γ corresponding multi-indices but other function systems are also possible (e.g. tensor products of sines and cosines). An element φ of this function space has a representation as a linear combination

$$\varphi(y) = \sum_{\gamma \in \mathcal{J}} \varphi_\gamma \Psi_\gamma(y) \tag{4.14}$$

of these basis functions. Let us suppose for the moment that Q is a scalar-valued random variable and the coefficients φ_γ are therefore scalar-valued, too. Minimising expression (4.11) for φ then becomes equivalent to solving

$$\frac{\partial}{\partial \varphi_\delta} \mathbb{E}[(Q - \sum_\gamma \varphi_\gamma \Psi_\gamma(Y))^2] = 0 \tag{4.15}$$

for all $\delta \in \mathcal{J}$. Using the linearity of the derivative operator and of the expectation leads to

$$\sum_\gamma \varphi_\gamma \mathbb{E}[\Psi_\gamma(Y)\, \Psi_\delta(Y)] = \mathbb{E}[Q\, \Psi_\delta(Y)]. \tag{4.16}$$

As this is a linear system of equations we can rewrite it in the compact form

$$\mathbf{A}\varphi = \mathbf{b} \tag{4.17}$$

with

$$[\mathbf{A}]_{\gamma\delta} = \mathbb{E}[\Psi_\gamma(Y)\, \Psi_\delta(Y)], \tag{4.18}$$

$$[\mathbf{b}]_\delta = \mathbb{E}[Q\, \Psi_\delta(Y)] \tag{4.19}$$

coefficients φ_γ collected in the vector φ. Note, that for the actual computation some linear ordering needs to be imposed on the indices $\gamma \in \mathcal{J}$, but this is not essential here and can be left to the implementation.

If the unknown Q and the measurements Y are given by polynomials – like, for example, a Wiener polynomial chaos expansion (PCE) or a generalized polynomial chaos expansion (GPC) [75, 192] – and the function space $\mathcal{V}_\varphi$ consists of polynomials, the expectations could in principle be computed exactly using the polynomial algebra. However, this is computationally very expensive and non-trivial to implement. More efficient is to approximate $\mathbf{A}$ and $\mathbf{b}$ by numerical integration via

$$\mathbb{E}[\Psi_\gamma(Y)\, \Psi_\delta(Y)] \approx \sum_k w_k \Psi_\gamma(Y(\xi_k))\, \Psi_\delta(Y(\xi_k)) \tag{4.20}$$

and

$$\mathbb{E}[Q\,\Psi_\delta(Y)] \approx \sum_k w_k Q(\xi_k)\,\Psi_\delta(Y(\xi_k)). \tag{4.21}$$

Choosing an integration rule of sufficient polynomial exactness these relations can also be made exact.

REMARK 1.2: Suppose Y and Q are polynomials of total degree p_Y and p_Q, respectively, and φ has total degree p_φ. Then the maximum degree in the expression for $\mathbf{A}$ will be $2p_Y p_\varphi$ and a Gauss integration rule of order $p_Y p_\varphi + 1$ will suffice. For the computation of $\mathbf{b}$ a rule of order $\lceil (p_Q + p_Y p_\varphi + 1)/2 \rceil$ will suffice for exactness of the integration. In practice, usually integration rules of lower degree of exactness have shown to work well.

In the case that Q is a vector-valued random variable, which it usually is, the component functions φ_i of φ in expression (4.11) approximating the components Q_i for $i \in [1 \ldots n]$ are completely independent. So, the problem of computing the minimiser in Equation (4.15) essentially factors into n independent problems and can be done component-wise. In order to compute the estimator $\hat{\varphi}$ now for a vector valued Q the vectors $\boldsymbol{\varphi}_i$ and $\mathbf{b}_i$ (defined by Equation (4.19) for each Q_i) can be collected into matrices and the whole system

$$\mathbf{A}[\boldsymbol{\varphi}_1, \cdots, \boldsymbol{\varphi}_n] = [\mathbf{b}_1, \cdots, \mathbf{b}_n] \tag{4.22}$$

solved at once, which makes the process often more efficient, especially if factorisation of the matrix $\mathbf{A}$ is involved.

4.2.2 Numerical examples

In this section, we present two examples for the numerical approximation of the conditional expectation via the MMSE. In the first we take two random variables for Q and Y – created arbitrarily via some multivariate polynomials with random coefficients – and approximate the MMSE estimator from $\mathcal{Y}$ to $\mathcal{Q}$. It will be shown that the approximations $\varphi_p(Y)$ converge to Q if the measurements Y give enough information about the underlying probability space. If the measurements are not sufficiently informative it can be seen that only the mean square error is better minimised with increasing approximation order p. In the second example we have Y given by a non-linear measurement function plus some additive noise, for which the analytical computation of the conditional expectation and Bayes' posterior is possible. This allows better comparison and exact computation of the errors made in the MMSE approximation.

Example 9 *In the following examples, the sample space Ω is $\mathbb{R}^d$ with a standard Gaussian product measure, which can be thought of simply as a collection of d independent standard Gaussian random variables. If the number of parameters is denoted n, then the random variable Q is a function from Ω to $\mathbb{R}^n$, which we create artificially as a vector of n polynomials of total degree p_q with randomly generated coefficients in the d standard Gaussian random variables each. The number of measurements is m and the random variable $Y : \Omega \to \mathbb{R}^m$ is generated likewise as m multivariate polynomials in d variables up to total degree p_y. The non-linear MMSE was then used to approximate the "unknown" random vector Q by the "measurements" Y.*

Figure 4.1 shows the non-linear MMSE for $d = m = n = 2$ and different values of p_φ, the polynomial degree of the estimator $\hat{\varphi}$. Since $d = m$ the estimator can be expected to converge for large values of p_φ, that is, $\lim_{p_\varphi \to \infty} \|Q - \hat{\varphi}(Y; p_\varphi)\| = 0$. This can be seen in the figures by noting *that the crosses (x), denoting the approximated values $\hat{Q} = \hat{\varphi}(Y; p_\varphi)$, are increasingly better centered in the circles (o), denoting the true values of Q, in the sequence of increasing p_φ.*

In the two leftmost graphs in Figure 4.2, one can see the MMSE estimation with two-dimensional sample and parameter space, like in the previous example, but only one measurement, that is, $m = 1$. As expected, the estimate can be only a one-dimensional subset, which is in the left figure for $p_\varphi = 1$, the linear estimator, a straight line. For the cubic estimator with $p_\varphi = 3$ in the middle figure, the estimate is non-linear, and matches better the shape of the original distribution. However, convergence cannot be achieved as there is just not enough information available in the measurements.

In the rightmost figure the parameters are $m = n = 3$, $d = 5$ and $p_\varphi = 4$. Even though, the number of measurements is the same as the number of parameters to estimate and the polynomial degree is relatively high, there is no apparent convergence, since the dimension of the sample space d is higher than m. In this setting, a measurement y is not sufficient to determine the exact event ω but only some subset of Ω that can have led to it, and thus the determination of the corresponding parameter q has still some remaining uncertainty. So, even for high value of p_φ no convergent approximation can be expected.

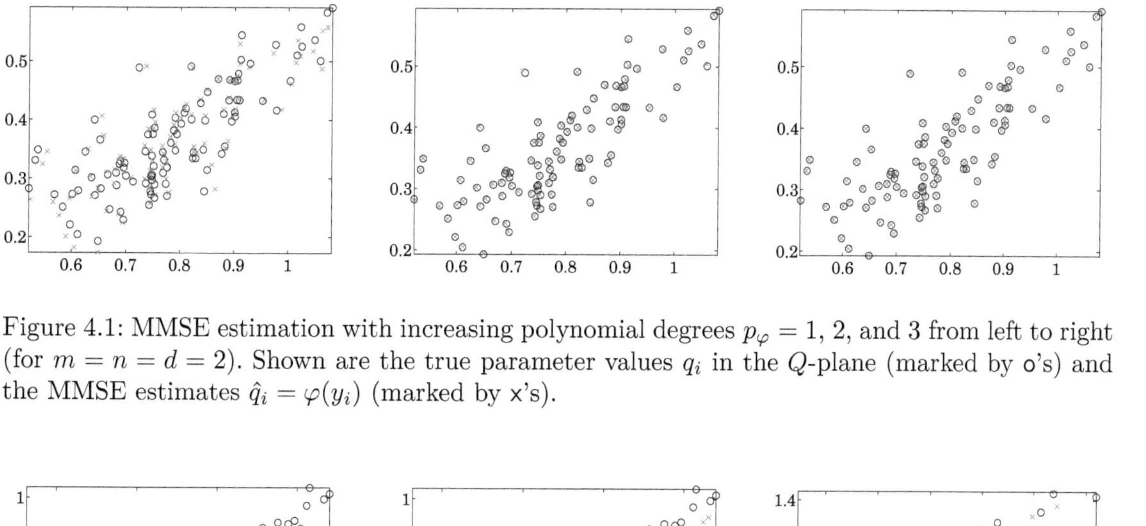

Figure 4.1: MMSE estimation with increasing polynomial degrees $p_\varphi = 1$, 2, and 3 from left to right (for $m = n = d = 2$). Shown are the true parameter values q_i in the Q-plane (marked by o's) and the MMSE estimates $\hat{q}_i = \varphi(y_i)$ (marked by x's).

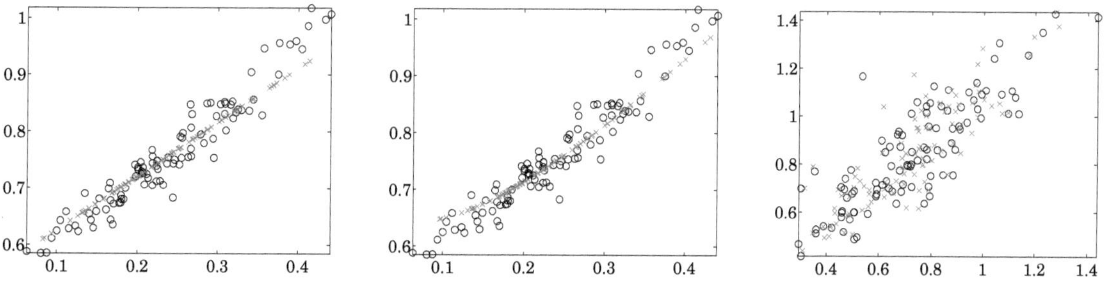

Figure 4.2: MMSE estimation with $m = 1, p_\varphi = 1$ (left), $m = 1, p_\varphi = 3$ (middle), and $m = 3, d = 5, p_\varphi = 4$ (right). True values Q are marked by o, and estimated values $\hat{Q} = \varphi(Y)$ are marked by x.

Example 10 *In this example, the measurement function is non-linear, but still conditional probabilities and conditional expectation can be computed analytically. Let Q have a uniform prior $Q \sim U[-1,1]$. Let E have a uniform distribution $E \sim U[-\delta,\delta]$ with $\delta = 0.5$. Let the system response G be given by $y = M(q) = \sin(q)$, such that $Y = \sin(Q) + E$. Then the MMSE is given by*

$$q = \hat{\varphi}(y) = \tfrac{1}{2}\left(\arcsin(\min(\sin(1), y+\delta)) + \arcsin(\max(-\sin(1), y-\delta))\right) \tag{4.23}$$

The MMSE and polynomial approximations of degree 1, 3, and 5 (even degree terms are zero since $\hat{\varphi}$ is an odd function) are shown in Figure 4.3 (left). The discontinuities in the prior and the error probability density functions introduce kinks in the MMSE at $\pm(\sin(1) - \delta)$, where the polynomial approximation is not very good. This can also be seen in Figure 4.3 (right), where the error is displayed, by reduced convergence at those points. This behaviour is mitigated for smooth prior and error distributions with much faster convergence of the polynomial approximations.

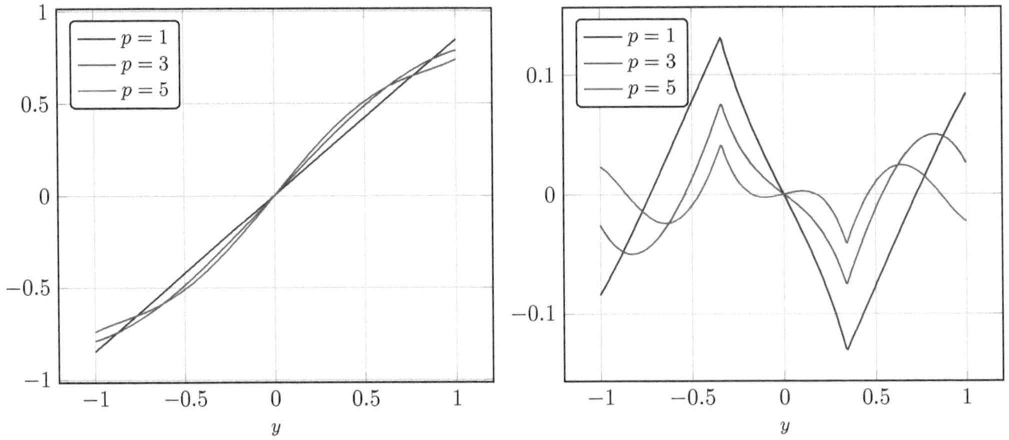

Figure 4.3: Approximation of the conditional expectation for different polynomial degrees (left) and difference to the true conditional expectation (right).

4.3 Parameter identification using the MMSE

The MMSE as derived in the last sections can be used directly for point estimates of the parameters q, that is, the posterior mean $q_m = \hat{\varphi}(y_m)$. However, the mean is often not informative enough, because it does not tell us anything about how certain this value is or how much trust we can put into it. In a Bayesian framework, we thus want to have a distribution which characterises the posterior density.

4.3.1 The MMSE filter

Suppose we have a parameter $q \in \mathcal{Q}$ and a corresponding measurement value $y = M(u;q) \in \mathcal{Y}$. From the section about the MMSE we know that $\hat{\varphi}(y)$ is the best estimate for q in the mean square sense. In terms of the random variables, we can therefore restate that as: for each ω, the conditional expectation $E(Q|Y)(\omega)$, or equivalently $\hat{\varphi}(Y(\omega))$, is the best estimate for Q given $Y(\omega)$. We can thus decompose Q into two components such that

$$\begin{aligned} Q &= \underbrace{(Q - Q_Y)} + Q_Y \\ &= \quad Q_Y^\top \quad + Q_Y \end{aligned}$$

where Q_Y is the conditional expectation and Q_Y^T is the residual part of Q. Q_Y and Q_Y^T are orthogonal, that is, uncorrelated, random variables. This can be easily seen, because Q_Y follows from the minimisation of $\mathbb{E}\|Q - \varphi(Y)\|^2$, and thus $Q_Y^T = Q - Q_Y$ must be orthogonal to $\varphi(Y)$ for every φ including $Q_Y = \hat{\varphi}(Y)$.[3]

Two uncorrelated random variables that are known to be (jointly) Gaussian are also independent. So, if Q_Y and Q_Y^T are Gaussian, then they are independent and thus Q_Y^T is also independent of Y and hence does not contain any information that can be inferred from the measurements Y. In this case Q_Y can be called the "predictable part" of the parameters Q by Y, and Q_Y^T the "unpredictable part", as there is no information in Y that can reduce our uncertainty in Q_Y. If the random variables are not Gaussian, then this is not strictly true, that is, Q_Y^T does contain information from Y. However, in many cases where the random variables are not too far from Gaussianity this assumption is a good approximation. Quantitative measures or estimates in this case are yet missing but currently under investigation.

The MMSE filter is now based on the following idea: the "predictable part" of Q is the component of the parameter uncertainty that can be removed by knowledge of an outcome of measurement Y. That means, if we have a concrete measurement y_m we can replace the Q_Y in Equation (4.24) below by the concrete prediction $q_m = \hat{\varphi}(y_m)$, the best estimate for the parameters. The new model Q' with reduced uncertainty is given by best estimate q_m plus remaining uncertainty Q_Y^T, that is,

$$Q' = q_m + Q_Y^T \tag{4.24}$$

This can be written as an update equation for Q in the form

$$Q' = Q + \hat{\varphi}(y_m) - \hat{\varphi}(Y) \tag{4.25}$$

or using conditional expectations

$$Q' = Q + \mathbb{E}[Q|Y = y_m] - \mathbb{E}[Q|Y], \tag{4.26}$$

which constitutes the so-called *MMSE filter.*

The implementation is straightforward given that the MMSE has been calculated beforehand. Depending on how the random variables Q and Y are represented in the code, the representation of Q' should be chosen accordingly. For example, if Q and Y are given as functions, then Q' could be defined as a new function that forwards its arguments to Q and Y respectively, like

$$Q' = \omega \mapsto \left(Q(\omega) + q_m - \sum_{\gamma \in \mathcal{J}} \varphi_\gamma \Psi_\gamma(Y(\omega)) \right) \tag{4.27}$$

[3]This can be visualised in the following way: Suppose in 3D space we have a point and a plane and we are looking for the point in the plane that minimises the distance to the original point. Then the residual, that is, the vector from the point to the minimiser is orthogonal to every vector in the plane. In our stochastic setting, the point corresponds to Q, the plane corresponds to the subspace in $\mathcal{Q}$ made up by all $\varphi(Q)$ and the Euclidean distance to the mean squared error. For a rigorous treatment in Hilbert spaces see, for example [123, Classical Projection Theorem].

If Q and Y are given via GPC expansions, then the GPC for Q' can be attained from the above, for example, by projection. Before we show numerical examples for the MMSE filter, we will first show that under certain conditions the filter is equivalent to the well-known *Kalman filter*, and can thus be regarded as a non-linear extension of the latter.

4.3.2 The Kalman filter

The Kalman filter is a numerical procedure for state estimation in dynamical systems [107]. It consists of a prediction step that describes how the distribution of the state estimate develops over one time step and a data assimilation step that incorporates new, but uncertain sensor data into the current state estimate. In the setting of parameter estimation, we are only interested in the data assimilation step, where the current state of the Kalman filter now corresponds to the parameter vector we want to estimate.

The model underlying the assimilation part of the Kalman filter can be summarised as follows: Let the current best estimate for the uncertain parameters be described by the random variable Q having a Gaussian distribution with mean q and variance C_Q, that is, $Q \sim N(q, C_Q)$. The observations shall be given by a measurement model $Y = HQ + E$ where H is the *observation matrix* and E is a mean-free Gaussian noise term with covariance C_E, i.e. $E \sim N(0, C_E)$. Then, after observing y_m, the best new estimate for the mean is given by

$$q' = q + K(y_m - Hq) \tag{4.28}$$

where $K = C_Q H^\top (C_E + H C_Q H^\top)^{-1}$ is the *Kalman gain*, and the variance of this updated estimate is given by

$$C_{Q'} = (I - KH)C_Q(I - KH)^\top + KC_E K^\top, \tag{4.29}$$

that is, then $Q' \sim N(q', C_{Q'})$. For a more extensive treatment we refer the interested reader to [123] and [130]. We show now that under the same conditions, that is, a linear observation operator and Gaussian uncertainties, the MMSE filter leads to exactly the same equations.

Since the Kalman filter is a linear filter, the only functions in the basis will be the constant and linear polynomials that we assign to the basis Ψ_i for $(i = 0 \dots m)$, that is,

$$\Psi_0(Y) = 1, \Psi_1(Y) = Y_1, \dots, \Psi_m(Y) = Y_m. \tag{4.30}$$

Since we assume Q to be vector valued, we have trial functions of the form

$$\varphi_i(Y) = \alpha_i + \beta_{i1}Y_1, \dots, \beta_{im}Y_m \tag{4.31}$$

with $i = 0 \dots n$. Collecting the α_i in a vector $\alpha = (\alpha_1, \dots, \alpha_n)^\top$ and the β_{ij} in a matrix K (the naming will become clear later) such that $(K)_{ij} = \beta_{ij}$, we can write this as

$$\varphi(Y) = \alpha + KY \tag{4.32}$$

In this setting, Equation (4.22) becomes

$$\begin{bmatrix} \mathbb{E}[1] & \mathbb{E}[Y^\top] \\ \mathbb{E}[Y] & \mathbb{E}[YY^\top] \end{bmatrix} \begin{bmatrix} \alpha^\top \\ K^\top \end{bmatrix} = \begin{bmatrix} \mathbb{E}[Q^\top] \\ \mathbb{E}[YQ^\top] \end{bmatrix}. \tag{4.33}$$

Using the expressions

$$\mathbb{E}[YY^\top] = C_Y + C_E + \mu_Y \mu_Y^\top \tag{4.34}$$

$$\mathbb{E}[YQ^\top] = C_{YQ} + \mu_Y \mu_Q^\top, \tag{4.35}$$

where $C_Y = \mathbb{E}[(Y - \mu_Y)(Y - \mu_Y)^\top]$ is the covariance matrix of Y and $C_{YQ} = \mathbb{E}[(Y - \mu_Y)(Q - \mu_Q)^\top]$ is the cross covariance matrix between Y and Q, we get

$$\begin{bmatrix} 1 & \mu_Y^\top \\ \mu_Y & C_Y + C_E + \mu_Y \mu_Y^\top \end{bmatrix} \begin{bmatrix} \alpha^\top \\ K^\top \end{bmatrix} = \begin{bmatrix} \mu_Q^\top \\ C_{YQ} + \mu_Y \mu_Q^\top \end{bmatrix}. \tag{4.36}$$

From the first row of Equation (4.36), we can express α as

$$\alpha = \mu_Q - K\mu_Y \tag{4.37}$$

and inserting this into the second row we attain

$$K = C_{QY}(C_Y + C_E)^{-1}, \tag{4.38}$$

whose expression corresponds to the K Kalman gain. The estimator then reads:

$$\hat{\varphi}(Y) = \mu_Q + K(Y - \mu_Y) \tag{4.39}$$

and the MMSE filter with the orthogonal decomposition becomes

$$Q' = Q - \hat{\varphi}(Y) + \hat{\varphi}(y_m) = Q + K(y_m - Y). \tag{4.40}$$

If we compute the mean and the variance of both sides of the last equation we get the usual update equations for the Kalman filter

$$\mu_Q' = \mu_Q + K(y_m - \mu_Y) \tag{4.41}$$

$$C_Q' = C_Q + KC_Y K^\top. \tag{4.42}$$

This means, in the linear case, the MMSE filter reduces to the Kalman filter, and the former can thus be seen as a non-linear generalization of the latter. In Section 4.3.3, a comparison of the performance between the MMSE and the Kalman filter for a simple non-linear example will be shown.

4.3.3 Numerical examples

In the following, we show two numerical examples for the MMSE filter with quadratic non-linearities. The examples were chosen such that the action of the filter can well be studied. For a realistic case study, the reader is referred to Chapter 8.

Example 11 *The one-dimensional case is well-suited for comparison to the Kalman filter. Let the system model be given be the quadratic relation*

$$A(u; q) = u - \alpha(q - q_0)^2 = 0 \tag{4.43}$$

with $q_0 = -3$ *and* $\alpha = 0.03$ *and a measurement operator given by the identity*

$$M(u; q) = u. \tag{4.44}$$

Thus, the relation between the parameters and the measurements is given by the response surface

$$G(q) = \alpha(q - q_0)^2. \tag{4.45}$$

As a prior, we assume a normal distribution with standard deviation 2, that is, $Q \sim N(0, 2^2)$*. The true value of the parameter is* $q_{\text{true}} = 3$*, of course taken to be unknown, and the measured value* $y_m = G(q_{\text{true}}) = 1.08$*, where no measurement error has been added.*

As the Bayes posterior is a combination of our prior belief and the information given by the data, it should be somewhere between the maximum of the prior (wider curve) and the black vertical line (the true parameter value) in Figure 4.4. The MMSE posterior for $p_\varphi = 1, 2, 3,$ *and* 4 *is shown as the narrower curve in Figure 4.4. It can be observed that in the linear case that corresponds to the Kalman filter, the posterior overshoots and lies even further away from the truth than the*

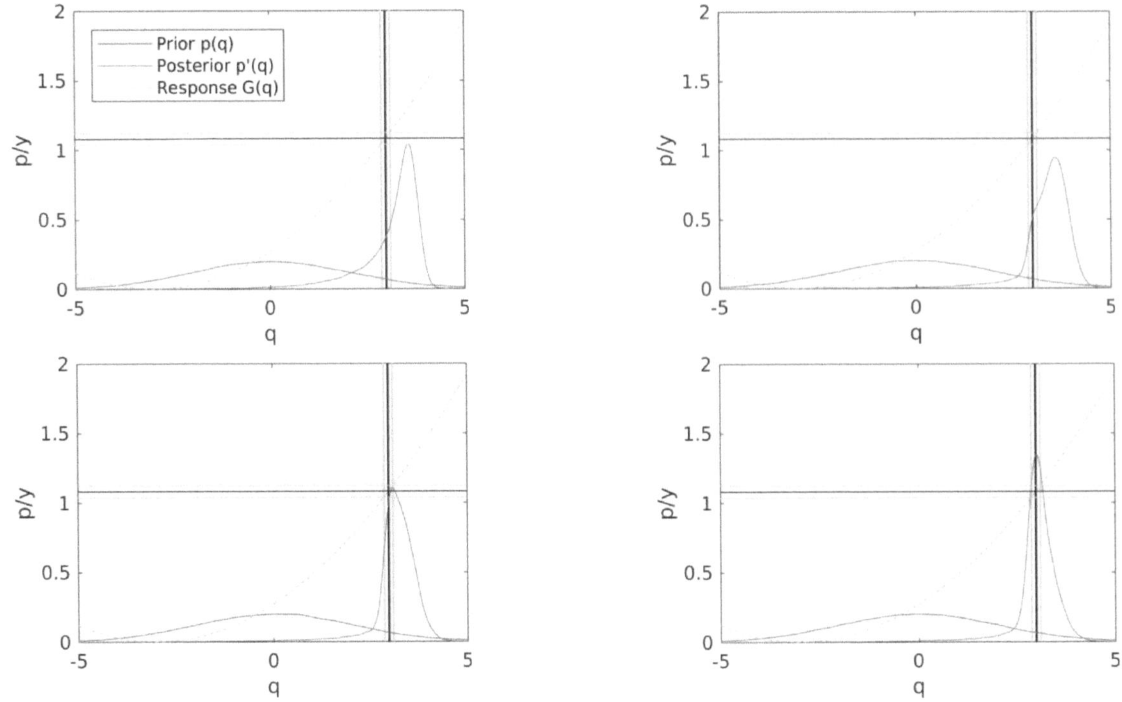

Figure 4.4: MMSE filter with different polynomial orders. The system response $G(q)$ is given by the dashed line, the prior by the wider curve and the MMSE filter posterior for different polynomials orders, that is, $p_\varphi = 1$, $p_\varphi = 2$, $p_\varphi = 3$, and $p_\varphi = 4$ from left to right and top to bottom, by the narrower curve. The measurement y_m is indicated by the horizontal black line, and plus and minus one standard deviation by the parallel grey lines. The corresponding parameter value is indicated by the vertical black line; again, the horizontal lines indicate one standard deviation.

prior would indicate. The reason is that the slope for the linear estimator is determined by sampling of the response surface where the prior is large. Since the response is relatively flat there, the inverse mapping has a steep slope and so the estimator will overestimate the true MMSE estimates. The higher-order estimators, taking non-linear terms into account, produce apparently much better results, especially for $p_\varphi = 3$ *and* $p_\varphi = 4$.

Example 12 *In the second example, we use a two-dimensional response surface and a Gaussian prior, such that depending on their parameters, the posterior can take very different shapes.*

The prior on Q *is given by normal distribution with mean at the center* $(0,0)^\top$ *and variance one, depicted by the coloured plane at* $z = 0$ *in Figure 4.5. The system model is given by*

$$A(u;q) = u - \alpha\|q - q_0\|^2 \tag{4.46}$$

with $q_0 = (1,3)^\top$ *and* $\alpha = 1.1$ *and the measurement operator again, like in the last example, by the identity* $M(u;q) = u$. *Thus the relation between the parameters and the response is*

$$G(q) = \alpha\|q - q_0\|^2. \tag{4.47}$$

The measurement was assumed to be $y_m = 12$ *here. Depending on the location of* q_0, *the value of* α *and the measurement* y_m, *very different shapes of the posterior distribution can be achieved: everything from circular shapes, over nearly Gaussian-like bumps, to very long, thin and straight shapes are possible. For the parameters described here, the resulting posterior is a slightly bent, "banana"-shaped bump, which can be seen in the left plot of Figure 4.5. The center plot shows sample points generated by a Metropolis-Hastings MCMC method, which follows the Bayes posterior very well.*

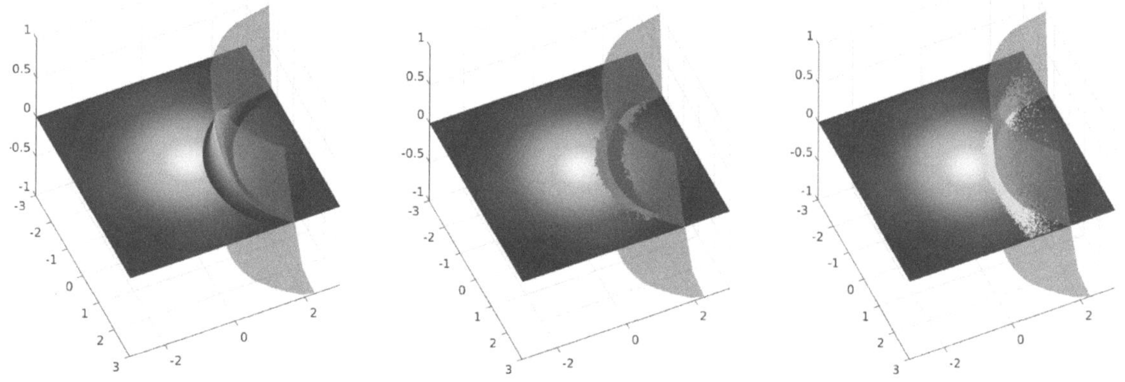

Figure 4.5: Comparison of the MMSE filter with $p_\varphi = 3$ (right) with a true Bayes posterior (left) and an MCMC simulation (center) for a two-dimensional example. The prior density is a standard normal indicated on the $z = 0$ plane. The response surface is given by a paraboloid shown by the light-blue transparent surface, shifted such a way that the measurement coincides with the zero plane. That way the set of parameters that are in agreement with the measurement are given by the intersection of those two surfaces, and the posterior should be close to this set (a circle) and to the maximum of prior density.

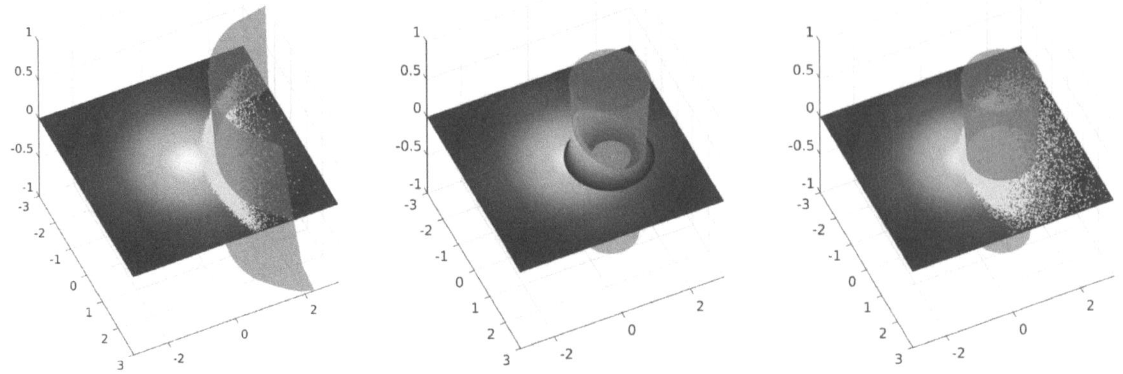

Figure 4.6: Continuation of Fig. 4.5. On the left the MMSE filter with $p_\varphi = 1$. In the center, the true posterior density for a Bayes posterior that is more difficult to approximate and on the right samples form the MMSE filter with $p_\varphi = 3$.

The plot on the right show samples generated from the cubic MMSE filter posterior, which also captures the center region well, however, it deviates at the bent edges of posterior. In Figure 4.6, on the left, samples from the linear MMSE filter are displayed. Here, the part of the posterior close to the center of the prior is also matched quite well; however, there are too many samples in the center of the response paraboloid than there should be.

The example in the center and on the right of Figure 4.6 shows results for the MMSE filter for a set of parameters ($\alpha = 3.3, q_0 = (0.3, 0.9)^\top, y_m = 3$) which gives rise to a more complicated, circular posterior. Here, the filter even with higher orders cannot capture the structure of the posterior density well. This behaviour that is, that the MMSE filter captures the Bayes posterior around the conditional mean very well, but deviates at the tails if the distribution is strongly curved or multi-modal, could be observed in other examples as well.

4.4 Conclusion

The MMSE filter has shown good performance in a variety of problems. Its advantage being that it is a deterministic method with calculable runtime, in contrast to, for example, Monte Carlo methods, with a non-predictable number of iterations for burn-in and convergence. The performance is generally superior to Kalman filters that use only linearisations (e.g. the EKF) as the MMSE filter can also take non-linearities into account. However, the performance can suffer during strongly non-linear mappings. Here, either different basis functions (like e.g. trigonometric functions or rational basis functions) should be used or approaches like the ensemble Kalman filter or MCMC methods could be considered. If only the conditional mean or the conditional variance is of interest, there are now also more efficient methods available (see e.g. [191]), that do not construct the MMSE as a function but directly compute its value for a given measurement.

Part II

Engineering Applications

5

Sparse Bayesian Learning and its Application in Bayesian System Identification

Yong Huang[1,*] and *James L. Beck*[2]

5.1 Introduction

In this chapter, a hierarchical sparse Bayesian learning (SBL) methodology is presented to perform sparse stiffness change inference for system identification and damage assessment purposes, based on the identified modal parameters from motion sensor data. The method aims to treat the inherent difficulties in model-based damage identification approaches, which involve comparing structural models identified from sets of measured modal data (natural frequencies and mode shapes) from a structure before and after possible damage.

One of the main difficulties is that existing model-based damage detection methods often require measurement information at locations corresponding to a large number of degrees of freedom (DOF), whereas, in reality, sensors are typically installed at only a limited number of locations, so the spatial distribution of the structural motion is not known completely. In addition, there are always some other modelling uncertainties involved, which may arise from many sources: simplifying approximations to develop the structural models; sensor noise; thermally-induced daily variations in structural stiffness, etc. Therefore, if the inverse problems in structural system identification and damage assessment are treated deterministically, they are typically ill-conditioned and ill-posed when using noisy, incomplete data; the uniqueness, existence, and robustness to noise of an inverse solution is not guaranteed. To deal with the ill-conditioning and ill-posedness in inverse problems, a common approach is to employ a regularised least-squares approach by adding a regularisation term to the data-matching term in the objective function to be optimised, which is often called Tikhonov Regularisation [181]. A regularised least-squares approach usually leads to a well-conditioned and well-posed deterministic optimisation problem; however, the relationship of the unique solution to the solution of the original inverse problem is uncertain.

The presence of substantial modelling uncertainties suggest that when solving inverse problems in structural system identification and damage assessment, one should not just search for a single "optimal" parameter vector to specify the structural model but rather attempt to describe the family of all plausible values of the model parameter vector that are consistent with both the observations and prior information. This leads us to consider inverse problems in structural system

[1] Harbin Institute of Technology (China).
[2] California Institute of Technology (USA).
* Corresponding author: huangyongthere@outlook.com

identification from a full Bayesian perspective, which provides a robust and rigorous framework due to its ability to account for model uncertainties. By treating the problem within a framework of plausible inference in the presence of incomplete information, the posterior probability distribution of the model parameters obtained from Bayes' theorem is used to quantify the plausibility of all models within a specified set of models, based on the available data. Therefore, the framework provides a promising way to locate structural damage, which may occur away from the sensor locations or be hidden from sight. Furthermore, being able to quantify the uncertainties of the structural model parameters accurately and appropriately is essential for a robust prediction of the future safety and reliability of the structure. The Bayesian framework for damage detection and assessment has been available for nearly two decades [189, 23, 200].

There is a fundamental trade-off in system identification and damage assessment due to the mathematical nature of the under-determined structural inverse problem based on the constraints imposed by the available data. This trade-off is between the spatial resolution of the damage locations and the reliability of the probabilistically-inferred damaged state. Ideally, we would like to treat each structural member as a substructure so that we can infer from the dynamic data which, if any, members have been damaged by a severe loading event or environmental deterioration. However, the information available from the structure's local network of sensors will generally be insufficient to support a member-level resolution of stiffness loss from damage, so larger substructures consisting of assemblages of structural members may be necessary in order to reduce the number of model parameters. In this case, defining a proper threshold to determine whether the damage features shift from their healthy state is also very important to alleviate false positive (false alarm) and false negative (missed alarm) detection; however, it is very challenging to establish a reliable threshold value in a rigorous manner for issuing a timely damage alarm [171].

A general strategy to alleviate this problem in structural system identification and damage assessment is to incorporate as much prior knowledge as possible in a Bayesian formulation in order to constrain the set of plausible solutions. For example, the engineer may wish to exploit the prior knowledge that structural stiffness change from damage typically occurs at a limited number of locations in a structure in the absence of its collapse. In this chapter, by exploiting this prior knowledge about the spatial sparseness of damage, the applicability of Bayesian methods is extended to produce reliable damage assessment even for high-dimensional model parameter spaces (higher-resolution damage localisation). This is accomplished by exploiting recent developments in SBL [183]. A specific hierarchical Bayesian model and the corresponding evidence maximisation strategy for Bayesian inference is utilised to promote spatial sparseness automatically in the inferred stiffness changes. The approach is consistent with the Bayesian Ockham's razor [22]. The sparseness of the inferred structural stiffness loss allows more robust damage localisation with higher resolution and the method issues a damage alarm without having to set any stiffness loss thresholds.

In Section 5.2, we present a general framework for SBL together with the Bayesian Ockham's razor, with a discussion of why they induce sparseness during Bayesian updating. In Section 5.3, a specific hierarchical Bayesian model is introduced that employs system modal parameters of the structure as latent variables. The Bayesian model is then used in an evidence maximisation procedure, yielding a Bayesian inference framework that lends itself to reliable, efficient and scale-invariant computation of the most plausible values of spatially-sparse stiffness reductions together with their uncertainty. The method exploits the information in the experimentally identified modal parameters from the current unknown state and the original healthy state. The effectiveness of the proposed method is illustrated in Section 5.4 where it is used to update a structural model during a calibration stage for an undamaged building (Example 13), and to accurately infer stiffness losses in a monitoring stage (Example 14), based on noisy, incomplete modal data.

5.2 Sparse Bayesian learning

5.2.1 General formulation of sparse Bayesian learning with the ARD prior

Given a set of I/O (input/output) data $\mathcal{D} = \{\hat{\mathbf{u}}, \hat{\mathbf{t}}\}$ for some system, suppose that the model prediction of the measured output is $\mathbf{t} = \boldsymbol{f}(\hat{\mathbf{u}}) + \mathbf{e} + \mathbf{m} \in \mathbb{R}^{N_o}$ involving a deterministic function $\mathbf{f}$ of the input vector $\hat{\mathbf{u}}$, along with uncertain model prediction error $\mathbf{e}$ and measurement noise $\mathbf{m}$. Assume that the function $\mathbf{f}$ is chosen as a weighted sum of N_p functions $\{\boldsymbol{\Theta}_j(\hat{\mathbf{u}})\}_{j=1}^{N_p}$:

$$\boldsymbol{f}(\hat{\mathbf{u}}) = \sum_{j=1}^{N_p} w_j \boldsymbol{\Theta}_j(\hat{\mathbf{u}}) = \boldsymbol{\Theta}(\hat{\mathbf{u}})\,\mathbf{w} \tag{5.1}$$

where $\boldsymbol{\Theta} = [\boldsymbol{\Theta}_{1,\ldots,}\boldsymbol{\Theta}_{N_p}]$ is a general $N_o \times N_p$ design matrix with column vectors $\{\boldsymbol{\Theta}_j(\hat{\mathbf{u}})\}$ and $\mathbf{w} = (w_1, \ldots, w_{N_p})^T$ is a corresponding coefficient parameter vector. Analysis of this model is facilitated by the adjustable parameters (or weights) $\mathbf{w}$ appearing linearly. The objective of the sparse Bayesian learning framework is to apply Bayes' theorem to infer the posterior probability density function (PDF) $p(\mathbf{w}|\hat{\mathbf{t}})$ such that $\mathbf{f}(\hat{\mathbf{u}})$ is a 'good' approximation of the system I/O relation based on data $\mathcal{D}$ and the parameter vector $\mathbf{w}$ is sparse, that is, many of its components are zero, thereby suppressing terms in Equation (5.1) that are inferred to be not needed.

The sparse Bayesian learning framework encodes a preference for sparser parameter vectors by making a special choice for the prior distribution for the parameter vector $\mathbf{w}$ that is known as the automatic relevance determination prior (ARD prior) [125, 183]:

$$p(\mathbf{w}|\boldsymbol{\alpha}) = \prod_{j=1}^{N_p} p(w_j|\alpha_j) = \prod_{j=1}^{N_p} \mathcal{N}(w_j|0,\alpha_j) = \prod_{j=1}^{N_p} \left[(2\pi)^{-1/2}\alpha_j^{-1/2}\exp\left\{-\frac{1}{2}\alpha_j^{-1}w_j^2\right\}\right] \tag{5.2}$$

where the key to the model sparseness is the utilisation of the N_p independent hyper-parameters that are the components of vector $\boldsymbol{\alpha}$ where α_j is the prior variance for $w_j's$, thereby moderating the strength of the regularisation constraint from the Gaussian prior on w_j. Note that a very small value of α_j implies that the corresponding terms in Equation (5.1) with coefficient w_j has an insignificant contribution to the modelling of the measurements $\hat{\mathbf{t}}$, because it produces essentially a Dirac delta-function at zero for the prior for w_j, and so for its posterior.

Based on the principle of maximum information entropy [100, 102], the combination of the prediction error $\mathbf{e}$ and measurement noise $\mathbf{m}$ is modelled as a zero-mean Gaussian vector with co-variance matrix $\beta^{-1}\mathbf{I}_{N_o}$. This probability model gives the largest uncertainty for the combination of the prediction error $\mathbf{e}$ and measurement noise $\mathbf{m}$ subject to the first two moment constraints: $\mathbf{E}[e_k + m_k] = 0, \mathbf{E}\left[(e_k + m_k)^2\right] = \beta^{-1}, k = 1, \ldots, N_o$. Thus, one gets a Gaussian likelihood function for the I/O data

$$p\left(\hat{\mathbf{t}}|\hat{\mathbf{u}},\mathbf{w},\beta\right)=\left(2\pi\beta^{-1}\right)^{-\frac{N_o}{2}}\exp\left(-\frac{\beta}{2}\|\hat{\mathbf{t}}-\boldsymbol{\Theta}\left(\hat{\mathbf{u}}\right)\mathbf{w}\|_2^2\right)=\prod_{j=1}^{N_p}\mathcal{N}\left(\hat{\mathbf{t}}|\boldsymbol{\Theta}\left(\hat{\mathbf{u}}\right)\mathbf{w},\beta^{-1}\mathrm{I}_{N_o}\right) \tag{5.3}$$

The likelihood function expresses the probability of getting the measurements $\hat{\mathbf{t}}$ based on the prediction from the model for specified parameters $\mathbf{w}$ and β. For $(\boldsymbol{\alpha},\beta)\in\mathbb{R}_+^{N_p+1}$, a set of candidate model classes $\mathcal{M}(\boldsymbol{\alpha},\beta)$ is defined by the likelihood function in (5.3) and the prior PDF on $\mathbf{w}$ given by Equation (5.2) [22].

The posterior parameter distribution $p\left(\mathbf{w}|\hat{\mathbf{t}},\hat{\mathbf{u}},\boldsymbol{\alpha},\beta\right)$ given by model class $\mathcal{M}(\boldsymbol{\alpha},\beta)$ is computed via Bayes' theorem

$$p\left(\mathbf{w}|\hat{\mathbf{t}},\hat{\mathbf{u}},\boldsymbol{\alpha},\beta\right)=p\left(\hat{\mathbf{t}}|\hat{\mathbf{u}},\mathbf{w},\beta\right)p\left(\mathbf{w}|\boldsymbol{\alpha}\right)/p\left(\hat{\mathbf{t}}|\hat{\mathbf{u}},\boldsymbol{\alpha},\beta\right) \tag{5.4}$$

where

$$p\left(\hat{\mathbf{t}}|\hat{\mathbf{u}},\boldsymbol{\alpha},\beta\right)=\int p\left(\hat{\mathbf{t}}|\hat{\mathbf{u}},\mathbf{w},\beta\right)p\left(\mathbf{w}|\boldsymbol{\alpha}\right)d\mathbf{w} \tag{5.5}$$

is the evidence function of the model class $\mathcal{M}(\boldsymbol{\alpha},\beta)$. Since both the prior and likelihood for $\mathbf{w}$ are Gaussian and the likelihood mean $\boldsymbol{\Theta}(\hat{\mathbf{u}})\,\mathbf{w}$ is linear in $\mathbf{w}$, the posterior PDF can be expressed analytically as a multivariate Gaussian distribution $p\left(\mathbf{w}|\hat{\mathbf{t}},\hat{\mathbf{u}},\boldsymbol{\alpha},\beta\right)=\mathcal{N}\left(\mathbf{w}|\boldsymbol{\mu},\boldsymbol{\Sigma}\right)$ with

$$\begin{aligned}\boldsymbol{\Sigma}&=\left(\beta\boldsymbol{\Theta}^T\boldsymbol{\Theta}+\mathbf{A}^{-1}\right)^{-1}\\ \boldsymbol{\mu}&=\beta\boldsymbol{\Sigma}\boldsymbol{\Theta}^T\hat{\mathbf{t}}\end{aligned} \tag{5.6}$$

where the matrix $\mathbf{A}=diag\left(\alpha_j,\ldots,\alpha_{N_p}\right)$. The marginal posterior PDF $p\left(\mathbf{w}|\hat{\mathbf{t}},\hat{\mathbf{u}}\right)$ can be computed by integrating out the posterior uncertainty in $\boldsymbol{\alpha}$ and β as below. Based on the assumption that the posterior $p\left(\boldsymbol{\alpha},\beta|\hat{\mathbf{t}},\hat{\mathbf{u}}\right)$ is highly peaked at the MAP (maximum *a posteriori*) value $\left\{\tilde{\boldsymbol{\alpha}},\tilde{\beta}\right\}$, we treat $\{\boldsymbol{\alpha},\beta\}$ as 'nuisance' parameters and integrate them out by applying Laplace's asymptotic approximation [26]:

$$p\left(\mathbf{w}|\hat{\mathbf{t}},\hat{\mathbf{u}}\right)=\int p\left(\mathbf{w}|\hat{\mathbf{t}},\hat{\mathbf{u}},\boldsymbol{\alpha},\beta\right)p\left(\boldsymbol{\alpha},\beta|\hat{\mathbf{t}},\hat{\mathbf{u}}\right)d\boldsymbol{\alpha}d\beta\approx p\left(\mathbf{w}|\hat{\mathbf{t}},\hat{\mathbf{u}},\tilde{\boldsymbol{\alpha}},\tilde{\beta}\right) \tag{5.7}$$

where:

$$\left\{\tilde{\boldsymbol{\alpha}},\tilde{\beta}\right\}=\arg\max_{[\boldsymbol{\alpha},\beta]}p\left(\boldsymbol{\alpha},\beta|\hat{\mathbf{t}},\hat{\mathbf{u}}\right)=\arg\max_{[\boldsymbol{\alpha},\beta]}\left\{p\left(\hat{\mathbf{t}}|\hat{\mathbf{u}},\boldsymbol{\alpha},\beta\right)p\left(\boldsymbol{\alpha}\right)p\left(\beta\right)\right\} \tag{5.8}$$

If flat, non-informative prior PDFs are assigned on hyper-parameters $\boldsymbol{\alpha}$ and β, the optimisation over $\boldsymbol{\alpha}$ and β is equivalent to maximising the evidence function $p\left(\hat{\mathbf{t}}|\hat{\mathbf{u}},\boldsymbol{\alpha},\beta\right)$. Two optimisation algorithms have been proposed to find the MAP values $\tilde{\boldsymbol{\alpha}}$ and $\tilde{\beta}$. One is Tipping's original iterative algorithm [183] , which we call the "Top-down" algorithm [96]. Another one is Tipping and Faul's "Fast Algorithm" [184], which we call the "Bottom-up" algorithm. It is found that the maximisation in (5.8) causes many hyper-parameters α_j to approach infinity during the learning process, thereby producing a sparse model $\mathbf{w}$. This is an instance of the Bayesian Ockham's razor [81, 125, 22], which will be discussed further in Section 5.2.2.

Having found the MAP estimates $\{\tilde{\alpha}, \tilde{\beta}\}$, our approximation to the robust posterior predictive distribution of the system response $\mathbf{t}^* \in \mathbb{R}^{N_o{}^*}$ for a new input $\mathbf{u}^*$ is:

$$\begin{aligned} p(\mathbf{t}^*|\hat{\mathbf{t}}, \hat{\mathbf{u}}, \mathbf{u}^*) &= \int p(\mathbf{t}^*, \mathbf{w}, \alpha, \beta|\hat{\mathbf{t}}, \hat{\mathbf{u}}, \mathbf{u}^*) d\mathbf{w} d\alpha d\beta \\ &= \int p(\mathbf{t}^*|\mathbf{u}^*, \mathbf{w}, \alpha, \beta) p(\mathbf{w}, \alpha, \beta|\hat{\mathbf{t}}, \hat{\mathbf{u}}) d\mathbf{w} d\alpha d\beta \\ &= \int p(\mathbf{t}^*|\mathbf{u}^*, \mathbf{w}, \alpha, \beta) p(\mathbf{w}|\hat{\mathbf{t}}, \hat{\mathbf{u}}, \alpha, \beta) p(\alpha, \beta|\hat{\mathbf{t}}, \hat{\mathbf{u}}) d\mathbf{w} d\alpha d\beta \\ &\approx \int p(\mathbf{t}^*|\mathbf{u}^*, \mathbf{w}, \tilde{\alpha}, \tilde{\beta}) p(\mathbf{w}|\hat{\mathbf{t}}, \hat{\mathbf{u}}, \tilde{\alpha}, \tilde{\beta}) d\mathbf{w} \end{aligned} \tag{5.9}$$

This can be analytically computed and corresponds to a Gaussian distribution,

$$p(\mathbf{t}^*|\hat{\mathbf{t}}, \hat{\mathbf{u}}, \mathbf{u}^*) = \mathcal{N}(\mathbf{t}^*|\mu^*, \mathbf{\Sigma}^*) \tag{5.10}$$

where the mean and covariance matrix can be expressed in terms of the posterior mean and covariance matrix of $\mathbf{w}$ in Equation (5.6) as

$$\begin{aligned} \mu^* &= \mathbf{\Theta}(\mathbf{u}^*)\mu(\tilde{\alpha}, \tilde{\beta}) \\ \mathbf{\Sigma}^* &= \tilde{\beta}^{-1}\mathbf{I}_{N_o{}^*} + \mathbf{\Theta}(\mathbf{u}^*)^T \mathbf{\Sigma}(\tilde{\alpha}, \tilde{\beta})\mathbf{\Theta}(\mathbf{u}^*) \end{aligned} \tag{5.11}$$

This robust predictive PDF takes into account the posterior uncertainties of the model parameter vector $\mathbf{w}$ when making a probabilistic prediction for output $\mathbf{t}^*$ corresponding to input $\mathbf{u}^*$.

5.2.2 Bayesian Ockham's razor implementation in sparse Bayesian learning

Finding the optimal values $(\tilde{\alpha}, \tilde{\beta})$ of the hyper-parameters as in Equation (5.8) is a procedure corresponding to Bayesian model class selection [27] because it chooses the MAP model class among the set of candidates. For this optimal model class $\mathcal{M}(\tilde{\alpha}, \tilde{\beta})$, many of the terms in the linear-in-the-parameters expansion in Equation (5.1) are suppressed. This procedure automatically implements an elegant and powerful version of Ockham's razor, known as the Bayesian Ockham's razor. A recent interesting information-theoretic interpretation [22] shows that the evidence $p(\hat{\mathbf{t}}|\hat{\mathbf{u}}, \alpha, \beta)$ in Equation (5.5) explicitly builds in a trade-off between a data-fit measure for the model class and an information-theoretic measure of its complexity that quantifies the amount of information that the model class extracts from the data $\mathbf{t}$. This result is based on using Equation (5.4) in the expression for the normalisation of the posterior PDF

$$\begin{aligned} \log\left[p\left(\hat{\mathbf{t}}|\hat{\mathbf{u}}, \alpha, \beta\right)\right] &= \int \log\left[\frac{p\left(\hat{\mathbf{t}}|\hat{\mathbf{u}}, \mathbf{w}, \beta\right) p\left(\mathbf{w}|\alpha\right)}{p\left(\mathbf{w}|\hat{\mathbf{t}}, \hat{\mathbf{u}}, \alpha, \beta\right)}\right] p\left(\mathbf{w}|\hat{\mathbf{t}}, \hat{\mathbf{u}}, \alpha, \beta\right) d\mathbf{w} \\ &= \int \log\left[p\left(\hat{\mathbf{t}}|\hat{\mathbf{u}}, \mathbf{w}, \alpha, \beta\right)\right] p\left(\mathbf{w}|\hat{\mathbf{t}}, \hat{\mathbf{u}}, \alpha, \beta\right) d\mathbf{w} \\ &- \int \log\left[\frac{p\left(\mathbf{w}|\hat{\mathbf{t}}, \hat{\mathbf{u}}, \alpha, \beta\right)}{p\left(\mathbf{w}|\alpha\right)}\right] p\left(\mathbf{w}|\hat{\mathbf{t}}, \hat{\mathbf{u}}, \alpha, \beta\right) d\mathbf{w} \\ &= \mathbf{E}\left[\log\left(p\left(\hat{\mathbf{t}}|\hat{\mathbf{u}}, \mathbf{w}, \alpha, \beta\right)\right)\right] - \mathbf{E}\left[\log\left[\frac{p\left(\mathbf{w}|\hat{\mathbf{t}}, \hat{\mathbf{u}}, \alpha, \beta\right)}{p\left(\mathbf{w}|\alpha\right)}\right]\right] \end{aligned} \tag{5.12}$$

where the expectations $\mathbf{E}[\cdot]$ are taken with respect to the posterior PDF $p(\mathbf{w}|\hat{\mathbf{t}}, \hat{\mathbf{u}}, \boldsymbol{\alpha}, \beta)$. The first term in Equation (5.12) is the posterior mean of the log likelihood function, which is a measure of the average data-fit of the model class $\mathcal{M}(\boldsymbol{\alpha}, \beta)$. The second term in (5.12) is the Kullback-Leibler information, or relative entropy, of the posterior relative to the prior, which is a measure of the model complexity (the amount of information gain about $\mathbf{w}$ from the data by $\mathcal{M}(\boldsymbol{\alpha}, \beta)$) and is always non-negative so it serves as a penalty term for the log evidence. The merit of Equation (5.12) is that it shows rigorously, without introducing *ad hoc* concepts, that the log evidence for $\mathcal{M}(\boldsymbol{\alpha}, \beta)$ explicitly builds in a trade-off between the data-fit of the model class and its information-theoretic complexity. As seen in expression (5.8), the evidence is the data-based controlling factor for the posterior probability of $\mathcal{M}(\boldsymbol{\alpha}, \beta)$. Therefore, the MAP model class $\mathcal{M}(\tilde{\boldsymbol{\alpha}}, \tilde{\beta})$ gives the optimal trade-off between the data-fit and information complexity among all model classes $\mathcal{M}(\boldsymbol{\alpha}, \beta)$.

It is found in [96] that the data-fit measure (the first term) decreases with a lower specified prediction accuracy (smaller β), and this is associated with sparser models (more α_j's tend to zero during the optimisation). This reduction is because smaller β allows more of the data misfit to be treated as prediction errors. At the same time, smaller β, with the associated smaller α_j's, produce a smaller Kullback-Leibler information (the second term in Equation (5.12)) between the posterior PDF and the prior PDF, indicating that less information is extracted from the measurements $\hat{\mathbf{t}}$ by the updated model and that the data-fit term is penalised less by the positive second term in Equation (5.12). On the other hand, larger β produces a model that fits the measurements with smaller error (larger data-fit measure in Equation (5.12)) but the model is under-sparse (more nonzero terms in Equation (5.1)) and so its relative entropy is large, that is, the second term in Equation (5.12) penalises data-fit more. In both cases, smaller and larger β, the models give a trade-off between data-fitting and model complexity (more sparseness corresponds to less model complexity) that may not be the optimal one that maximises the log evidence in Equation (5.12). The learning of hyper-parameters $\boldsymbol{\alpha}$ and β by maximising the evidence function $p(\hat{\mathbf{t}}|\hat{\mathbf{u}}, \boldsymbol{\alpha}, \beta)$ produces the optimal trade-off that causes many hyper-parameters α_j to approach zero with a reasonably large value of β, giving a model $\mathbf{w}$ that is both sufficiently sparse and fits the data vector $\hat{\mathbf{t}}$ well, that is, it gives the best balance between data fitting and model complexity. We can also say that sparse Bayesian learning automatically penalises models that "under-fit" or "over-fit" the associated data $\hat{\mathbf{t}}$. This is important in system identification applications, since over-fitting of the measured data often leads to unreliable response predictions since they depend too much on the detailed information of the specific data, that is, measurement noise and environmental effects.

5.3 Applying sparse Bayesian learning to system identification

5.3.1 Hierarchical Bayesian model class for system identification

Suppose that N_s sets of measured vibration time histories are available from a structure and N_m modes of the structural system have been identified by applying Bayesian modal identification [15] for each set of time histories so that we have a vector of identified (taking the MAP values) system natural frequencies $\hat{\boldsymbol{\omega}}^2 = \left[\hat{\omega}_{1,1}^2, \ldots, \hat{\omega}_{1,N_m}^2, \hat{\omega}_{2,1}^2, \ldots, \hat{\omega}_{N_s,N_m}^2\right]^T \in \mathbb{R}^{N_s N_m \times 1}$ and mode shapes $\hat{\boldsymbol{\Psi}} = \left[\hat{\boldsymbol{\Psi}}_{1,1}^T, \ldots, \hat{\boldsymbol{\Psi}}_{1,N_m}^T, \hat{\boldsymbol{\Psi}}_{2,1}^T, \ldots, \hat{\boldsymbol{\Psi}}_{N_s,N_m}^T\right]^T \in \mathbb{R}^{N_s N_m N_o \times 1}$, where $\hat{\boldsymbol{\Psi}}_{r,i} \in \mathbb{R}^{N_o}$ gives the identified components of the system mode shape of the mth mode ($m = 1, \ldots, N_m$) at the N_o observed degrees of freedom (DOFs) from the rth data segment ($r = 1, \ldots, N_s$). In real applications, the number of observed DOFs N_o is usually smaller that N_d, the number of DOFs of an appropriate structural model. To represent the actual underlying modal parameters of the structural system, system natural frequencies $\boldsymbol{\omega}^2 = \left[\omega_1^2, \ldots, \omega_{N_m}^2\right]^T \in \mathbb{R}^{N_m \times 1}$ and system mode shapes $\boldsymbol{\phi} = \left[\boldsymbol{\phi}_1^T, \ldots, \boldsymbol{\phi}_{N_m}^T\right]^T \in \mathbb{R}^{N_d N_m \times 1}$ are introduced at the same N_d DOFs as the structural model, but they are distinct from the model modal parameters, as will be seen clearly later. The introduction of the concept of system modal parameters is also useful in methods for solving the model updating problem because it avoids mode matching [23, 200]. A set of parameterised linear structural models with classical damping are chosen to produce normal modes of vibration where each model has the same known mass matrix $\mathbf{M} \in \mathbb{R}^{N_d \times N_d}$, which is assumed to be inferred with sufficient accuracy from structural drawings. Taking an appropriate sub-structuring (perhaps focusing on likely damage locations), the uncertain stiffness matrix $\mathbf{K} \in \mathbb{R}^{N_d \times N_d}$ is parameterised as a linear combination of $(N_\theta + 1)$ substructure stiffness matrices $\mathbf{K}_j, j = 0, 1, \ldots N_\theta$:

$$\mathbf{K}(\boldsymbol{\theta}) = \mathbf{K}_0 + \sum_{j=1}^{N_\theta} \theta_j \mathbf{K}_j \tag{5.13}$$

where $\mathbf{K}_j \in \mathbb{R}^{N_d \times N_d}, j = 1, \ldots, N_\theta$, is the prior choice of the jth substructure stiffness matrix, representing the nominal contribution of the jth substructure to the overall stiffness matrix $\mathbf{K}$. The corresponding stiffness scaling parameter $\theta_j, j = 1, \ldots, N_\theta$, is a factor that allows modification of the nominal jth substructure stiffness so it is more consistent with the real structure behaviour. The stiffness scaling parameter vector $\boldsymbol{\theta} = \left[\theta_1, \ldots, \theta_{N_\theta}\right]^T \in \mathbb{R}^{N_\theta}$ represents the structural model parameter vector to be updated by dynamic data. The stiffness matrices $\mathbf{K}_j$ could come from a finite-element model of the structure, then it would be appropriate to choose all $\theta_j = 1$ to give the most probable value *a priori* for the stiffness scaling parameter vector $\boldsymbol{\theta}$. For system identification and damage detection purposes, we will exploit the prior knowledge that damage-induced stiffness reductions typically occur in a limited number of locations in the absence of structural collapse, such as the connections of steel members where local buckling or weld fracture can occur. Suppose that there is a "calibration" value $\widehat{\boldsymbol{\theta}}$ available, which is either derived theoretically (e.g. from a finite-element model) or is identified from previous experimental data from the calibration (undamaged) state, then the potential change $\Delta\boldsymbol{\theta} = \boldsymbol{\theta} - \widehat{\boldsymbol{\theta}}$ is expected to be a *sparse vector* with relatively few non-zero components.

It is not assumed that the system modal parameters $\boldsymbol{\omega}^2$ and $\boldsymbol{\phi}$ satisfy exactly the eigenvalue problem of any given structural model specified by stiffness scaling parameters $\boldsymbol{\theta}$. There will always be modelling errors $\mathbf{e} = \left[\mathbf{e}_1^T, \ldots, \mathbf{e}_{N_m}^T\right] \in \mathbb{R}^{N_m N_d}$, where

$$\mathbf{e}_m = \left(\mathbf{K}(\boldsymbol{\theta}) - \omega_m^2 \mathbf{M}\right) \boldsymbol{\phi}_m, m = 1, \ldots, N_m \tag{5.14}$$

for the ith system mode and the structural model specified by $\boldsymbol{\theta}$. The uncertain eigenvalue equation error $\mathbf{e}$ therefore provides a bridge between the behaviours of the real system and a structural model. Equation (5.14) is used to create the following joint prior PDF for system modal parameters $\boldsymbol{\omega}^2$ and $\boldsymbol{\phi}$ and stiffness scaling parameters $\boldsymbol{\theta}$ [94]:

$$p\left(\boldsymbol{\omega}^2, \boldsymbol{\phi}, \boldsymbol{\theta} | \beta\right) = c_0 \left(\frac{2\pi}{\beta}\right)^{-N_m N_d / 2} \exp\left\{-\frac{\beta}{2} \sum_{m=1}^{N_m} \| \left(\mathbf{K}(\boldsymbol{\theta}) - \omega_m^2 \mathbf{M}\right) \boldsymbol{\phi}_m \|^2\right\} \tag{5.15}$$

where c_0 is a normalising constant and $\|\cdot\|$ denotes the Euclidean vector norm, so $\|\boldsymbol{x}\|^2 = \boldsymbol{x}^T\boldsymbol{x}$. This prior PDF is based on choosing a zero-mean Gaussian PDF with covariance matrix $\beta^{-1}\mathrm{I}_d = diag\left(\beta^{-1}, \ldots, \beta^{-1}\right)$ as a probability model for the eigenvalue equation error $\mathbf{e}$. This joint probability model for $\mathbf{e}_1, \ldots, \mathbf{e}_{N_m}$ maximises Shannon's information entropy (i.e. it gives the largest uncertainty) for the equation errors subject to the moment constraints: $\mathbf{E}\left[(\mathbf{e}_m)_k\right] = 0, \mathbf{E}\left[(\mathbf{e}_m)_k^2\right] = \beta^{-1}, k = 1, \ldots, N_d, m = 1, \ldots, N_m$ [100, 102]. The finite value of the equation-error precision parameter β in the joint prior PDF provides a soft constraint for the eigen-equation and it allows for the explicit control of how closely the system and model modal parameters agree. Notice that as $\beta \to \infty$, the system modal parameters $\boldsymbol{\omega}^2$ and $\boldsymbol{\phi}$ become tightly clustered around the modal parameters corresponding to the structural model specified by $\boldsymbol{\theta}$, which are given by Equation (5.14) as $\left(\mathbf{K}(\boldsymbol{\theta}) - \omega_m^2\mathbf{M}\right)\boldsymbol{\phi}_m = \mathbf{0}, m = 1, \ldots, N_m$. Note also that if $\boldsymbol{\theta}$ is specified then these modal parameters are always the most plausible prior values of the system modal parameters.

Using the probability product rule, the joint prior PDF $p\left(\boldsymbol{\phi}, \boldsymbol{\omega}^2, \boldsymbol{\theta}|\beta\right)$ in Equation (5.15) can be decomposed into the product of the conditional PDF for the model parameter vector $\boldsymbol{\theta}$ that is conditional on the system modal parameter vectors $\boldsymbol{\omega}^2$ and $\boldsymbol{\phi}$, and a marginal PDF for $\boldsymbol{\omega}^2$ and $\boldsymbol{\phi}$:

$$p\left(\boldsymbol{\phi}, \boldsymbol{\omega}^2, \boldsymbol{\theta}|\beta\right) = p\left(\boldsymbol{\theta}|\boldsymbol{\phi}, \boldsymbol{\omega}^2, \beta\right) p\left(\boldsymbol{\phi}, \boldsymbol{\omega}^2|\beta\right) \tag{5.16}$$

It is seen that the exponent in Equation (5.15) is a quadratic in the model parameter vector $\boldsymbol{\theta}$ and so the overall equation can be analytically integrated with respect to $\boldsymbol{\theta}$ to get the marginal prior PDFs for the system modal parameter vectors $\boldsymbol{\omega}^2$ and $\boldsymbol{\phi}$:

$$\begin{aligned} p\left(\boldsymbol{\omega}^2, \boldsymbol{\phi}|\beta\right) &= \int p\left(\boldsymbol{\omega}^2, \boldsymbol{\phi}, \boldsymbol{\theta}|\beta\right) d\boldsymbol{\theta} \\ &= c\left(\frac{2\pi}{\beta}\right)^{\frac{(N_\theta - N_d N_m)}{2}} \left|\mathbf{H}^T\mathbf{H}\right|^{\frac{-1}{2}} \exp\left\{-\frac{\beta}{2}\left(\mathbf{b}^T\mathbf{b} - \mathbf{b}^T\mathbf{H}\left(\mathbf{H}^T\mathbf{H}\right)^{-1}\mathbf{H}^T\mathbf{b}\right)\right\} \end{aligned} \tag{5.17}$$

where

$$\mathbf{H} = \begin{bmatrix} \mathbf{K}_1\boldsymbol{\phi}_1 & \cdots & \mathbf{K}_{N_\theta}\boldsymbol{\phi}_1 \\ \vdots & \ddots & \vdots \\ \mathbf{K}_1\boldsymbol{\phi}_{N_m} & \cdots & \mathbf{K}_{N_\theta}\boldsymbol{\phi}_{N_m} \end{bmatrix}_{N_m N_d \times N_\theta}, \mathbf{b} = \begin{bmatrix} \left(\omega_1^2\mathbf{M} - \mathbf{K}_0\right)\boldsymbol{\phi}_1 \\ \vdots \\ \left(\omega_{N_m}^2\mathbf{M} - \mathbf{K}_0\right)\boldsymbol{\phi}_{N_m} \end{bmatrix}_{N_m N_d \times 1} \tag{5.18}$$

Then the Gaussian conditional prior PDF for $\boldsymbol{\theta}$ can be derived from Equations (5.15), (5.16), and (5.17) as:

$$\begin{aligned} p\left(\boldsymbol{\theta}|\boldsymbol{\phi}, \boldsymbol{\omega}^2, \beta\right) &= \left(\frac{2\pi}{\beta}\right)^{\frac{-N_\theta}{2}} \left|\mathbf{H}^T\mathbf{H}\right|^{\frac{1}{2}} \exp\left\{-\frac{\beta}{2}\left(\boldsymbol{\theta} - \left(\mathbf{H}^T\mathbf{H}\right)^{-1}\mathbf{H}^T\mathbf{b}\right)^T \mathbf{H}^T\mathbf{H}\left(\boldsymbol{\theta} - \left(\mathbf{H}^T\mathbf{H}\right)^{-1}\mathbf{H}^T\mathbf{b}\right)\right\} \\ &= \mathcal{N}\left(\boldsymbol{\theta}|\left(\mathbf{H}^T\mathbf{H}\right)^{-1}\mathbf{H}^T\mathbf{b}, \left(\beta\mathbf{H}^T\mathbf{H}\right)^{-1}\right) \end{aligned} \tag{5.19}$$

Note that one complication in Bayesian system identification using modal data is that the model for the modal parameters characterising the modal data $\widehat{\boldsymbol{\omega}}^2$ and $\widehat{\boldsymbol{\Psi}}$ is a nonlinear function of the stiffness scaling parameter vector $\boldsymbol{\theta}$. Rather than directly tackling this challenging nonlinear inverse problem, our formulation involves a series of coupled linear-in-the-parameter problems that allow analytical construction of the conditional posterior PDFs needed for Bayesian inference; so it provides an efficient way to perform Bayesian system identification.

Using the Principle of Maximum Information Entropy again, the combined prediction errors and measurement errors for the system modal parameters $\boldsymbol{\phi}$ and $\boldsymbol{\omega}^2$ are modelled independently as zero-mean Gaussian variables with unknown variances, so the likelihood functions for $\boldsymbol{\phi}$ and $\boldsymbol{\omega}^2$ based on the measured quantities $\widehat{\boldsymbol{\omega}}^2$ and $\widehat{\boldsymbol{\Psi}}$ are given by

$$p\left(\widehat{\boldsymbol{\Psi}}|\boldsymbol{\phi},\eta\right)=\mathcal{N}\left(\widehat{\boldsymbol{\Psi}}|\boldsymbol{\Gamma}\boldsymbol{\phi},\eta\mathrm{I}_{N_oN_sN_m}\right) \tag{5.20}$$

$$p\left(\widehat{\boldsymbol{\omega}}^2|\boldsymbol{\omega}^2,\rho\right)=\mathcal{N}\left(\widehat{\boldsymbol{\omega}}^2|\mathbf{T}\boldsymbol{\omega}^2,\rho\mathrm{I}_{N_sN_m}\right) \tag{5.21}$$

where the selection matrix $\boldsymbol{\Gamma}\in\mathbb{R}^{N_oN_sN_m\times N_dN_m}$ with "1s" and "0s" picks the observed DOF in the "measured" mode shape data set $\widehat{\boldsymbol{\Psi}}$ from the full system mode shapes $\boldsymbol{\phi}$; $\mathbf{T}=[\mathrm{I}_{N_m},\ldots,\mathrm{I}_{N_m}]^T\in\mathbb{R}^{N_sN_m\times N_m}$ is the transformation matrix which connects the vector of N_s sets of N_m identified natural frequencies $\widehat{\boldsymbol{\omega}}^2$ and the N_m system natural frequencies $\boldsymbol{\omega}^2$. Hyper-parameters η and ρ are prescribed variances for the predictions of the identified mode shapes $\widehat{\boldsymbol{\Psi}}$ and natural frequencies $\widehat{\boldsymbol{\omega}}^2$ from $\boldsymbol{\phi}$ and $\boldsymbol{\omega}^2$, respectively, which will be learned solely from the modal data. The hyper-priors for η and ρ are taken as locally non-informative ones that are uniform over large intervals $(0,\eta_{0,max})$ and $(0,\rho_{0,max})$.

For the structural stiffness scaling parameters $\boldsymbol{\theta}$, it is desirable to utilise the prior knowledge of sparseness in the change $\Delta\boldsymbol{\theta}=\boldsymbol{\theta}-\widehat{\boldsymbol{\theta}}$ to alleviate ill-conditioning in the stiffness inverse problem. For this purpose, the calibration value $\widehat{\boldsymbol{\theta}}$ is chosen as pseudo-data to define the likelihood function for $\boldsymbol{\theta}$ by

$$p\left(\widehat{\boldsymbol{\theta}}|\boldsymbol{\theta},\boldsymbol{\alpha}\right)=\prod_{j=1}^{N_\theta}\mathcal{N}\left(\hat{\theta}_j|\theta_j,\alpha_j\right) \tag{5.22}$$

Although the conventional strategy in SBL is to use an ARD Gaussian *prior* PDF [183] to model sparseness as shown in Equation (5.2), here the ARD concept is incorporated in the *likelihood function,* along with the prior on $\boldsymbol{\theta}$ in Equation (5.19). This likelihood function gives a measure of the plausibility of getting the calibration value $\widehat{\boldsymbol{\theta}}$ when the structural model is specified by the parameter vector $\boldsymbol{\theta}$. The hyper-parameter vector $\boldsymbol{\alpha}$ containing the variances α_j controls how for $\boldsymbol{\theta}$ departs from $\widehat{\boldsymbol{\theta}}$. It is learned from the pseudo-data $\widehat{\boldsymbol{\theta}}$ and modal data $\widehat{\boldsymbol{\omega}}^2$ and $\widehat{\boldsymbol{\Psi}}$ using a locally non-informative prior to $\boldsymbol{\alpha}$.

On the other hand, if a sparse stiffness change is not desired, then the pseudo-likelihood function in (5.22) is not required for the stiffness scaling parameter $\boldsymbol{\theta}$ and only the likelihood function $p\left(\widehat{\boldsymbol{\Psi}},\widehat{\boldsymbol{\omega}}^2|\boldsymbol{\phi},\boldsymbol{\omega}^2,\boldsymbol{\theta}\right)=p\left(\widehat{\boldsymbol{\Psi}}|\boldsymbol{\phi}\right)p\left(\widehat{\boldsymbol{\omega}}^2|\boldsymbol{\omega}^2\right)$ is used for Bayesian model updating. Theoretically, this choice corresponds to all $\alpha_j\to\infty$ in Equation (5.22), so $\widehat{\boldsymbol{\theta}}$ is irrelevant. It would be appropriate, for example, if: 1) sufficient modal data $\widehat{\boldsymbol{\omega}}^2$ and $\widehat{\boldsymbol{\Psi}}$ are available to allow a non-sparse set of stiffness changes to be reliably inferred, if needed; or 2) there is no reliable calibration value $\widehat{\boldsymbol{\theta}}$ available to impose sparsity on the change $\Delta\boldsymbol{\theta}$.

Finally, it remains to define a hyper-prior for hyper-parameter β. A widely used exponential prior PDF is taken

$$p\left(\beta|b_0\right)=\mathrm{Exp}\left(\beta|b_0\right)=b_0\exp\left(-b_0\beta\right) \tag{5.23}$$

which is the maximum entropy prior for β with the mean constraint $\mathbf{E}\left(\beta|b_0\right)=1/b_0$. The hyper-prior for b_0 is taken as a locally non-informative one.

Based on the presented Bayesian model, the posterior PDF of all uncertain model parameters θ can be inferred. To simplify the notation, the pseudo-data used for structural model updating is denoted as $\mathcal{D} = \left[\widehat{\Psi}^T, \left(\widehat{\omega}^2\right)^T, \widehat{\theta}^T\right]^T$ if the pseudo-likelihood function in Equation (5.22) is utilised, otherwise $\widehat{\theta}$ is dropped from $\mathcal{D}$. For system identification, our primary goal is to infer $p(\theta|\mathcal{D})$, the marginal posterior PDF of the stiffness scaling parameters, which is obtained by marginalising over all other uncertain parameters as 'nuisance' parameters in the joint posterior PDF $p\left(\omega^2, \phi, \theta, \beta, \rho, \eta, \alpha, b_0|\mathcal{D}\right)$. By combining all the stages of the hierarchical Bayesian model, the joint posterior PDF is given by Bayes' theorem as follows

$$p\left(\omega^2, \phi, \theta, \beta, \rho, \eta, \alpha, b_0|\mathcal{D}\right) = \frac{p\left(\widehat{\omega}^2|\omega^2, \rho\right) p\left(\widehat{\Psi}|\phi, \eta\right) p\left(\widehat{\theta}|\theta, \alpha\right) p\left(\omega^2, \phi, \theta|\beta\right) p(\rho)\, p(\eta)\, p(\alpha)\, p(\beta|b_0)\, p(b_0)}{p\left(\widehat{\omega}^2, \widehat{\Psi}, \widehat{\theta}\right)} \tag{5.24}$$

where PDFs $p\left(\widehat{\omega}^2|\omega^2, \rho\right), p\left(\widehat{\Psi}|\phi, \eta\right), p\left(\widehat{\theta}|\theta, \alpha\right), p\left(\omega^2, \phi, \theta|\beta\right)$, and $p(\beta|b_0)$ are defined in Equations (5.21), (5.20), (5.22), (5.15) and (5.23), respectively, and the prior PDFs $p(\rho), p(\eta), p(\alpha), p(\beta|b_0)$, and $p(b_0)$ are uniform. The acyclic graph of the hierarchical Bayesian model is shown in Figure 5.1, where each arrow denotes the conditional dependencies used in the joint probability model. The bidirectional arrow in the graph of the hierarchical Bayesian model represents the statistical interaction between structural model parameters ω^2 and ϕ and structural model parameters θ, which comes from the joint prior $p\left(\omega^2, \phi, \theta|\beta\right)$ in Equation (5.15).

The resulting expression in Equation (5.24) is intractable because the high-dimensional normalising integral for $p\left(\widehat{\omega}^2, \widehat{\Psi}, \widehat{\theta}\right)$ cannot be computed analytically. A fast SBL algorithm [95] is presented next to tackle this problem effectively. The algorithm characterises the posterior distribution of the stiffness scaling parameters θ by simply using the MAP values for all other model parameters learned from the available data $\left\{\widehat{\omega}^2, \widehat{\Psi}, \widehat{\theta}\right\}$.

5.3.2 Fast sparse Bayesian learning algorithm

The fast SBL algorithm focuses on an analytical derivation of the posterior PDF of the stiffness scaling parameter θ. We collect all uncertain parameters in Figure 5.1 except θ in the vector $\delta = \left[\left(\omega^2\right)^T, \rho, \phi^T, \eta, \alpha^T, \beta, b_0\right]^T$ and consider them as 'nuisance' parameters to be marginalised out of $p(\theta, \delta|\mathcal{D})$ by using Laplace's approximation method [26]. The stochastic model class $\mathcal{M}(\delta)$ for the stiffness scaling parameters θ is defined by the likelihood functions in Equations (5.20), (5.21), and(5.22), and the priors given by Equations (5.17), (5.19), and (5.23). The full posterior uncertainty in θ is explicitly incorporated when finding the MAP estimates of all parameters in δ using the evidence maximisation procedure.

5.3.2.1 Formulation

Using the probability product rule, the full posterior $p\left(\theta, \delta|\widehat{\omega}^2, \widehat{\Psi}, \widehat{\theta}\right)$ is rewritten as

$$p\left(\theta, \delta|\widehat{\omega}^2, \widehat{\Psi}, \widehat{\theta}\right) = p\left(\theta|\delta, \widehat{\omega}^2, \widehat{\Psi}, \widehat{\theta}\right) p\left(\delta|\widehat{\omega}^2, \widehat{\Psi}, \widehat{\theta}\right) \tag{5.25}$$

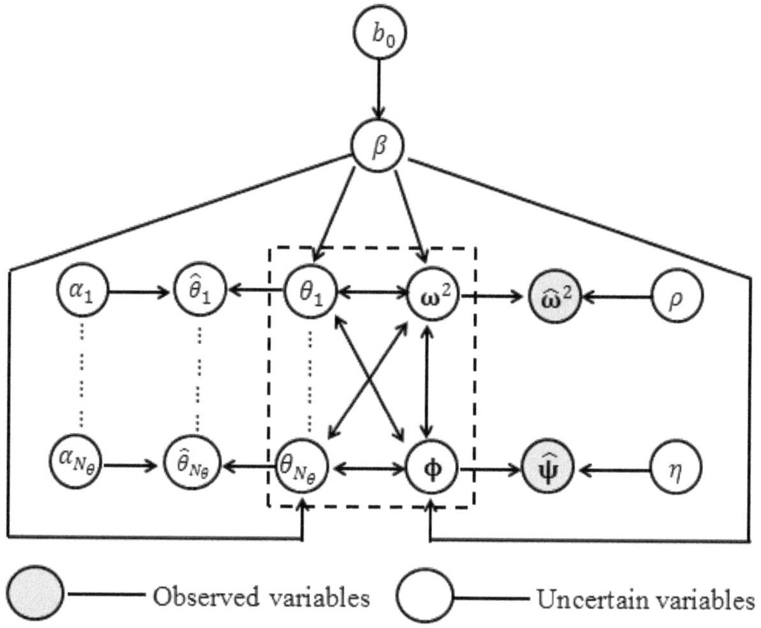

Figure 5.1: Acyclic graph representing the information flow in the hierarchical sparse Bayesian model.

The first factor in Equation (5.25) is the posterior PDF of the stiffness scaling parameter vector conditional on δ, which is given by

$$p\left(\theta|\delta,\hat{\omega}^2,\hat{\Psi},\hat{\theta}\right)=p\left(\theta|\delta,\hat{\theta}\right)=\frac{p\left(\hat{\theta}|\theta,\delta\right)p\left(\theta|\delta\right)}{p\left(\hat{\theta}|\delta\right)} \tag{5.26}$$

where the first equality holds because θ is independent of the measured modal parameters $\hat{\omega}^2$ and $\hat{\Psi}$, when conditional on the system modal parameters ω^2 and ϕ in δ (see Figure 5.1) and the second equality due to Bayes' theorem. As a consequence of combining a Gaussian prior (Eq. (5.19)) and a linear model within a Gaussian likelihood (Eq. (5.22)), the conditional posterior is also Gaussian

$$p\left(\theta|\delta,\hat{\omega}^2,\hat{\Psi},\hat{\theta}\right)=\mathcal{N}\left(\theta|\mu_\theta,\boldsymbol{\Sigma}_\theta\right) \tag{5.27}$$

where mean μ_θ and covariance matrix $\boldsymbol{\Sigma}_\theta$ are

$$\mu_\theta=\boldsymbol{\Sigma}_\theta\left(\beta\mathbf{H}^T\mathbf{b}+\mathbf{A}^{-1}\hat{\theta}\right) \tag{5.28a}$$

$$\boldsymbol{\Sigma}_\theta=\left(\beta\mathbf{H}^T\mathbf{H}+\mathbf{A}^{-1}\right)^{-1} \tag{5.28b}$$

where $\mathbf{A}=diag\left(\alpha_1,\ldots,\alpha_{N_\theta}\right)$. The normalising constant $p\left(\hat{\theta}|\delta\right)$ for the posterior PDF in Equation (5.26) is called the *pseudo-evidence* (or *pseudo-marginal likelihood*) *for the model class* $\mathcal{M}\left(\delta\right)$ given by pseudo-data $\widehat{\theta}$ and it can be analytically evaluated

$$\begin{aligned} p\left(\widehat{\theta}|\delta\right) &= \int p\left(\widehat{\theta}|\theta,\delta\right) p\left(\theta|\delta\right) d\theta \\ &= \int p\left(\widehat{\theta}|\theta,\alpha\right) p\left(\theta|\omega^2,\phi,\beta\right) d\theta \\ &= \mathcal{N}\left(\widehat{\theta}|\left(\mathbf{H}^T\mathbf{H}\right)^{-1}\mathbf{H}^T\mathbf{b}, \mathbf{A}+\beta^{-1}\left(\mathbf{H}^T\mathbf{H}\right)^{-1}\right) \end{aligned} \tag{5.29}$$

where the hierarchical structure exhibited in Figure 5.1 is utilised. This is a Gaussian distribution over the N_θ−dimensional pre-damage parameter vector θ_u evaluated at its MAP value $\widehat{\theta}$, and is readily evaluated for arbitrary values of δ.

The second factor in Equation (5.25) is given by Bayes' theorem

$$p\left(\delta|\widehat{\omega}^2,\widehat{\Psi},\widehat{\theta}\right) = \frac{p\left(\widehat{\omega}^2,\widehat{\Psi},\widehat{\theta}|\delta\right) p\left(\delta\right)}{p\left(\widehat{\omega}^2,\widehat{\Psi},\widehat{\theta}\right)} \tag{5.30}$$

Again, the posterior PDF $p\left(\delta|\widehat{\omega}^2,\widehat{\Psi},\widehat{\theta}\right)$ is always intractable due to the high-dimensional normalisation integral $p\left(\widehat{\omega}^2,\widehat{\Psi},\widehat{\theta}\right)$. One must turn to an approximation strategy where all parameters in δ are learned by MAP estimation from the PDF $p\left(\delta|\widehat{\omega}^2,\widehat{\Psi},\widehat{\theta}\right)$, which involves maximising the product of the likelihood $p\left(\widehat{\omega}^2,\widehat{\Psi},\widehat{\theta}|\delta\right)$ and the prior $p\left(\delta\right)$ in Equation (5.30). It is assumed that the model class $\mathcal{M}\left(\delta\right)$ is globally identifiable based on the modal data, meaning here that the likelihood $p\left(\widehat{\omega}^2,\widehat{\Psi},\widehat{\theta}|\delta\right)$ in Equation (5.30) has a unique global maximum over δ, and then so does the posterior at the MAP value $\widetilde{\delta}$. To get the marginal posterior PDF $p\left(\theta|\widehat{\omega}^2,\widehat{\Psi},\widehat{\theta}\right)$, δ is then treated as a 'nuisance' parameter vector and it is integrated out using Laplace's asymptotic approximation [26]:

$$p\left(\theta|\widehat{\omega}^2,\widehat{\Psi},\widehat{\theta}\right) = \int p\left(\theta|\delta,\widehat{\omega}^2,\widehat{\Psi},\widehat{\theta}_u\right) p\left(\delta|\widehat{\omega}^2,\widehat{\Psi},\widehat{\theta}\right) d\delta \approx p\left(\theta|\widetilde{\delta},\widehat{\omega}^2,\widehat{\Psi},\widehat{\theta}\right) \tag{5.31}$$

where $\widetilde{\delta} = \arg\max p\left(\delta|\widehat{\omega}^2,\widehat{\Psi},\widehat{\theta}\right) = \arg\max p\left(\widehat{\omega}^2,\widehat{\Psi},\widehat{\theta}|\delta\right) p\left(\delta\right)$, because the denominator in Equation (5.30) is independent of δ. By using the hierarchical structure exhibited in Figure 5.1 and substituting Equations (5.20), (5.21), (5.29), (5.17), and (5.23), the object function $J_1\left(\delta\right) = p\left(\widehat{\omega}^2,\widehat{\Psi},\widehat{\theta}|\delta\right) p\left(\delta\right)$ to be maximized is given by

$$J_1\left(\delta\right) = p\left(\widehat{\omega}^2|\omega^2,\rho\right) p\left(\widehat{\Psi}|\phi,\eta\right) p\left(\widehat{\theta}|\omega^2,\phi,\beta,\alpha\right) p\left(\omega^2,\phi|\beta\right) p\left(\beta|b_0\right) \tag{5.32}$$

To find the MAP value $\widetilde{\delta}$, explicit expressions can be obtained for iterative maximisation of $\log J_1\left(\delta\right)$ where all parameters in δ are updated successively, as shown in the following. By maximising $\log J_1\left(\delta\right)$ with respect to the system mode shapes ϕ and the associated hyper-parameter

η and letting all other parameters be fixed at their MAP values, the MAP estimates $\tilde{\phi}$ and $\tilde{\eta}$ are derived as

$$\begin{aligned}\tilde{\phi} = &\left(\tilde{\beta}\tilde{\mathbf{F}}^T\tilde{\mathbf{F}} + \tilde{\beta}\text{diag}\left(\text{tr}\left(\tilde{\boldsymbol{\Sigma}}_\theta\mathbf{T}_1\right), \ldots, \text{tr}\left(\tilde{\boldsymbol{\Sigma}}_\theta\mathbf{T}_{N_m N_d}\right)\right) + \tilde{\eta}\boldsymbol{\Gamma}^T\boldsymbol{\Gamma}\right)^{-1} \\ &\cdot\left(\tilde{\eta}\boldsymbol{\Gamma}^T\hat{\boldsymbol{\Psi}} - \tilde{\beta}\left[\text{tr}\left(\tilde{\boldsymbol{\Sigma}}_\theta\mathbf{U}_1\right), \ldots, \text{tr}\left(\tilde{\boldsymbol{\Sigma}}_\theta\mathbf{U}_{N_m N_d}\right)\right]^T\right)\end{aligned} \tag{5.33}$$

$$\tilde{\eta} = \frac{N_s N_m N_o}{\|\hat{\boldsymbol{\Psi}} - \boldsymbol{\Gamma}\tilde{\phi}\|^2} \tag{5.34}$$

where $\tilde{\boldsymbol{\Sigma}}_\theta$ is given by Equation (5.28b) using $\tilde{\beta}$, $\tilde{\alpha}$ and also $\tilde{\phi}$ in $\mathbf{H}$ in Equation (5.18),

$$\tilde{\mathbf{F}} = \begin{bmatrix} \mathbf{K}(\tilde{\mu}_\theta) - \tilde{\omega}_1^2\mathbf{M} & \cdots & \mathbf{0} \\ \vdots & \ddots & \vdots \\ \mathbf{0} & \cdots & \mathbf{K}(\tilde{\mu}_\theta) - \tilde{\omega}_1^2\mathbf{M} \end{bmatrix}_{N_m N_d \times N_m N_d} \tag{5.35}$$

where $\tilde{\mu}_\theta$ is given by Equation (5.28a) using $\tilde{\beta}$, $\tilde{\alpha}$ and also $\tilde{\phi}$, $\tilde{\omega}^2$ in $\mathbf{b}$ and $\mathbf{H}$ in (5.18), and for $q = 1, \ldots, N_m N_d$

$$\begin{aligned}\mathbf{T}_q &= \boldsymbol{\Pi}_q^T\boldsymbol{\Pi}_q \\ \mathbf{U}_q &= \boldsymbol{\Pi}_q^T \cdot \left(\sum_{s\neq q}^{N_m N_d} \tilde{\phi}_s\boldsymbol{\Pi}_s\right) \\ \boldsymbol{\Pi}_q &= \frac{\partial\mathbf{H}}{\partial\phi_q}\end{aligned} \tag{5.36}$$

Determining the MAP value $\tilde{\phi}$ requires repeated evaluation of the inverse of $\mathbf{X} = \left(\tilde{\beta}\tilde{\mathbf{F}}^T\tilde{\mathbf{F}} + \tilde{\eta}\boldsymbol{\Gamma}^T\boldsymbol{\Gamma} + \tilde{\beta}\text{diag}\left(\text{tr}\left(\tilde{\boldsymbol{\Sigma}}_\theta\mathbf{T}_1\right), \ldots, \text{tr}\left(\tilde{\boldsymbol{\Sigma}}_\theta\mathbf{T}_{N_m N_d}\right)\right)\right)$. The dimension $N_m N_d$ of ϕ will often be very large in applications (e.g. $N_m N_d > 100$). To facilitate the goal of efficient computation of the inverse of the matrix $\mathbf{X}$, it is rewritten in the block diagonal form: $\mathbf{X} = \text{diag}(\mathbf{X}_1 \ldots, \mathbf{X}_{N_m})$, where the $N_d \times N_d$ matrices $\mathbf{X}_i$ for $i = 1, \ldots, N_m$ are given by

$$\mathbf{X}_i = \tilde{\beta}\left(\mathbf{K}(\tilde{\mu}_\theta) - \tilde{\omega}_i^2\mathbf{M}\right)^2 + \tilde{\beta}\text{diag}\left(\text{tr}\left(\tilde{\boldsymbol{\Sigma}}_\theta\mathbf{T}_{(i-1)N_d+1}\right), \ldots, \text{tr}\left(\tilde{\boldsymbol{\Sigma}}_\theta\mathbf{T}_{iN_d}\right)\right) + \tilde{\eta}N_m\boldsymbol{\Gamma}_s^T\boldsymbol{\Gamma}_s$$

and then the MAP value $\tilde{\phi}$ is given by:

$$\tilde{\phi} = \text{diag}\left(\mathbf{X}_1^{-1}, \ldots, \mathbf{X}_{N_m}^{-1}\right)\left(\tilde{\eta}\boldsymbol{\Gamma}^T\hat{\boldsymbol{\Psi}} - \tilde{\beta}\left[\text{tr}\left(\tilde{\boldsymbol{\Sigma}}_\theta\mathbf{U}_1\right), \ldots, \text{tr}\left(\tilde{\boldsymbol{\Sigma}}_\theta\mathbf{U}_{N_m N_d}\right)\right]^T\right) \tag{5.37}$$

where $\boldsymbol{\Gamma}_s$ is the sub-matrix which contains the elements of the first N_s rows and N_d columns of $\boldsymbol{\Gamma}$. The estimation of $\tilde{\phi}$ now involves inversions of N_m square matrices $\mathbf{X}_i$ $(i = 1, \ldots, N_m)$ of order N_d, whereas the original estimation of $\tilde{\phi}$ in Equation (5.33) involves an inversion of one square matrix of order $(N_m N_d)$. The latter requires N_m^2 more computational effort than Equation (5.37).

Similarly, the MAP estimates of the system natural frequencies ω^2 and their associated hyper-parameters ρ are also derived as

$$\widetilde{\omega}^2 = \left[\tilde{\rho}^{-1}\mathbf{L}^T\mathbf{L} + \tilde{\beta}\widetilde{\mathbf{G}}^T\widetilde{\mathbf{G}}\right]^{-1}\left(\tilde{\rho}^{-1}\mathbf{L}^T\widehat{\omega}^2 + \tilde{\beta}\widetilde{\mathbf{G}}^T\tilde{\mathbf{c}}\right) \tag{5.38}$$

$$\tilde{\rho} = \frac{N_s N_m}{\|\widehat{\omega}^2 - \mathbf{T}\widetilde{\omega}^2\|^2} \tag{5.39}$$

where

$$\widetilde{\mathbf{G}} = \begin{bmatrix} \mathbf{M}\widetilde{\phi}_1 & \cdots & \mathbf{0} \\ \vdots & \ddots & \vdots \\ \mathbf{0} & \cdots & \mathbf{M}\widetilde{\phi}_{N_m} \end{bmatrix}_{N_d N_m \times N_m} \tag{5.40}$$

$$\tilde{\mathbf{c}} = \begin{bmatrix} \mathbf{K}(\widetilde{\mu}_\theta)\,\widetilde{\phi}_1 \\ \vdots \\ \mathbf{K}(\widetilde{\mu}_\theta)\,\widetilde{\phi}_{N_m} \end{bmatrix}_{N_d N_m \times 1} \tag{5.41}$$

It is shown in the Appendix that by maximising $\log J_1(\delta)$ with respect to each hyper-parameter α_j, the MAP estimates $\widetilde{\alpha}_j, j = 1, \ldots, N_\theta$, are given by

$$\widetilde{\alpha}_j = \left(\widetilde{\Sigma}_\theta\right)_{jj} + \left(\widehat{\theta} - \widetilde{\mu}_\theta\right)_j^2 \tag{5.42}$$

Finally, the MAP values $\tilde{\beta}$ and $\tilde{b}_0$ are derived as

$$\tilde{\beta} = \frac{N_d N_m - \sum_{j=1}^{N_\theta}\left(1 - \widetilde{\alpha}_j^{-1}\widetilde{\Sigma}_{jj}\right)}{\|\widetilde{\mathbf{H}}\widetilde{\mu}_\theta - \widetilde{\mathbf{b}}\|^2 + 2\tilde{b}_0^{-1}} \tag{5.43}$$

$$\tilde{b}_0 = \frac{\tilde{\beta}}{2} \tag{5.44}$$

A nice feature of this formulation is that $\tilde{\beta}$ is inversely dependent on the square of the scaling of the mode shape and the stiffness and mass matrices, which leads to the desirable property that the inferred stiffness reduction results are scale-invariant, since the terms $\beta\mathbf{H}^T\mathbf{H}$ and $\beta\mathbf{H}^T\mathbf{b}$ in Equation (5.28a) and Equation (5.28b) are automatically independent of these scale selections. The optimisation problems above are coupled, that is, the expressions for the MAP estimates for each parameter depend on the MAP values of other parameters. In order to calculate the overall MAP value $\widetilde{\delta}$, the proposed algorithm goes through the sequence of MAP estimation Equation (5.37), (5.34), (5.38), (5.39), (5.42), (5.43), and (5.44) and then updates the mean and covariance of the Gaussian posterior for θ in (5.27) by substituting $\widetilde{\delta}$ into Equations (5.28a) and (5.28b) before repeating this process until convergence occurs. As shown in the Appendix, the maximisation of the logarithm function of the product of pseudo-evidence $p\left(\widehat{\theta}|\omega^2, \phi, \beta, \alpha\right)$ and PDF $p\left(\omega^2, \phi|\beta\right)$ in the objective function $J_1(\delta)$ in Equation (5.32) automatically produces the optimal trade-off between data fitting and model complexity that causes many hyper-parameters α_j to approach zero with a reasonably large value of β. This procedure gives a model for θ that has both a sufficiently small number of components θ_j differing from those in $\widehat{\theta}$ and fits the system modal parameters ω^2 and ϕ well. This is an instance of the Principle of Model Parsimony or the Bayesian Ockham's razor [22].

5.3.2.2 Proposed fast SBL algorithm for stiffness inversion

The above formulation leads to the following algorithm for probabilistic inference of structural stiffness changes. For the calibration stage, model sparseness of changes of the stiffness scaling parameters $\boldsymbol{\theta}_u$ is not expected and hence, the hyper-parameter vector $\boldsymbol{\alpha}$ is not optimised; instead, all components α_j in $\boldsymbol{\alpha}$ are fixed with some very large values. For the monitoring stage, components α_j in $\boldsymbol{\alpha}$ are estimated to ensure the inference of sparse changes of the stiffness scaling parameters $\boldsymbol{\theta}_d$ compared with the calibration values $\widehat{\boldsymbol{\theta}}$.

Algorithm 9: Fast SBL algorithm for stiffness inversion

INPUTS:

▷ N_s sets of identified modal data $\widehat{\boldsymbol{\omega}}^2 = \widehat{\boldsymbol{\omega}}_u^2$ (if calibration stage), or $\widehat{\boldsymbol{\omega}}_d^2$ (if monitoring stage), and $\widehat{\boldsymbol{\Psi}} = \widehat{\boldsymbol{\Psi}}_u$ (if calibration stage), or $\widehat{\boldsymbol{\Psi}}_d$ (if monitoring stage), for N_m modes from the calibration stage ($N_s \geq 3$).

▷ **If** calibration stage, a chosen nominal value $\boldsymbol{\theta}_0$ of $\boldsymbol{\theta}$ (e.g. from a finite-element model) and associated prior covariance matrix $\boldsymbol{\Sigma}_0 = \sigma_0^2 \mathbf{I}_{N_\theta}$ with large variance σ_0^2.

▷ **If** monitoring stage, MAP estimate $\widehat{\boldsymbol{\theta}}$ and the associated posterior covariance matrix $\boldsymbol{\Sigma}_u$ from the calibration stage.

▷ Choose initial values for parameters $\tilde{\eta}, \tilde{\rho}, \tilde{b}_0$, and $\tilde{\beta}$.

ALGORITHM:

Initialise $\boldsymbol{\mu}_\theta = \boldsymbol{\theta}_0$ (if calibration stage), or $\boldsymbol{\mu}_\theta = \widehat{\boldsymbol{\theta}}$ (if monitoring stage), $\boldsymbol{\Sigma}_\theta = \boldsymbol{\Sigma}_0$ (if calibration stage), or $\boldsymbol{\Sigma}_\theta = \boldsymbol{\Sigma}_u$ (if monitoring stage), and $\widetilde{\boldsymbol{\omega}}^2 = \sum_{r=1}^{N_s} \widehat{\boldsymbol{\omega}}_r^2 / N_s$.

If calibration stage, fix all components in $\widetilde{\boldsymbol{\alpha}}$ with large values (e.g. $\widetilde{\alpha}_j = 10^9$).

If monitoring stage, initialise $\widetilde{\alpha}_j = N_\theta^2, j = 1, \ldots, N_\theta$.

While convergence criterion on $\boldsymbol{\theta}$ is not met **(Outer loop)**.

While convergence criterion on β is not met **(Inner loop)**.

Update MAP $\widetilde{\boldsymbol{\phi}}$ using Eq. (5.37), and then update $\tilde{\eta}$ using Eq. (5.34).

Update MAP $\widetilde{\boldsymbol{\omega}}^2$ using Eq. (5.38), and then update $\tilde{\rho}$ using Eq. (5.39).

If at the monitoring stage, update $\widetilde{\boldsymbol{\alpha}}$ using Eq. (5.42); Set $\widetilde{\alpha}_j = 0$ if α_j $(j = 1, \ldots, N_\theta)$ becomes smaller than $\alpha_{min} (e.g.\ \alpha_{min} = 10^{-9})$.

Update MAP $\tilde{\beta}$ using Eq. (5.43).

Calculate the conditional posterior mean $\boldsymbol{\mu}_\theta$ and covariance matrix $\boldsymbol{\Sigma}_\theta$ for $\boldsymbol{\theta}$ using Eqs. (5.28a) and Eq. (5.28b).

End while (the ratio of change of β between the current iteration and the previous iteration of the inner loop is sufficiently small (e.g. smaller than 0.001)).

Update $\tilde{b}_0$ using Eq. (5.44).

End while (the ratio of change in all components of $\boldsymbol{\mu}_\theta$ between the current iteration and the previous iteration of the outer loop is sufficiently small (e.g. smaller than 0.01)).

OUTPUTS:

▷ MAP estimates of all uncertain parameters in $\boldsymbol{\delta}$.
▷ MAP estimates $\widehat{\boldsymbol{\theta}} = \boldsymbol{\mu}_{\theta}$ (if at the calibration stage), or $\widetilde{\boldsymbol{\theta}}_d = \boldsymbol{\mu}_{\theta}$ (if at the monitoring stage) and covariance matrix $\boldsymbol{\Sigma}_u = \boldsymbol{\Sigma}_{\theta}$ (if at the calibration stage), or $\boldsymbol{\Sigma}_d = \boldsymbol{\Sigma}_{\theta}$ (if at the monitoring stage) of the stiffness scaling parameters $\boldsymbol{\theta}$.

There are several implementation details for the algorithm as follows:

1. *Initialisation of hyper-parameters*: Using Equations (5.34), (5.39), (5.43), and (5.44) along with some approximations, the initial parameter values for hyper-parameters $\bar{\eta}, \bar{\rho}, \bar{b}_0$, and $\bar{\beta}$ are initialised as:

$$\bar{\eta} = \frac{10 N_s N_m N_o}{\|\widehat{\boldsymbol{\Psi}}\|^2} \tag{5.45}$$

$$\bar{\rho} = \frac{10 N_s N_m}{\|\widehat{\boldsymbol{\omega}}^2\|^2} \tag{5.46}$$

$$\bar{b}_0 = \frac{2 N_s N_o}{\sum_r^{N_s} \| \left(\mathbf{K}(\boldsymbol{\theta}_0) - \widehat{\omega}_1^2 \mathbf{M}\right) \widehat{\boldsymbol{\Psi}}_{r,1} \|^2} \tag{5.47}$$

$$\bar{\beta} = \frac{4 N_s N_o}{\sum_r^{N_s} \| \left(\mathbf{K}(\boldsymbol{\theta}_0) - \widehat{\omega}_1^2 \mathbf{M}\right) \widehat{\boldsymbol{\Psi}}_{r,1} \|^2} \tag{5.48}$$

For the initial value of $\overline{\alpha}_j$, we use $\overline{\alpha}_j = N_{\theta}^2,\ j = 1, \ldots, N_{\theta}$, which is inspired by [183].

2. *Optimisation of the hyper-parameter* b_0: For stiffness inversion with real data where there are large modelling errors, it was sometimes observed during finding the iterative solution for the MAP estimates $\widetilde{\boldsymbol{\delta}}$ that the components of the estimated system mode shape $\widetilde{\boldsymbol{\phi}}$ kept decreasing towards zero while the equation-error precision parameter $\widetilde{\beta}$ tended to infinity. To avoid this occurring, $\widetilde{b}_0$ is updated separately in an outer loop until convergence of the optimisation occurs for the other parameters in $\boldsymbol{\delta}$.

3. *Determination of the effective dimension of the structural model parameter vector* $\boldsymbol{\theta}_d$: For the monitoring stage, the algorithm starts by considering all substructures as possibly damaged (the reason for the choice $\widetilde{\alpha}_j = N_{\theta}^2,\ j = 1, \ldots, N_{\theta}$, at the first iteration) and then causes the "inactive" components θ_j of $\boldsymbol{\theta}_d$, which have $\widetilde{\alpha}_j < 10^{-9}$, to be exactly equal to $\widehat{\theta}_j$ from the calibration stage when optimising over the hyper-parameters α_j. After convergence, there are only a few "active" $\theta_j{}'s$ that are changed from their calibration values $\widehat{\theta}_j$ and their corresponding substructures are considered to be damaged.

5.3.2.3 Damage assessment

Using Algorithm 9, the posterior PDFs of the stiffness scaling parameters $\boldsymbol{\theta}_d$ and $\boldsymbol{\theta}_u$ for the current monitoring stage (where damage is possible) and the undamaged calibration stage, respectively, are obtained based on the modal data $\left[\widehat{\boldsymbol{\omega}}_d^2, \widehat{\boldsymbol{\Psi}}_d\right]$ and $\left[\widehat{\boldsymbol{\omega}}_u^2, \widehat{\boldsymbol{\Psi}}_u\right]$, respectively. The posterior means $\widetilde{\boldsymbol{\theta}}_d$ and $\widetilde{\boldsymbol{\theta}}_u$ and covariance matrices $\boldsymbol{\Sigma}_d$ and $\boldsymbol{\Sigma}_u$ can be used to compute the probability that a given substructure stiffness parameter $\theta_{d,j}$ in the current monitoring stage has been reduced by more than a specified fraction of its value $\theta_{u,j}$ in the initial calibration stage of the structure [189, 23, 200]. An asymptotic Gaussian approximation [26] is used for the integrals involved to give

$$\begin{aligned} P_j^{dam}(f) &= P\left(\theta_{d,j} < (1-f)\,\theta_{u,j} | \widehat{\omega}_u^2, \widehat{\Psi}_u, \widehat{\omega}_d^2, \widehat{\Psi}_d\right) \\ &= \int P\left(\theta_{d,j} < (1-f)\,\theta_{u,j} | \theta_{u,j}, \widehat{\omega}_u^2, \widehat{\Psi}_u, \widehat{\omega}_d^2, \widehat{\Psi}_d\right) p\left(\theta_{u,j} | \widehat{\omega}_u^2, \widehat{\Psi}_u\right) d\theta_{u,j} \\ &\approx \Phi\left[\frac{(1-f)\,\widetilde{\theta}_{u,j} - \widetilde{\theta}_{d,j}}{\sqrt{(1-f)^2\,\sigma_{d,j}^2 + \sigma_{u,j}^2}}\right] \end{aligned} \tag{5.49}$$

where $\Phi(\cdot)$ is the standard Gaussian cumulative distribution function; $\widetilde{\theta}_{d,j}$ and $\widetilde{\theta}_{u,j}\left(=\widehat{\theta}_j\right)$ denote the MAP value of the stiffness scaling parameters $\theta_{d,j}$ and $\theta_{u,j}$, respectively, estimated from Equation (5.28a); $\sigma_{d,j}$ and $\sigma_{u,j}$ are the corresponding posterior standard deviations of the stiffness scaling parameters $\theta_{d,j}$ and $\theta_{u,j}$, respectively, which are the square root of the diagonal elements of the posterior covariance matrix $\boldsymbol{\Sigma}_\theta$ given in Equation (5.28b).

5.4 Case studies

Example 13 *The first example considers a twelve-story shear building with linear dynamics and applies the algorithm only for calibration; no sparseness of structural stiffness changes from their prior values is imposed. The lumped mass of each floor is taken to be 90 metric tons, and the inter-story stiffness is* $k_0 = 225.30\,\mathrm{MN/m}$ *for all stories. This gives the first five modal frequencies as* 1.00, 2.98, 4.92, 6.78, *and* 8.53 Hz. *For system identification, one stiffness scaling parameter* θ_j *is used for each story,* $j = 1, \cdots, 12$, *where* $\boldsymbol{K}_j = \theta_j \overline{\boldsymbol{K}}_j$ *is the uncertain contribution of the jth story to the global stiffness matrix* $\boldsymbol{K}$, *as in Equation* (5.13), *where* $\boldsymbol{K}_0 = \mathbf{0}$ *and* $\overline{\boldsymbol{K}}_j$ *is the 'nominal' contribution, which, for convenience in assessing results, is taken as the exact contribution in the structural model used to generate the data. Thus,* $\theta_j = 1$ *gives the exact value for* $\boldsymbol{K}_j$ *at the calibration (undamaged) stage. To simulate the results of model identification, zero-mean Gaussian noise is added to the exact modal parameters with a standard deviation of* 2% *of the exact modal frequencies and mode shapes. The goal of this example is to test the performance of the proposed iterative method described in Algorithm 9 for identifying the structural model parameters* $\boldsymbol{\theta}$ *at the calibration stage. The initial value of each* θ_j *for starting the iterations is selected randomly from a uniform distribution within the interval between* 1 *and* 3 *to demonstrate the robustness to the initial choice.*

In the first set of experiments, we study the effect of different choices of equation-error precision β *on the stiffness identification performance. First, the results of the identified MAP values of the stiffness scaling parameters* θ_j *using three choices of hyper-parameter* β *and the first four measured modes* ($N_m = 4$) *identified from three data segments* ($N_s = 3$) *of complete measurements* ($N_o = 12$) *are tabulated in Table 5.1, where the three fixed values of* β *are obtained by varying the* $\bar{\beta}$ *values calculated from Equation* (5.48) *by factors 1, 5, and 20. The associated posterior coefficients of variation (c.o.v), which are calculated from the ratio of the square root of the diagonal elements of the posterior covariance matrix* $\boldsymbol{\Sigma}_\theta$ *in Equation* (5.28b) *to the corresponding MAP values* $\boldsymbol{\mu}_\theta$ *in*

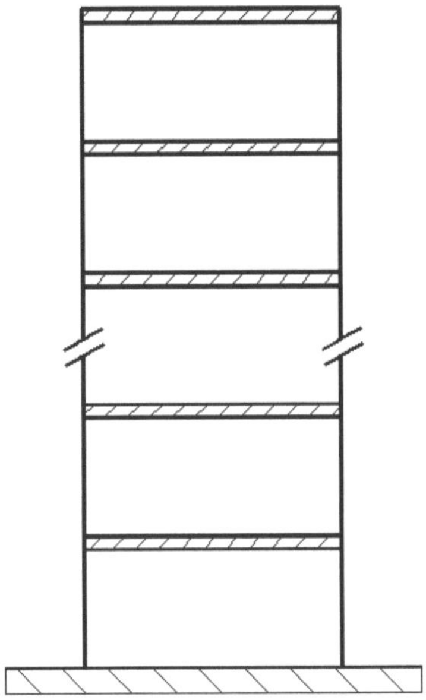

Figure 5.2: A 12-storey shear building structure.

Equation (5.28a), *are also presented in Table 5.1. All components in the MAP vector* $\tilde{\alpha}$ *are fixed at large values when implementing the pseudo-codes in Table 5.1. It is seen that the various values of* β *produce different MAP identification results and hence proper selection of the hyper-parameter* β *is important for identification accuracy. It is further observed that the associated posterior uncertainty is highly dependent on the selected values of* β*, which can give a misleading confidence in the accuracy of the MAP estimates. Then, the algorithm is run again with* β *optimised but varying the value* $\bar{\beta}$ *from Equation* (5.48) *by factors 1, 5, and 20 to get three choices of initial values. Four measured modes* ($N_m = 4$) *are identified from three data segments* ($N_s = 3$)*. No matter what initial values of* β *is chosen, all runs converge to the same MAP value of* β *and the same MAP vector and associated c.o.v of the stiffness scaling parameters* θ*. The results are presented at the last column of Table 5.1. The identification errors for the obtained results are generally smaller than those with* β *fixed that are tabulated in the other columns of Table 5.1. Therefore, the use of the optimisation scheme for the selection of the hyper-parameter* β *in the proposed algorithm gives it the ability to accurately identify the stiffness scaling parameters* θ.

The iteration histories for convergence of the MAP values of the stiffness scaling parameters are shown in Figure 5.3, which correspond to the results in Table 5.1. When β *is fixed at a small value, that is,* $\beta = \bar{\beta}$ *(Figure 5.3 (a)(i)), the convergence for the algorithm is very fast, occurring in the first few iterations, but the final identified MAP values have larger errors than the case with* β *optimised (Table 5.1). Larger* β *should be selected for more accurate identification results. However, when the* β *selected is too large, that is,* $\beta = 20\bar{\beta}$ *(Figure 5.3 (c)(i)), the convergence is very slow, requiring more than 200 iterations. In contrast, the algorithm with* β *optimised produces accurate identification results in the first few iterations, no matter what the initial* β *is, as shown in Figure 5.3 (ii), which shows the advantage of learning*

Table 5.1: Identification results with four measured modes ($N_m = 4$) and three data segments ($N_s = 3$), and β fixed at different values of $\bar{\beta}$, $5\bar{\beta}$, and $20\bar{\beta}$, and optimised with different initial values of $\bar{\beta}$, $5\bar{\beta}$, and $20\bar{\beta}$ (Example 13).

β		Fixed at $\bar{\beta}$		Fixed at $5\bar{\beta}$		Fixed at $20\bar{\beta}$		Optimised	
Param.	Init. values	MAP	c.o.v.(%)	MAP	c.o.v.(%)	MAP	c.o.v.(%)	MAP	c.o.v.(%)
θ_1	1.702	0.994	2.91	0.992	1.51	0.991	0.51	0.999	0.17
θ_2	2.878	0.994	4.02	0.997	2.07	0.999	0.70	1.008	0.23
θ_3	2.752	1.015	3.40	1.013	1.77	1.011	0.60	1.007	0.20
θ_4	2.100	0.995	2.32	0.990	1.20	0.991	0.41	0.993	0.13
θ_5	2.245	1.022	2.46	1.018	1.28	1.017	0.43	1.015	0.14
θ_6	2.174	1.004	3.00	0.994	1.56	0.992	0.53	1.003	0.17
θ_7	1.415	0.993	2.40	0.990	1.24	0.989	0.42	0.998	0.14
θ_8	1.602	0.996	2.35	0.993	1.22	0.991	0.41	1.000	0.14
θ_9	1.941	1.013	3.09	0.988	1.61	0.989	0.54	0.991	0.18
θ_{10}	1.461	1.008	2.57	1.008	1.33	1.006	0.45	1.008	0.15
θ_{11}	2.688	0.990	2.15	0.991	1.12	0.989	0.38	0.997	0.13
θ_{12}	1.389	1.016	2.63	1.016	1.36	1.015	0.46	1.013	0.15

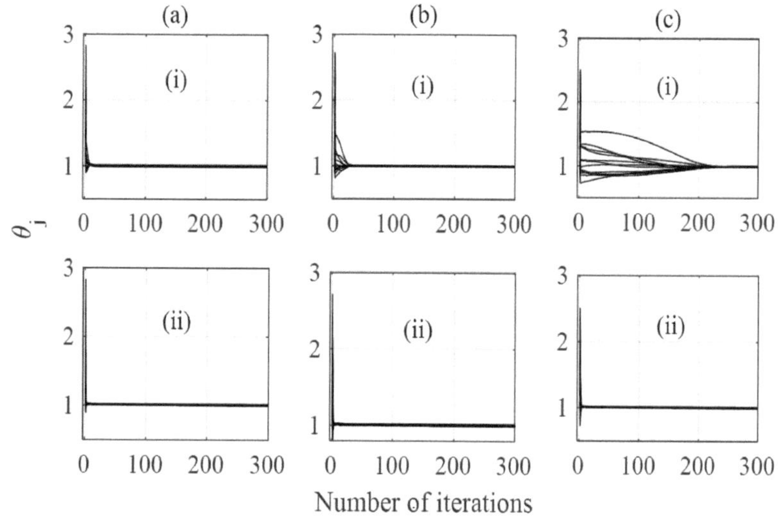

Figure 5.3: Iteration histories for the MAP values of the twelve stiffness scaling parameters ($N_m = 4, N_s = 3$): (i) using the algorithm with β fixed at different values: (a) $\bar{\beta}$; (b) $5\bar{\beta}$; (c) $20\bar{\beta}$, (ii) using the algorithm with β optimised starting from different initial values: (a) $\bar{\beta}$; (b) $5\bar{\beta}$; (c) $20\bar{\beta}$, where $\bar{\beta}$ is estimated from Equation (5.48).

Table 5.2: Identification results using different numbers of measured modes (N_m) but the same number of data segments ($N_s = 3$) (Example 13).

		3 modes ($N_m = 3$)		5 modes ($N_m = 5$)		6 modes ($N_m = 5$)	
Param.	Init. values	MAP	c.o.v.(%)	MAP	c.o.v.(%)	MAP	c.o.v.(%)
θ_1	1.702	0.994	0.44	1.010	0.09	1.005	0.05
θ_2	2.878	1.000	0.47	0.986	0.15	0.999	0.08
θ_3	2.752	1.031	0.67	0.996	0.08	0.999	0.04
θ_4	2.100	1.022	0.47	1.001	0.08	1.012	0.06
θ_5	2.245	0.965	0.36	1.004	0.09	1.004	0.04
θ_6	2.174	0.993	0.35	1.006	0.07	0.998	0.05
θ_7	1.415	0.981	0.42	0.998	0.09	1.006	0.04
θ_8	1.602	0.976	0.50	1.007	0.08	1.013	0.05
θ_9	1.941	0.998	0.39	1.014	0.07	1.001	0.05
θ_{10}	1.461	0.992	0.34	1.000	0.10	0.990	0.05
θ_{11}	2.688	1.009	0.35	0.992	0.07	1.005	0.05
θ_{12}	1.389	0.956	0.47	1.010	0.07	1.004	0.04
β	$\overline{\beta}^{(a)}$	29.670	-	67.689	-	100.851	-

β from the modal data based on the hierarchical Bayesian model in Figure 5.1. The algorithm is also run using from three to five measured modes ($N_m = 3-6$) identified from three data segments ($N_m = 3$), with all hyper-parameters learned using the pseudo-code in Algorithm 9. The final identified MAP values and their associated c.o.v are presented in Table 5.2. The results show, as expected, that using more measured modes results in smaller identification errors and smaller associated uncertainty than the results when using fewer modes; while the identified MAP value $\tilde{\beta}$ gets larger.

We now consider different numbers of data segments to give multiple estimates of the identified modal parameters, where the number of measured modes is $N_m = 4$. The stiffness identification results are presented in Tables 5.3 and 5.4, for full-sensor and partial-sensor scenarios, respectively. For the full-sensor scenario, measurements from all twelve floors are available ($N_o = 12$) while for the partial-sensor scenario, only measurements from six sensors are utilised which are located on the first, fourth, fifth, seventh, tenth, and top floors ($N_o = 6$). It is not surprising to see that the MAP estimates $\tilde{\boldsymbol{\theta}}$ become closer and closer to their actual value with the increase in the number of the data segments and observed DOFs. Moreover, the corresponding uncertainty also decreases, implying higher confidence in the identification results. All of these benefits come from more information in the data that is available for constraining the parameter updating. Notice that when the number of data segments is selected to be fifty for the full-sensor scenario, the identification errors for the MAP estimates of all stiffness scaling parameters θ_j become smaller than 0.6%, which is accurate enough for the MAP values $\tilde{\boldsymbol{\theta}}$ to be utilised as pseudo-data for the likelihood function in Equation (5.22) in a subsequent structural health monitoring stage.

Example 14 *In the second example, the applicability of the proposed methodology to identify sub-structure stiffness reductions is illustrated with a three-dimensional five-story, two-bay by two-bay*

Table 5.3: Identification results using various numbers of data segments for the *full-sensor* scenario ($N_o = 12$) (Example 13).

		5 segments		10 segments		50 segments	
Param.	Init. values	MAP	c.o.v.(%)	MAP	c.o.v.(%)	MAP	c.o.v.(%)
θ_1	1.702	0.991	0.18	0.994	0.18	0.997	0.17
θ_2	2.878	0.987	0.25	0.988	0.24	0.996	0.23
θ_3	2.752	1.009	0.22	0.992	0.21	0.994	0.20
θ_4	2.100	0.993	0.18	0.992	0.14	1.000	0.13
θ_5	2.245	1.006	0.16	0.991	0.15	0.993	0.14
θ_6	2.174	0.981	0.19	0.983	0.18	0.998	0.17
θ_7	1.415	1.003	0.15	1.005	0.14	1.003	0.14
θ_8	1.602	0.992	0.15	0.994	0.14	0.999	0.14
θ_9	1.941	0.987	0.20	0.996	0.19	1.005	0.18
θ_{10}	1.461	0.990	0.16	0.994	0.16	0.998	0.15
θ_{11}	2.688	0.996	0.14	0.993	0.13	0.995	0.13
θ_{12}	1.389	0.996	0.17	0.998	0.16	1.000	0.15
β	$\overline{\beta}^{(a)}$	63.212	-	57.176	-	68.178	-

Table 5.4: Identification results using various number of data segments for the *partial-sensor* scenario ($N_o = 6$) (Example 13).

		5 segments		10 segments		50 segments	
Param.	Init. values	MAP	c.o.v.(%)	MAP	c.o.v.(%)	MAP	c.o.v.(%)
θ_1	1.702	0.967	0.23	0.982	0.20	0.997	0.17
θ_2	2.878	0.915	0.32	0.936	0.28	0.988	0.23
θ_3	2.752	1.048	0.27	1.067	0.22	0.999	0.20
θ_4	2.100	1.063	0.18	1.010	0.15	1.001	0.13
θ_5	2.245	1.037	0.19	1.025	0.16	0.997	0.14
θ_6	2.174	1.052	0.24	1.076	0.20	1.019	0.17
θ_7	1.415	0.990	0.19	0.972	0.16	0.995	0.14
θ_8	1.602	1.122	0.19	1.065	0.16	1.005	0.14
θ_9	1.941	0.944	0.24	0.975	0.20	0.999	0.18
θ_{10}	1.461	0.953	0.21	0.959	0.18	0.984	0.15
θ_{11}	2.688	1.041	0.17	1.021	0.15	0.998	0.13
θ_{12}	1.389	0.927	0.21	0.934	0.17	0.993	0.15
β	$\overline{\beta}^{(a)}$	36.471	-	50.662	-	67.562	-

steel-braced frame structure, which is based on the IASC-ASCE Phase II experimental SHM Benchmark model. The studied structure has a 3m× 3m plan and is 5.5m tall. There are four columns for each floor, one at each corner, and each face in each floor is stiffened by braces. A diagram for the structure is depicted in Figure 5.2 along with its dimensions, in which the x-direction is the strong direction of the columns. A structural model with 150 DOF is used to represent the real structure and to generate the modal data before and after damage. The damage is simulated by reducing the Young's moduli of certain braces in the structural model.

For braced-frame structures, structural damage usually occurs in braces and beam-column connections due to buckling and fracture, respectively. Ideally, we would like to treat each structural member (e.g. brace) as a substructure so that we can infer from the dynamic modal data which, if any, members have been damaged by the event. However, the information available from the structure's local network of sensors will generally be insufficient to support a member-level resolution of stiffness loss from damage, so larger substructures consisting of assemblages of structural members may be necessary in order to reduce the number of model parameters in θ. *A trade-off is therefore required between the number of substructures (and hence the resolution of the damage locations) and the reliability of the probabilistically-inferred damage state.*

For the brace damage inference procedure, a 3-D 15-DOF shear-building model with rigid floors and three DOFs per floor (translations parallel to the x- *and* y-*axes and rotation about the* z-*axis) is employed. Each column contributes an inter-story stiffness of 15 and 20 MN/m in the* x *and* y *direction, respectively. For each face, the stiffness in the braces and beam in each floor is taken to be 20MN/m. As a result, the inter-story stiffness is 100 and 120MN/m in the* x *and* y *direction, respectively. The floor mass is taken to be 12 metric tons for each floor so that the first twelve natural frequencies of the structure are the values tabulated in the first row in Table 5.5 (Configuration 1 is the undamaged calibration case). In order to locate the faces sustaining brace damage, the damage inference model has four stiffness parameters for each story to give a stiffness scaling parameter vector* θ *with twenty components, corresponding to twenty substructures (four faces of five stories). The stiffness matrix* $\boldsymbol{K}$ *is parameterised as:*

$$\boldsymbol{K}(\theta) = \boldsymbol{K}_0 + \Sigma_u \Sigma_v \theta_{uv} \overline{\boldsymbol{K}}_{uv} \tag{5.50}$$

where $u = 1, \ldots, 5$ *refers to the story number and* $v = +x, -x, +y, -y$ *indicates the direction of the outward normal of each face at each floor. The "nominal" stiffness matrices* $\overline{\boldsymbol{K}}_{uv}$ *in Equation* (5.50) *are defined to make the nominal value of each stiffness scaling parameter* θ_{uv} *to be 1.0 and* $\boldsymbol{K}_0$ *is the nominal stiffness matrix contribution from the columns and beams that is not updated. Three damage patterns are defined as follows: 1) Configuration 2: 8.3% reduction in* $\theta_{1,+x}, \theta_{3,+x}$ *and* $\theta_{5,+x}$ *; 2) Configuration 3: 8.3% reduction in* $\theta_{1,+x}, \theta_{3,+x}$ *and* $\theta_{5,+x}$ *; 12% reduction in* $\theta_{1,-y}, \theta_{3,-y}$ *and* $\theta_{5,-y}$ *; 3) Configuration 4: 8.3% reduction in* $\theta_{1,+x}$ *; 12% reduction in* $\theta_{1,-y}$.

In practice, when performing modal identification, some lower modes might not be excited sufficiently to be able to be detected and also the order of the modes might switch when damage occurs, but this is not a problem for the proposed algorithm because it does not require matching the experimental and structural model modes. In this illustrative example, it is assumed that only the first four x- *and* y-*directional modes are measured for all damage patterns, which means that information is available about the 1st, 2nd, 4th, 5th, 7th, 8th, 10th, and 11th modes. Both the full-sensor and partial-sensor scenarios are considered for the identified natural frequencies and mode shape components. For the full-sensor scenario, measurements are available at the center of each side at each floor with the directions parallel to the side in either the positive* x *direction or* y *direction. For the partial-sensor scenario, only the subset of measurements corresponding to the third and the fifth floor are available. For the modal data, zero-mean Gaussian noise with standard deviation equal to 4% of* $\|\phi\|_2$ *is added to the "measured" components of the system mode shapes* ϕ *and the modal fre-*

Table 5.5: System modal frequencies for various damage patterns (Example 14).

Mode order	Frequency											
	1	2	3	4	5	6	7	8	9	10	11	12
Config. 1	4.14	4.53	7.51	12.07	13.22	19.03	20.84	21.93	24.44	26.78	27.88	30.54
Config. 2	4.14	4.47	7.46	12.07	13.04	19.03	20.45	21.76	24.44	26.54	27.88	30.22
Config. 3	4.06	4.47	7.39	11.83	13.04	18.50	20.45	21.54	24.13	26.53	27.46	30.22
Config. 4	4.09	4.49	7.44	11.96	13.14	18.90	20.75	21.76	24.37	26.72	27.86	30.52

quencies ω^2 of the identified modes, respectively. The system modal frequencies for various damage patterns are tabulated in Table 5.5. For the calibration stage, in order to get accurate inferred MAP estimates of the structural stiffness scaling parameters $\widehat{\theta}$, we utilise the proposed algorithm based on 100 independent sets of modal data ($N_s = 100$) from tests of the undamaged structure (Config. 1). The MAP values and the corresponding c.o.v. (coefficients of variation) are shown in Tables 5.6 and 5.7 for the two sensor scenarios. The initial value for each stiffness scaling parameter θ_{uv} is taken randomly from a uniform distribution within the interval between 1 and 3 , which over-estimates their values. It is evident from the tables that the MAP estimates of all stiffness scaling parameters θ_{uv} are accurate, that is, the errors are smaller than 1.7% and 2.5% for full-sensor and partial-sensor scenarios, respectively.

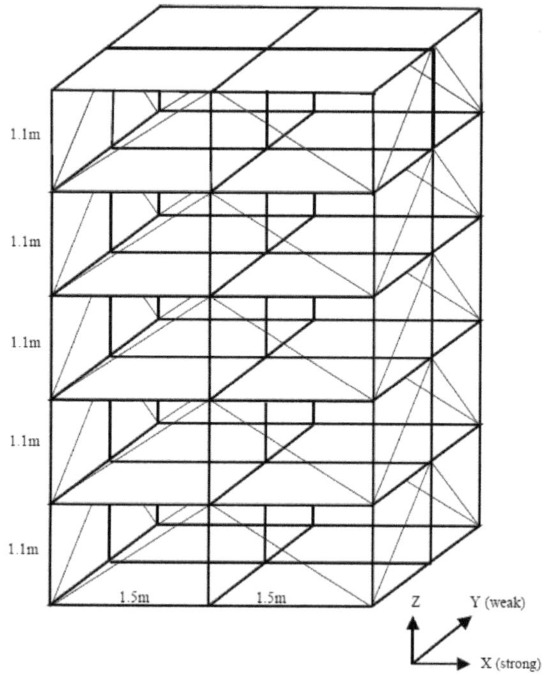

Figure 5.4: Scheme of the investigated structure.

Table 5.6: Identification results for the full-sensor scenario (Example 14).

Param.	Config. 1.fs* MAP $(\widehat{\theta})$	Config. 1.fs* c.o.v.(%)	Config. 2.fs MAP ratio $(\widetilde{\theta}/\widehat{\theta})$	Config. 2.fs c.o.v.(%)	Config. 3.fs MAP ratio $(\widetilde{\theta}/\widehat{\theta})$	Config. 3.fs c.o.v.(%)	Config. 4.fs MAP ratio $(\widetilde{\theta}/\widehat{\theta})$	Config. 4.fs c.o.v.(%)
$\theta_{1,+x}$	0.991	0.13	**0.920**	**0.08**	**0.927**	**0.08**	**0.902**	**0.08**
$\theta_{2,+x}$	1.008	0.08	1.000	0.00	1.000	0.00	1.000	0.00
$\theta_{3,+x}$	1.016	0.09	**0.921**	**0.06**	**0.913**	**0.06**	1.000	0.00
$\theta_{4,+x}$	0.987	0.07	1.000	0.00	1.000	0.00	1.000	0.00
$\theta_{5,+x}$	0.983	0.07	**0.913**	**0.04**	**0.914**	**0.04**	1.000	0.00
$\theta_{1,+y}$	0.989	0.19	1.000	0.00	1.000	0.00	1.000	0.00
$\theta_{2,+y}$	0.994	0.11	1.000	0.00	1.000	0.00	1.000	0.00
$\theta_{3,+y}$	1.006	0.09	1.000	0.00	1.000	0.00	1.000	0.00
$\theta_{4,+y}$	1.011	0.08	1.000	0.00	1.000	0.00	1.000	0.00
$\theta_{5,+y}$	1.008	0.10	1.000	0.00	1.000	0.00	1.000	0.00
$\theta_{1,-x}$	0.997	0.20	1.000	0.00	1.000	0.00	1.000	0.00
$\theta_{2,-x}$	0.989	0.15	1.000	0.00	1.000	0.00	1.000	0.00
$\theta_{3,-x}$	0.990	0.16	1.000	0.00	1.000	0.00	1.000	0.00
$\theta_{4,-x}$	1.010	0.11	1.000	0.00	1.000	0.00	1.000	0.00
$\theta_{5,-x}$	1.012	0.11	1.000	0.00	1.000	0.00	1.000	0.00
$\theta_{1,-y}$	0.998	0.17	1.000	0.00	**0.875**	**0.25**	**0.871**	**0.25**
$\theta_{2,-y}$	0.996	0.01	1.000	0.00	1.000	0.00	1.000	0.00
$\theta_{3,-y}$	0.988	0.08	1.000	0.00	**0.881**	**0.10**	1.000	0.00
$\theta_{4,-y}$	1.002	0.09	1.000	0.00	1.000	0.00	1.000	0.00
$\theta_{5,-y}$	0.994	0.10	1.000	0.00	**0.883**	**0.15**	1.000	0.00

*.fs denotes the full-sensor scenario.

During the monitoring stage, the MAP value $\widehat{\theta}$ from the calibration stage is used as pseudo-data for θ and the proposed algorithm is implemented based on ten sets of identified modal parameters $\widehat{\omega}_r^2$ and $\widehat{\Psi}_r$ $(r = 1, \dots, 10)$ as the primary data for all damage cases. The stiffness ratios of the MAP estimates of the stiffness scaling parameters $\widetilde{\theta}_{uv}$ with respect to those inferred from the calibration stage (Config. 1) and their associated c.o.v. are also tabulated in Tables 5.6 and 5.7 for the three damage patterns. The actual damaged locations are made bold for comparison. It is observed that most of the components with non-bold font have their stiffness ratio exactly equal to one, showing they are unchanged by the monitoring stage data, and the corresponding c.o.v. for each of these components is zero, which means these substructures have no stiffness reduction with full confidence (conditional on the modelling) compared with that of the calibration stage. This is a benefit of the proposed sparse Bayesian formulation which reduces the uncertainty of the unchanged components. It produces sparse models by learning the hyper-parameter α, where $\widetilde{\alpha}_{sv} \to 0$ implies that $\Sigma_{(sv)(sv)} \to 0$ and so $\widetilde{\theta}_{sv} \to \left(\widehat{\theta}\right)_{sv}$. For the posterior uncertainty quantification of those components with bold font,

Table 5.7: Identification results for the partial-sensor scenario (Example 14).

	Config. 1.ps*		Config. 2.ps		Config. 3.ps		Config. 4.ps	
Param.	MAP value ($\widehat{\theta}$)	c.o.v.(%)	MAP ratio ($\widetilde{\theta}/\widehat{\theta}$)	c.o.v.(%)	MAP ratio ($\widetilde{\theta}/\widehat{\theta}$)	c.o.v.(%)	MAP ratio ($\widetilde{\theta}/\widehat{\theta}$)	c.o.v.(%)
$\theta_{1,+x}$	0.987	0.17	**0.916**	**0.07**	**0.923**	**0.08**	**0.911**	**0.08**
$\theta_{2,+x}$	1.008	0.08	1.000	0.00	1.000	0.00	1.000	0.00
$\theta_{3,+x}$	1.025	0.09	**0.921**	**0.06**	**0.926**	**0.06**	1.000	0.00
$\theta_{4,+x}$	0.977	0.07	1.000	0.00	1.000	0.00	1.000	0.00
$\theta_{5,+x}$	0.979	0.07	**0.917**	**0.04**	**0.911**	**0.04**	1.000	0.00
$\theta_{1,+y}$	0.985	0.22	1.000	0.00	1.000	0.00	1.000	0.00
$\theta_{2,+y}$	0.994	0.11	1.000	0.00	0.975	0.12	1.000	0.00
$\theta_{3,+y}$	1.004	0.09	1.000	0.00	1.000	0.00	1.000	0.00
$\theta_{4,+y}$	1.009	0.08	1.000	0.00	1.024	0.09	1.000	0.00
$\theta_{5,+y}$	1.007	0.08	1.000	0.00	1.000	0.00	1.000	0.00
$\theta_{1,-x}$	0.998	0.19	1.000	0.00	1.000	0.00	1.000	0.00
$\theta_{2,-x}$	0.988	0.14	1.000	0.00	1.000	0.00	1.000	0.00
$\theta_{3,-x}$	0.980	0.15	1.000	0.00	1.000	0.00	1.000	0.00
$\theta_{4,-x}$	1.016	0.12	1.000	0.00	1.000	0.00	1.000	0.00
$\theta_{5,-x}$	1.014	0.12	1.011	0.07	1.000	0.00	1.000	0.00
$\theta_{1,-y}$	0.999	0.18	1.000	0.00	**0.878**	**0.25**	**0.896**	**0.25**
$\theta_{2,-y}$	0.997	0.10	1.000	0.00	1.000	0.00	1.000	0.00
$\theta_{3,-y}$	0.991	0.08	1.000	0.00	**0.874**	**0.11**	1.000	0.00
$\theta_{4,-y}$	1.005	0.08	1.000	0.00	0.978	0.10	1.000	0.00
$\theta_{5,-y}$	0.995	0.07	1.000	0.00	**0.895**	**0.15**	1.000	0.00

*.ps denotes the partial-sensor scenario.

the c.o.v. values in Table 5.7 for the partial sensor scenario are generally higher as compared with those in Table 5.6 for a full-sensor scenario, since the sensor data for the partial-sensor scenario provide less information to constrain the updated parameter values.

If we issue a damage alarm for a substructure when the corresponding stiffness ratio is smaller than one (stiffness loss is larger than zero), it is seen that no false or missed damage indications are produced for the full-sensor scenario. While for the partial-sensor scenario, two false detections are observed for the Config. 3.ps case in the substructures corresponding to $\theta_{2,+y}$ *and* $\theta_{4,-y}$ *with 2.5% and 2.2% stiffness reductions, respectively. However, the value for no damage is around two standard deviations of 1% from the MAP values for* $\theta_{2,+y}$ *and* $\theta_{4,-y}$ *and so the probability of no damage is high. There are also two stiffness scaling parameters,* $\theta_{5,-x}$ *for Config. 2.ps case and* $\theta_{4,+y}$ *for Config. 3.ps case, where the stiffness ratio is larger than one; however, they show a very small increase in stiffness of 1% and 2%, respectively. For the identified damage extent, we observe that the identified stiffness ratios are close to their actual values (0.917 and 0.880) for both the full and partial sensor scenarios, and it is not surprising that the identified values are more accurate for the full sensor scenario. Therefore, the damage patterns are reliably detected in both qualitative and quantitative ways.*

A more complete picture of substructure damage is given by the probabilities of damage, $P_{sv}^{dam}(f)$*, with different fractional stiffness losses* $f \in [0,1]$*, calculated using Equation (5.49). In Figure 5.5, the probability curves of damage are shown for the twenty* θ_{sv}*'s for various damage patterns from applying the proposed algorithm, for both the full- and partial-sensor scenarios. The non-zero probabilities for negative fractional stiffness losses* f *are a result of the Gaussian approximations used in the calculation of the curves. It is clear in all cases that the substructures with real*

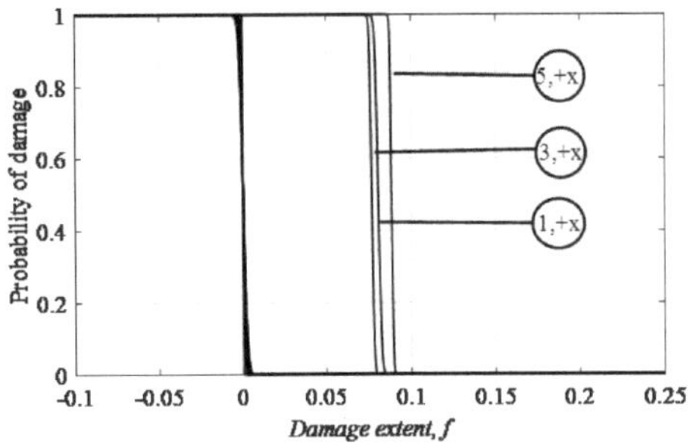

(a)(i)Config. 2.fs

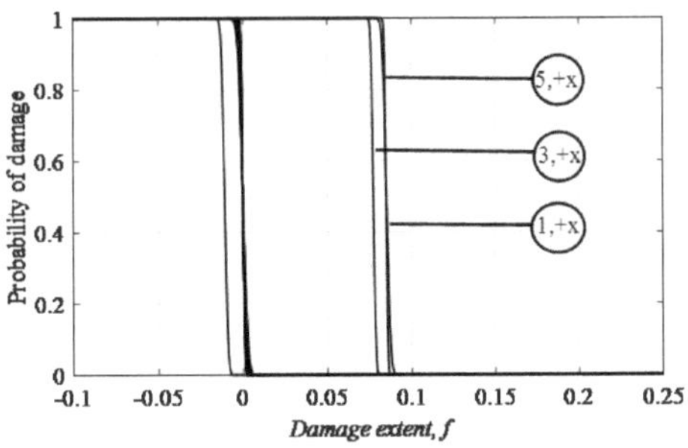

(a)(ii)Config. 2.ps

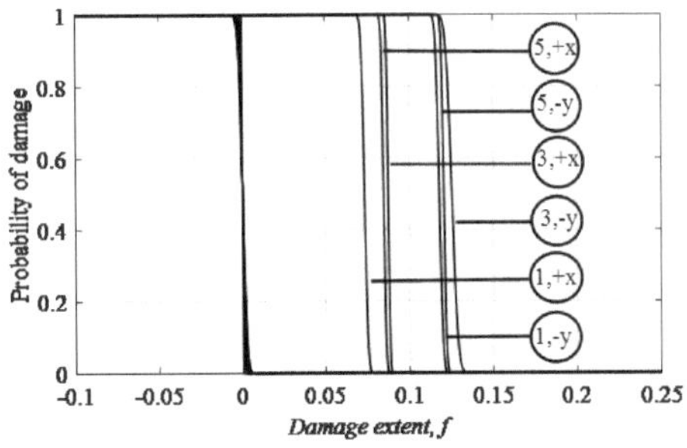

(b)(i)Config. 3.fs

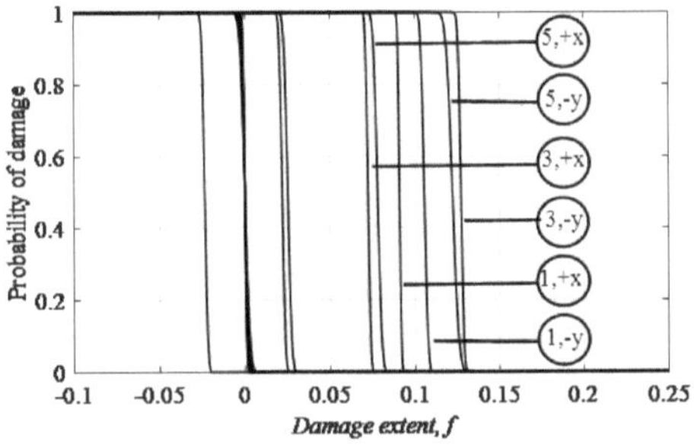

(b)(ii)Config. 3.ps

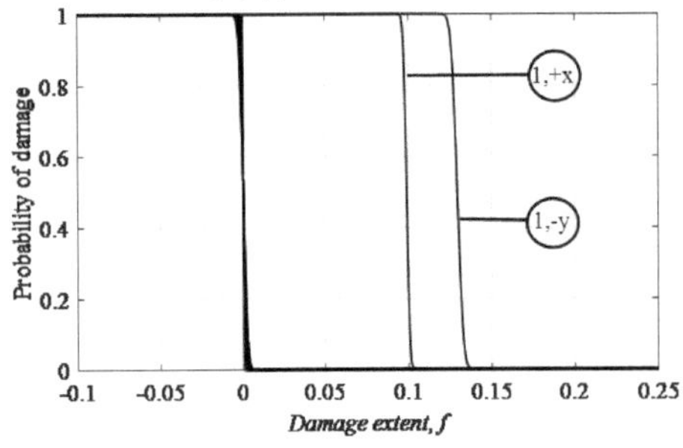

(c)(i)Config. 4.fs

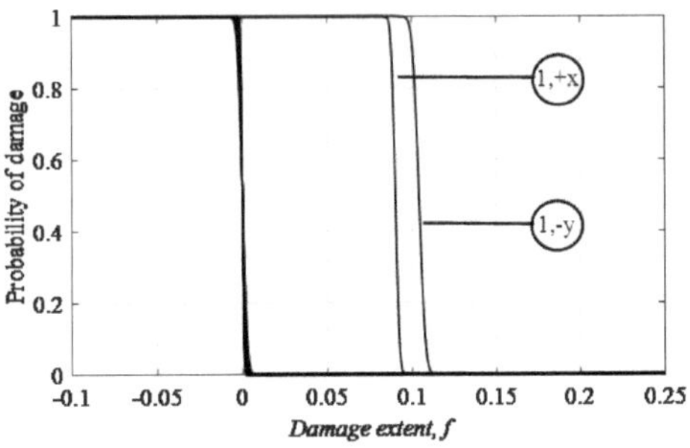

(c)(ii)Config. 4.ps

Figure 5.5: Probability of substructure damage exceeding f for damage patterns: (a) Config. 2, (b) Config. 3, (c) Config. 4, and two sensor scenarios: (i) full-sensor scenario, (ii) partial-sensor scenario.

stiffness losses have a damage probability of almost unity with a large damage extent. Consider the $-y$ face of the first story as an example. For the full-sensor cases Config. 3.fs and Config. 4.fs, this substructure has a damage probability of almost unity for a stiffness loss ratio less than 11.5% and a probability of almost zero that the damage exceeds 13.5% (its actual value is 12%). However, for the partial-sensor cases Config. 3.ps and Config. 4.ps, this substructure is inferred to have a damage probability of almost unity for a stiffness loss ratio less than 9.5% and a probability of almost zero that the damage exceeds 11.5%, showing an underestimation of the damage extent.

For the substructures that are undamaged, most of the posterior median fractional stiffness losses corresponding to a damage probability of 0.5 *are close to zero and probabilities decrease to zero when the damage exceeds* 0.5%*, showing a very small plausibility that damage has happened in these substructures. The exceptions are the two curves for the case of Config. 3.ps with partial-sensor scenario (Figure 5.5(b)(ii)), which have probabilities close to 1.0 for damage fractions up to 2.5%, corresponding to the two false detections shown in Table 5.7. Compared with traditional Bayesian updating methods without model sparseness imposed (e.g. [200]), the confidence for correct damage indication is high in the proposed method. The reduced posterior uncertainty for the inferred stiffness scaling parameter* θ *from the proposed hierarchical sparse Bayesian learning framework helps suppress the occurrence of false and missed damage indications and increases the confidence of correct damage detection.*

5.5 Concluding remarks

A new hierarchical sparse Bayesian learning methodology is presented to perform sparse stiffness loss inference for damage identification purposes based on identified noisy incomplete modal parameters from structural sensor data. For each chosen substructure, not only the most probable estimate of the spatially-sparse stiffness changes is inferred from the identified modal parameters but also the associated posterior uncertainties of the stiffness parameters are quantified, including the probability of substructure damage of various amounts. The approach has four important benefits: first, no matching of model and experimental modes is needed; second, rather than directly solving the challenging nonlinear inverse problem related to a structural model eigenvalue equation, the proposed formulation applies an efficient iterative procedure that involves a series of coupled linear regression problems and provides a tractable form for sparse Bayesian learning. Third, the identified stiffness reductions are independent of the scaling of the stiffness and mass matrices, as well as the scaling of the identified mode shapes. Fourth, the new algorithm estimates all uncertain model hyper-parameters solely from the data, giving an algorithm for which no user-intervention is needed.

The illustrative examples confirm the effectiveness and robustness of the new approach. The first example demonstrates the ability of the proposed method to update a structural model during a calibration stage for an undamaged building, showing the benefit of the hierarchical Bayesian modelling and learning of the hyper-parameters. The second example using a three-dimensional five-story braced frame structure shows that for all cases, the simulated damage under study is reliably detected with the high accuracy of the identified stiffness reduction achieved by exploiting damage sparseness. The occurrence of false-positive and false-negative damage detection in the presence of modelling errors is effectively suppressed by the proposed method. The method also has an important advantage for actual applications; it can be used to update the structural stiffness efficiently based on the information in the modal data from dynamic testing without knowing if any significant modes are missing in the modal data set, or whether the ordering of the modes switches due to damage.

The method has general applicability for Bayesian system identification. It can be used, for example, to find sparse adjustments of the stiffness distribution specified by a finite-element structural model so that it becomes more consistent with the measured response of the structure. Alternatively, provided sufficient data are available, the method allows a set of stiffness changes to be inferred without imposing sparseness. This is relevant for the calibration stage where many structural stiffness parameters of a nominal (e.g. finite-element) model may need to be changed.

Appendices

Appendix A: Derivation of MAP estimation equations for α and β

The objective function for the derivation of MAP estimation equations of $\boldsymbol{\alpha}$ and β is defined as:

$$\begin{aligned}
J(\boldsymbol{\alpha},\beta) =& \log\left[p\left(\widehat{\boldsymbol{\theta}}|\boldsymbol{\omega}^2,\boldsymbol{\phi},\beta,\boldsymbol{\alpha}\right)p\left(\boldsymbol{\omega}^2,\boldsymbol{\phi}|\beta\right)p\left(\beta|b_0\right)\right]\\
=&\frac{1}{2}\left(N_dN_m-N_\theta\right)\log\beta-\frac{1}{2}\log\left|\mathbf{A}+\left(\beta\mathbf{H}^T\mathbf{H}\right)^{-1}\right|\\
&-\frac{1}{2}\left(\widehat{\boldsymbol{\theta}}-\left(\mathbf{H}^T\mathbf{H}\right)^{-1}\mathbf{H}^T\mathbf{b}\right)^T\left(\mathbf{A}+\left(\beta\mathbf{H}^T\mathbf{H}\right)^{-1}\right)^{-1}\left(\widehat{\boldsymbol{\theta}}-\left(\mathbf{H}^T\mathbf{H}\right)^{-1}\mathbf{H}^T\mathbf{b}\right)\\
&-\frac{\beta}{2}\left(\mathbf{b}^T\mathbf{b}-\mathbf{b}^T\mathbf{H}\left(\mathbf{H}^T\mathbf{H}\right)^{-1}\mathbf{H}^T\mathbf{b}\right)+\log b_0-b_0\beta+c
\end{aligned}\tag{A1}$$

where c is a constant independent of $\boldsymbol{\alpha}$ and β. Following [94], we can simplify the following two terms in (A1):

$$\log\left|\mathbf{A}+\left(\beta\mathbf{H}^T\mathbf{H}\right)^{-1}\right| = -N_\theta\log\beta-\log\left|\mathbf{H}^T\mathbf{H}\right|+\log|\mathbf{A}|+\log\left|\boldsymbol{\Sigma}^{-1}\right|\tag{A2}$$

and

$$\begin{aligned}
&\left(\widehat{\boldsymbol{\theta}}-\left(\mathbf{H}^T\mathbf{H}\right)^{-1}\mathbf{H}^T\mathbf{b}\right)^T\left(\mathbf{A}+\left(\beta\mathbf{H}^T\mathbf{H}\right)^{-1}\right)^{-1}\left(\widehat{\boldsymbol{\theta}}-\left(\mathbf{H}^T\mathbf{H}\right)^{-1}\mathbf{H}^T\mathbf{b}\right)\\
&+\beta\left(\mathbf{b}^T\mathbf{b}-\mathbf{b}^T\mathbf{H}\left(\mathbf{H}^T\mathbf{H}\right)^{-1}\mathbf{H}^T\mathbf{b}\right)\\
&=\left(\widehat{\boldsymbol{\theta}}-\boldsymbol{\mu}_\theta\right)^T\mathbf{A}^{-1}\left(\widehat{\boldsymbol{\theta}}-\boldsymbol{\mu}_\theta\right)+\beta\left(\boldsymbol{\mu}_\theta-\left(\mathbf{H}^T\mathbf{H}\right)^{-1}\mathbf{H}^T\mathbf{b}\right)^T\mathbf{H}^T\mathbf{H}\left(\boldsymbol{\mu}_\theta-\left(\mathbf{H}^T\mathbf{H}\right)^{-1}\mathbf{H}^T\mathbf{b}\right)\\
&\quad+\beta\left(\mathbf{b}^T\mathbf{b}-\mathbf{b}^T\mathbf{H}\left(\mathbf{H}^T\mathbf{H}\right)^{-1}\mathbf{H}^T\mathbf{b}\right)\\
&=\left(\widehat{\boldsymbol{\theta}}-\boldsymbol{\mu}_\theta\right)^T\mathbf{A}^{-1}\left(\widehat{\boldsymbol{\theta}}-\boldsymbol{\mu}_\theta\right)+\beta\|\mathbf{H}\boldsymbol{\mu}_\theta-\mathbf{b}\|^2
\end{aligned}$$

Then, the partial derivatives of the objective function in (A1) with respect to α_j and β are given by

$$\begin{aligned}
\frac{\partial J(\boldsymbol{\alpha},\beta)}{\partial\alpha_j}&=\frac{1}{2}\frac{\partial}{\partial\alpha_j}\left[-\log\left|\mathbf{A}+\left(\beta\mathbf{H}^T\mathbf{H}\right)^{-1}\right|-\left(\widehat{\boldsymbol{\theta}}-\boldsymbol{\mu}_\theta\right)^T\mathbf{A}^{-1}\left(\widehat{\boldsymbol{\theta}}-\boldsymbol{\mu}_\theta\right)\right]\\
&=\frac{1}{2}\left[-\alpha_j^{-1}+\alpha_j^{-2}\left(\boldsymbol{\Sigma}_\theta\right)_{jj}+\alpha_j^{-2}\left(\widehat{\boldsymbol{\theta}}-\boldsymbol{\mu}_\theta\right)_j^2\right]
\end{aligned}\tag{A4}$$

$$\begin{aligned}
\frac{\partial J(\boldsymbol{\alpha},\beta)}{\partial\beta}&=\frac{1}{2}\frac{\partial}{\partial\beta}\left[N_dN_m\log\beta-\log\left|\mathbf{A}+\left(\beta\mathbf{H}^T\mathbf{H}\right)^{-1}\right|-\beta\|\mathbf{H}\boldsymbol{\mu}_\theta-\mathbf{b}\|^2-\left(\widehat{\boldsymbol{\theta}}-\boldsymbol{\mu}_\theta\right)^T\mathbf{A}^{-1}\left(\widehat{\boldsymbol{\theta}}-\boldsymbol{\mu}_\theta\right)-2b_0^{-1}\beta\right]\\
&=\frac{1}{2}\beta^{-1}N_dN_m-\frac{1}{2}\beta^{-1}\sum_{j=1}^{N_\theta}\left(1-\alpha_j^{-1}\left(\boldsymbol{\Sigma}_\theta\right)_{jj}\right)-\frac{1}{2}\|\mathbf{H}\boldsymbol{\mu}_\theta-\mathbf{b}\|^2-b_0^{-1}
\end{aligned}\tag{A5}$$

Setting the derivatives of $J(\boldsymbol{\alpha}, \beta)$ to zero leads to the update formulae for α_j and β given previously in Equations (5.42) and (5.43), respectively. It is seen that the MAP estimation equations of $\boldsymbol{\alpha}$ and β in (5.42) and (5.43) involve the posterior covariance matrix $\boldsymbol{\Sigma}_{\boldsymbol{\theta}}$, which is independent of the hyper-parameters ρ, η, and b_0. This is because the full posterior uncertainty in the stiffness scaling parameters $\boldsymbol{\theta}$ is explicitly incorporated when finding the MAP estimates of $\boldsymbol{\alpha}$ and β by maximising the logarithm function of the product of the pseudo-evidence $p\left(\widehat{\boldsymbol{\theta}}|\boldsymbol{\delta}\right) = p\left(\widehat{\boldsymbol{\theta}}|\boldsymbol{\omega}^2, \boldsymbol{\phi}, \beta, \boldsymbol{\alpha}\right)$ with prior PDFs that depend on $\boldsymbol{\alpha}$ and β. For the logarithm function of the product of pseudo-evidence $p\left(\widehat{\boldsymbol{\theta}}|\boldsymbol{\omega}^2, \boldsymbol{\phi}, \beta, \boldsymbol{\alpha}\right)$ and PDF $p\left(\boldsymbol{\omega}^2, \boldsymbol{\phi}|\beta\right)$ in $J(\boldsymbol{\alpha}, \beta)$ in Equation (A1), the information-theoretical interpretation of the trade-off between data fitting and model complexity [22] can be demonstrated as follows, where we use Bayes' theorem and the hierarchical model structure in Figure 5.1:

$$
\begin{aligned}
&\log\left[p\left(\widehat{\boldsymbol{\theta}}|\boldsymbol{\omega}^2, \boldsymbol{\phi}, \beta, \boldsymbol{\alpha}\right) p\left(\boldsymbol{\omega}^2, \boldsymbol{\phi}|\beta\right)\right] \\
&= \int \log\left[p\left(\widehat{\boldsymbol{\theta}}|\boldsymbol{\omega}^2, \boldsymbol{\phi}, \beta, \boldsymbol{\alpha}\right) p\left(\boldsymbol{\omega}^2, \boldsymbol{\phi}|\beta\right)\right] p\left(\boldsymbol{\theta}|\widehat{\boldsymbol{\theta}}, \boldsymbol{\omega}^2, \boldsymbol{\phi}, \beta, \boldsymbol{\alpha}\right) d\boldsymbol{\theta} \\
&= \int \log\left[\frac{p\left(\boldsymbol{\theta}|\boldsymbol{\omega}^2, \boldsymbol{\phi}, \beta\right) p\left(\boldsymbol{\omega}^2, \boldsymbol{\phi}|\beta\right) p\left(\widehat{\boldsymbol{\theta}}|\boldsymbol{\theta}, \boldsymbol{\alpha}\right)}{p\left(\boldsymbol{\theta}|\widehat{\boldsymbol{\theta}}, \boldsymbol{\omega}^2, \boldsymbol{\phi}, \beta, \boldsymbol{\alpha}\right)}\right] p\left(\boldsymbol{\theta}|\widehat{\boldsymbol{\theta}}, \boldsymbol{\omega}^2, \boldsymbol{\phi}, \beta, \boldsymbol{\alpha}\right) d\boldsymbol{\theta} \\
&= \int \log\left[p\left(\boldsymbol{\omega}^2, \boldsymbol{\phi}, \boldsymbol{\theta}|\beta\right)\right] p\left(\boldsymbol{\theta}|\widehat{\boldsymbol{\theta}}, \boldsymbol{\omega}^2, \boldsymbol{\phi}, \beta, \boldsymbol{\alpha}\right) d\boldsymbol{\theta} - \int \log\left[\frac{p\left(\boldsymbol{\theta}|\widehat{\boldsymbol{\theta}}, \boldsymbol{\omega}^2, \boldsymbol{\phi}, \beta, \boldsymbol{\alpha}\right)}{p\left(\widehat{\boldsymbol{\theta}}|\boldsymbol{\theta}, \boldsymbol{\alpha}\right)}\right] p\left(\boldsymbol{\theta}|\widehat{\boldsymbol{\theta}}, \boldsymbol{\omega}^2, \boldsymbol{\phi}, \beta, \boldsymbol{\alpha}\right) d\boldsymbol{\theta} \\
&= \int \log\left[p\left(\boldsymbol{\omega}^2, \boldsymbol{\phi}, \boldsymbol{\theta}|\beta\right)\right] p\left(\boldsymbol{\theta}|\widehat{\boldsymbol{\theta}}, \boldsymbol{\omega}^2, \boldsymbol{\phi}, \beta, \boldsymbol{\alpha}\right) d\boldsymbol{\theta} - \int \log\left[\frac{p\left(\boldsymbol{\theta}|\widehat{\boldsymbol{\theta}}, \boldsymbol{\omega}^2, \boldsymbol{\phi}, \beta, \boldsymbol{\alpha}\right)}{p\left(\boldsymbol{\theta}|\widehat{\boldsymbol{\theta}}, \boldsymbol{\alpha}\right)}\right] p\left(\boldsymbol{\theta}|\widehat{\boldsymbol{\theta}}, \boldsymbol{\omega}^2, \boldsymbol{\phi}, \beta, \boldsymbol{\alpha}\right) d\boldsymbol{\theta}
\end{aligned}
\tag{A6}
$$

where the pseudo likelihood $p\left(\widehat{\boldsymbol{\theta}}|\boldsymbol{\theta}, \boldsymbol{\alpha}\right) = \mathcal{N}\left(\widehat{\boldsymbol{\theta}}|\boldsymbol{\theta}, \mathbf{A}\right) = \mathcal{N}\left(\boldsymbol{\theta}|\widehat{\boldsymbol{\theta}}, \mathbf{A}\right) = p\left(\boldsymbol{\theta}|\widehat{\boldsymbol{\theta}}, \boldsymbol{\alpha}\right)$. Equation (A6) shows that the logarithm function, which is to be maximised, is the difference between the posterior mean of the log joint prior PDF $p\left(\boldsymbol{\omega}^2, \boldsymbol{\phi}, \boldsymbol{\theta}|\beta\right)$ (the first term) and the relative entropy (or Kullback-Leibler information) of the posterior PDF $p\left(\boldsymbol{\theta}|\widehat{\boldsymbol{\theta}}, \boldsymbol{\omega}^2, \boldsymbol{\phi}, \beta, \boldsymbol{\alpha}\right)$ of $\boldsymbol{\theta}$ with respect to the PDF $p\left(\boldsymbol{\theta}|\widehat{\boldsymbol{\theta}}, \boldsymbol{\alpha}\right)$ conditional only on $\boldsymbol{\alpha}$ and the MAP vector $\widehat{\boldsymbol{\theta}}$, which is obtained from the calibration stage (the second term). The first term quantifies the ability of the modal parameters corresponding to the structural model specified by $\boldsymbol{\theta}$ to match the system modal parameters $\boldsymbol{\omega}^2$ and $\boldsymbol{\phi}$; it is maximised if the model modal parameters become tightly clustered around the system modal parameters $\boldsymbol{\omega}^2$ and $\boldsymbol{\phi}$, that is, the equation error precision parameter $\beta \to \infty$. The second term reflects the amount of information extracted from the system modal parameters $\boldsymbol{\omega}^2$ and $\boldsymbol{\phi}$, and so from the "measured" modal data $\widehat{\boldsymbol{\omega}}^2$ and $\widehat{\boldsymbol{\Psi}}$, as implied by the observation of the hierarchical model structure exhibited in Figure 5.1. It penalises the models that have more parameter components θ_j differing from those in $\widehat{\boldsymbol{\theta}}$, and therefore forces the model updating to extract less information from the system modal parameters $\boldsymbol{\omega}^2$ and $\boldsymbol{\phi}$. Over-extraction of information from the system modal parameters $\boldsymbol{\omega}^2$ and $\boldsymbol{\phi}$ will produce a structural model vector $\boldsymbol{\theta}$ with too large of a difference from $\widehat{\boldsymbol{\theta}}$ that is overly sensitive to the details of the information in the specified system modal parameters $\boldsymbol{\omega}^2$ and $\boldsymbol{\phi}$, and therefore in the "measured" modal parameters $\widehat{\boldsymbol{\omega}}^2$ and $\widehat{\boldsymbol{\Psi}}$. In other words, the measurement noise and other environmental effects may not be "smoothed out" so they may have an excessive effect on the damage detection performance.

In summary, the maximisation of (A6) automatically produces the optimal trade-off between data fitting and model complexity that causes many hyper-parameters α_j to approach zero with a reasonably large value of β, giving a model $\boldsymbol{\theta}$ that has both a sufficiently small number of components θ_j differing from those in $\widehat{\boldsymbol{\theta}}$ and fits the system modal parameters $\boldsymbol{\omega}^2$ and $\boldsymbol{\phi}$ well. This is an instance of the Principle of Model Parsimony or the Bayesian Ockham's razor. The Bayesian procedure effectively implements Ockham's razor by assigning lower probabilities to a structural model whose parameter vector $\boldsymbol{\theta}$ has too large or too small differences from $\widehat{\boldsymbol{\theta}}$ obtained at the calibration stage (too few or too many $\alpha_j \to 0$), thereby suppressing the occurrence of false and missed damage detection alarms.

6

Ultrasonic Guided-waves Based Bayesian Damage Localisation and Optimal Sensor Configuration

Sergio Cantero-Chinchilla,[1,*] *Juan Chiachío*,[2] *Manuel Chiachío*[2] and *Dimitrios Chronopoulos*[1]

Structural health monitoring is comprised of different stages, which range from the optimisation of a structural system to the provision of future predictions about its integrity. In this chapter, both the localisation of damage and the optimisation of a structural health monitoring system are rigorously addressed and illustrated for ultrasonic guided-waves. The complexities related to the measurement and post-processing process of the ultrasonic signals lead to uncertainties that need to be quantified for unbiased diagnostics and fully informed decision-making. The framework for damage localisation consists of two hierarchical Bayesian inverse problems, whereby both model class selection of several candidate signal post-processing models and the damage position inference are carried out simultaneously. Furthermore, the optimal sensor configuration (number and location) is addressed by maximising the value instead of the amount of information, whereby a trade-off between cost and damage-related information is obtained. Several examples show the rigour and accuracy of these Bayesian methods for structural health monitoring.

6.1 Introduction

Structural health monitoring (SHM) is a core element of condition-based maintenance, whereby maintenance actions are planned based on the actual condition of the structural asset, in contrast to schedule-based maintenance. Among the existing SHM techniques available [198, 19, 174], this chapter focuses on ultrasonic guided-waves, due to its potential in exploring large thin-walled structures with relatively low attenuation [177, 32]. These characteristics have led safety-critical industries, such as aerospace, to focus on guided-waves as an SHM technique. Nonetheless, the ultrasonic measurements convey uncertainties coming from different sources: (1) electronic noise generated in the generation and acquisition ultrasonic system; (2) material properties variability; and (3) the lack of knowledge (epistemic uncertainty) about post-processing techniques and wave propagation. Accounting for these uncertainty sources is crucial for obtaining rigorous information about the structural health, and therefore to carry out fully informed maintenance decision-making.

[1] University of Nottingham (UK).
[2] University of Granada (Spain).
[*] Corresponding author: sergio.canterochinchilla@bristol.ac.uk

The effectiveness and reliability of a SHM system depend heavily on the configuration (i.e. number, type, and number) of the sensors. In order to provide a reliable, yet efficient, configuration for a SHM system, the adoption of an optimal sensor configuration is a key aspect in any SHM system, since it enables a balance to be struck between the performance of the SHM system and its related cost [138]. Theoretically, an infinitely reliable system would require an infinite amount of information; however, such a theoretical rule finds an exception in practice when features related to the complexity of the monitoring system, such as the cost or the weight, are taken into consideration. The latter suggests a trade-off between reliability and complexity for SHM design that needs to be assessed rigorously for optimal SHM functionality. Authors have investigated these aspects and the most relevant approaches can be broadly categorised into two groups, namely, approaches using the *value of information* as an optimality criterion [91, 165] and those using *cost-benefit* analysis [115]. The first approach provides the sensor and/or actuator configuration that holds the best balance between the cost of the SHM system and the amount of information gain by the system [35]. Alternatively, the second one typically uses performance indices (e.g. the probability of detection [138] or the Lindley information measure [122]) and cost-related features associated within the layout of sensors of the SHM system [33], which are simultaneously maximised and minimised respectively using a multi-objective type of optimisation algorithm.

Furthermore, the localisation of damage is particularly relevant for industries such as the aerospace and the wind power industries, given that some of the structures used in aircraft or wind turbines are very large, and some of them are out of range for other non-destructive testing techniques such as visual inspection. To address such a challenging scenario, the damage localisation stage can be addressed using guided-wave based SHM along with post-processing techniques applied to ultrasonic measurements [1]. Two general approaches are typically adopted at this stage for damage assessment: (1) model-based inverse problems [41], whereby complete damage information can be obtained at a considerable computational cost; and (2) inverse problems based on post-processed signal features, whereby other relevant information can be obtained more efficiently. With regard to the model-based inverse problem type of identification, a physics-based wave propagation model is required. Alternatively, the second approach uses simpler models (such as the time-of-flight) whose outcomes are comparable with the ones provided by post-processing techniques (e.g. time-frequency transforms) instead of raw experimental data [34]. Note that an excessively high level of information is not usually required at the damage localisation stage, given that depending on the location of the damage, this might have different levels of relevance.

This chapter illustrates the use of uncertainty quantification approaches for ultrasonic guided-wave-based SHM in two different monitoring stages: (1) the localisation of damage, and (2) the optimisation of the sensor configuration. Considering that the optimisation of the SHM system is designed so that the preciseness of the damage localisation is maximized, we firstly describe the localisation methodology in Section 6.2. Then, the optimisation of sensor configurations using the value of information for an optimal reconstruction of the damage position is shown in Section 6.3.

6.2 Damage localisation

The reconstruction of the damage position in plate-like structures is addressed in this section by using ultrasonic guided-wave measurements and a two-level Bayesian inverse problem.

6.2.1 Time-frequency model selection

The damage localisation problem based on signal features requires the selection of a signal post-processing technique that extracts such features. Here, time-frequency (TF) models are used, whereby the frequency and energy content of the ultrasonic signal at each instant of time is obtained. The energy representation is then used to obtain the time of flight of the scattered signal, that is, the wave generated at the damage position, by which the Bayesian damage localisation can be addressed. The selection of a particular TF model among a set of candidate TF models [51, 93, 50, 20] is typically based on previous experience or specific characteristics. However, a more rigorous decision would be based on probability-logic assumptions, whereby the most plausible model among a set of candidates would be obtained.

In order to address the TF model selection problem, a finite set of N_m candidate models may be selected in $\mathbf{M} = \{\mathcal{M}_1^{(k)}, \ldots, \mathcal{M}_{N_m}^{(k)}\}$, where $\mathcal{M}_j^{(k)}$ denotes the jth *model class* applied to the signal acquired in the kth PZT sensor. Given that these models are just different alternatives to mimic the same reality, based on different hypotheses and assumptions, different outcomes can be expected from a variety of models. Thus, for a particular model, the validity of such hypotheses depends on the adopted values of certain model parameters. In this case, the parameters that control the output of the TF models are fixed and assumed to have *perfect information* (i.e. no uncertainty is related to them). The uncertain model parameter is selected to be the dispersion parameter, once the stochastic embedding has been introduced. Thus, to simultaneously identify both the plausibility of each TF model and the values of the model parameters that better suit the information coming from the raw ultrasonic data, a Bayesian inverse problem (BIP) is proposed here. It is worth mentioning that this BIP addresses only the data acquired in one PZT sensor, hence offering the possibility to run several BIPs in parallel for the different sensors available in a structure. Thus, uncertainties related to the sensors, such as different working environments or manufacturing defects, are taken into account.

6.2.1.1 Stochastic embedding of TF models

Let us consider a candidate TF model mathematically defined by the relationship $g_j : \mathbb{R}^n \to \mathbb{R}$ between a discrete signal $\mathcal{D}^{(k)} \in \mathbb{R}^n$ acting as input and the model output $g_j \in \mathbb{R}$, where k denotes the kth sensor in the structure. Next, let $\hat{d}_j^{(k)} \in \mathbb{R}$ be the first energy peak observed in the scattered ultrasound signal, that is, the time of flight, so that $\hat{d}_j^{(k)} = g_j(\mathcal{D}^{(k)})$. This representation could lead to biased results should no disagreement between measurement and model output be taken into account; thus, such a signal feature would be more rigorously represented as an uncertain variable, as follows

$$\tilde{d}^{(k)} = g_j\left(\mathcal{D}^{(k)}\right) + \epsilon \tag{6.1}$$

where ϵ is an error term which accounts for the discrepancy between $\hat{d}_j^{(k)}$ and $\tilde{d}^{(k)}$, namely the modelled and measured values for the first energy peak, respectively. This type of error term also is used to capture the uncertainty coming from the data and both the model and its parameters. Following the Principle of Maximum Information Entropy [22, 100], this error can be conservatively assumed to be modelled as a zero-mean Gaussian distribution with standard deviation σ_ϵ, that is, $\epsilon \sim \mathcal{N}(0, \sigma_\epsilon)$. Thus, following Equation (6.1), a probabilistic description of the TF model can be obtained as

$$p\left(\tilde{d}^{(k)}|\mathcal{M}_j^{(k)},\sigma_\epsilon\right) = \left(2\pi\sigma_\epsilon^2\right)^{-\frac{1}{2}} \exp\left(-\frac{1}{2}\left(\frac{\tilde{d}^{(k)} - g_j^{-1}\left(\mathcal{D}^{(k)}\right)}{\sigma_\epsilon}\right)^2\right) \tag{6.2}$$

This probabilistic model, associated to the *prior* PDF of the model parameter σ_ϵ, $p(\sigma_\epsilon|\mathcal{M}_j^{(k)})$, corresponds to the previously mentioned model class $\mathcal{M}_j^{(k)}$. This prior PDF represents the initial degree of belief of the values of σ_ϵ within a set of possible values $\mathbf{\Theta} \subseteq \mathbb{R}$ before the information from measurements is incorporated through Bayesian updating, as explained further below. For all the sensors in the structure, the stochastic model is defined independently, thus accounting for different potential sources of errors and uncertainties.

6.2.1.2 Model parameters estimation

In a previous sensitivity study, the dispersion parameter σ_ϵ was revealed as the most critical parameter, that is, the variation of it resulted in significant alterations in the model class assessment. Thus, a first stage of the BIP is conceived to obtain the set of most plausible values for σ_ϵ given a set of data $\hat{\mathbf{d}}^{(k)} = \{\hat{d}_1^{(k)}, \ldots, \hat{d}_{N_m}^{(k)}\}$, which corresponds to a set of N_m times of flight observed by each TF model class. To this end, the posterior PDF $p(\sigma_\epsilon|\hat{\mathbf{d}}^{(k)}, \mathcal{M}_j^{(k)})$ of the dispersion parameter σ_ϵ given the jth TF model class $\mathcal{M}_j^{(k)}$, is provided using Bayes' theorem, as follows

$$p\left(\sigma_\epsilon|\hat{\mathbf{d}}^{(k)},\mathcal{M}_j^{(k)}\right) = c^{-1} p\left(\hat{\mathbf{d}}^{(k)}|\sigma_\epsilon,\mathcal{M}_j^{(k)}\right) p\left(\sigma_\epsilon|\mathcal{M}_j^{(k)}\right) \tag{6.3}$$

where c is a normalising constant, so that

$$\int_{\mathbf{\Theta}} p\left(\sigma_\epsilon|\hat{\mathbf{d}}^{(k)},\mathcal{M}_j^{(k)}\right) d\sigma_\epsilon = \int_{\mathbf{\Theta}} c^{-1} p\left(\hat{\mathbf{d}}^{(k)}|\sigma_\epsilon,\mathcal{M}_j^{(k)}\right) p\left(\sigma_\epsilon|\mathcal{M}_j^{(k)}\right) d\sigma_\epsilon = 1 \tag{6.4}$$

In Equation (6.3), $p(\hat{\mathbf{d}}^{(k)}|\sigma_\epsilon, \mathcal{M}_j^{(k)})$ is the likelihood function, which expresses how likely the data $\hat{\mathbf{d}}^{(k)}$ are reproduced by the stochastic model in Equation (6.2) if model class $\mathcal{M}_j^{(k)}$ is adopted. Figure 6.1 depicts an example of four likelihood functions of their corresponding TF model classes with different dispersion parameters. Observe that the σ_ϵ parameter drives the maximum possible probability density value, therefore it plays a crucial role in the model class assessment, that is, the lower the dispersion parameter is, the higher the probability density is obtained, hence making such a model class more likely to be selected among the set of candidates if a uniform prior distribution is assumed. This likelihood function can be obtained by substitution of the values of $\hat{\mathbf{d}}^{(k)}$ as the output of the stochastic model, given the data independence, as follows

$$p\left(\hat{\mathbf{d}}^{(k)}|\sigma_\epsilon,\mathcal{M}_j^{(k)}\right) = \prod_{\ell=1}^{N_m} p\left(\hat{d}_\ell^{(k)}|\sigma_\epsilon,\mathcal{M}_j^{(k)}\right) \tag{6.5}$$

Therefore, Equation (6.3) can be rewritten as

$$p\left(\sigma_\epsilon|\hat{\mathbf{d}}^{(k)},\mathcal{M}_j^{(k)}\right) \propto \left\{\prod_{\ell=1}^{N_m} p\left(\hat{d}_\ell^{(k)}|\sigma_\epsilon,\mathcal{M}_j^{(k)}\right)\right\} p\left(\sigma_\epsilon|\mathcal{M}_j^{(k)}\right) \tag{6.6}$$

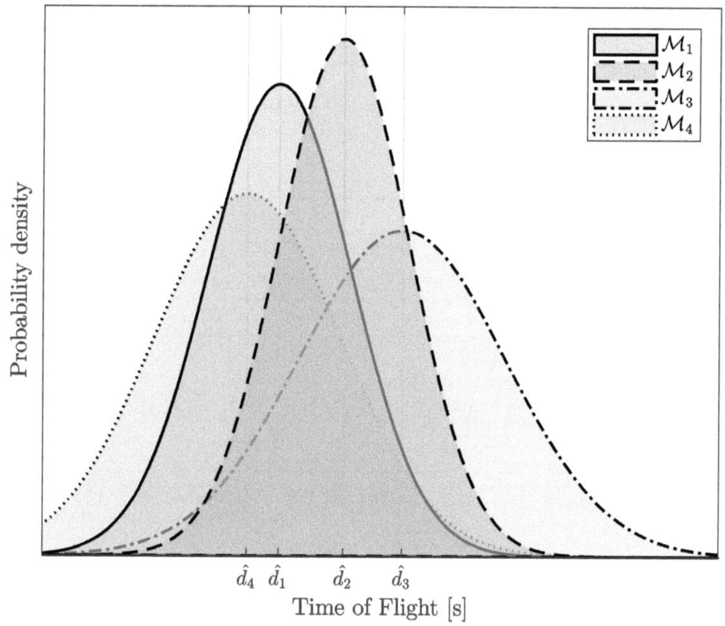

Figure 6.1: Likelihood functions derived from each time-of-flight $(\hat{d}_j^{(k)})$. The standard deviation of the proposed model ranking is expected to have different values in each model class. The time-of-flight data are then substituted in the likelihood function $p(\hat{\mathbf{d}}^{(k)}|\sigma_\epsilon, \mathcal{M}_j^{(k)})$.

Furthermore, note that the evaluation of the normalising constant c in Equation (6.3) cannot usually be evaluated analytically except for special cases based upon linear models and Gaussian uncertainties [9]. However, stochastic simulation based on MCMC methods [131, 90] can be used to obtain samples from the posterior avoiding the evaluation of c.

6.2.1.3 Model class assessment

The probabilistic approach for model class assessment is motivated by the uncertainty about the TF model based on the assumed hypotheses and simplifications adopted for its formulation [22, 100]. Once the posterior $p(\sigma_\epsilon|\hat{\mathbf{d}}^{(k)}, \mathcal{M}_j^{(k)})$ is obtained, the plausibility of the model class $\mathcal{M}_j^{(k)}$ can be obtained by applying the Total Probability Theorem as

$$P\left(\mathcal{M}_j^{(k)}|\hat{\mathbf{d}}^{(k)}\right) = \int P\left(\mathcal{M}_j^{(k)}|\hat{\mathbf{d}}^{(k)}, \sigma_\epsilon\right) p\left(\sigma_\epsilon|\hat{\mathbf{d}}^{(k)}\right) d\sigma_\epsilon =$$
$$= \int \underbrace{\frac{p(\hat{\mathbf{d}}^{(k)}|\mathcal{M}_j^{(k)}, \sigma_\epsilon)P(\mathcal{M}_j^{(k)}|\mathbf{M})}{\sum_{i=1}^{N_m} p(\hat{\mathbf{d}}^{(k)}|\mathcal{M}_i^{(k)}, \sigma_\epsilon)P(\mathcal{M}_i^{(k)}|\mathbf{M})}}_{\text{(Bayes' theorem)+(Total Probability theorem)}} p\left(\sigma_\epsilon|\hat{\mathbf{d}}^{(k)}\right) d\sigma_\epsilon \tag{6.7}$$

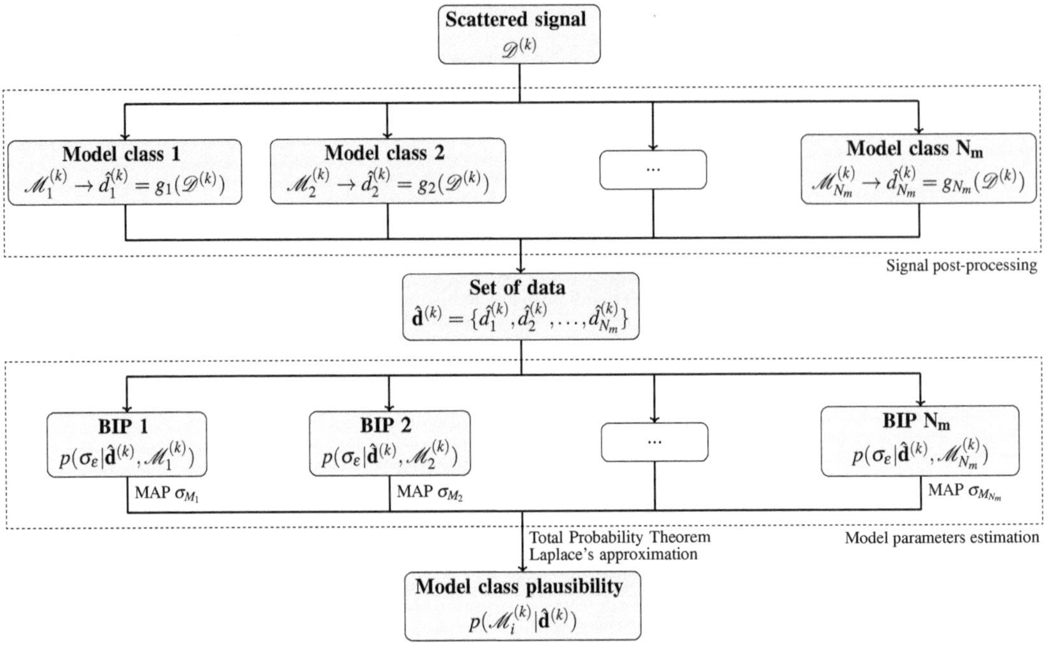

Figure 6.2: Flowchart describing the ultrasonic guided-waves-based model class assessment problem for one arbitrary scattered signal.

where $p\left(\sigma_\epsilon|\hat{\mathbf{d}}^{(k)}\right)$ denotes the posterior PDF. Note that the conditioning on $\mathbf{M}$ has been suppressed for simplicity. Equation (6.7) can be simplified by applying the asymptotic Laplace's approximation [22] as follows

$$P\left(\mathcal{M}_j^{(k)}|\hat{\mathbf{d}}^{(k)}\right) \approx \frac{p(\hat{\mathbf{d}}^{(k)}|\mathcal{M}_j^{(k)}, \sigma_{M_j})P(\mathcal{M}_j^{(k)})}{\sum_{i=1}^{N_m} p(\hat{\mathbf{d}}^{(k)}|\mathcal{M}_i^{(k)}, \sigma_{M_j})P(\mathcal{M}_i^{(k)})} \tag{6.8}$$

where σ_{M_j} is the maximum a posteriori (MAP) value of the posterior PDF $p(\sigma_\epsilon|\hat{\mathbf{d}}^{(k)}, \mathcal{M}_j^{(k)})$. Due to the use of the asymptotic Laplace's approximation, the computation of such TF model class probability is straightforward given the posterior PDF of the uncertain parameters $p(\sigma_\epsilon|\hat{\mathbf{d}}^{(k)}, \mathcal{M}_j^{(k)})$. This simplicity makes the proposed model class assessment specially suitable for real world applications, such as for the aerospace industry, which would be interested in introducing ultrasonic guided-wave based SHM devices, where on-board systems require efficient data post-processing. Figure 6.2 depicts the workflow of the proposed model class assessment for the ultrasonic data obtained in the kth sensor, $\mathcal{D}^{(k)}$.

Example 15 *Let's assume a rectangular thin-walled structure (2 mm thickness) with four PZT sensors arbitrarily placed and a central actuator, as observed in Figure 6.3a. For convenience, the experiments are virtually carried out by Abaqus/Explicit®. The actuator displacement is modelled as a perpendicular boundary condition, while the structure is modelled using squared shell elements (S4R). Thus, among the most used guided-wave modes used in SHM applications, considering that the excitation is carried out at* 300 *kHz, just the anti-symmetric 0 (A0) mode, also known as*

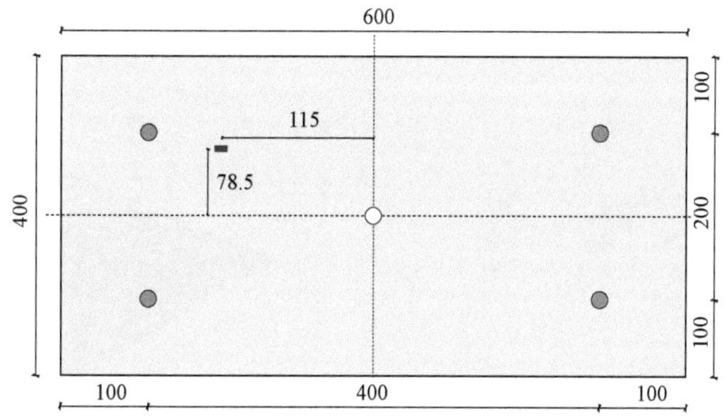

(a) Sensors and actuator layout. The numbering of the sensors starts with S1 on the top-left sensor and continues clockwise.

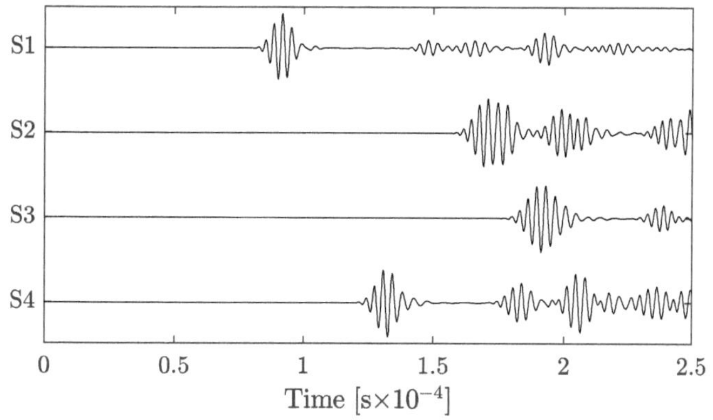

(b) Scattered signals obtained in each sensor, S1 through S4.

Figure 6.3: Flat aluminium plate along with the sensor's layout in (a), in the grey circles, and the actuator position, in the white circle. The damage is indicated in the rectangular darker area placed in the top-left area of the plate. In addition, the scattered signals obtained at the sensor positions are depicted in (b) in the time domain.

bending mode, is excited. The advantages of using just one mode rely on avoiding post-processing complexities caused by mode mixture, given that different modes travel at different speeds provided the same frequency. The mesh size is determined so that, at least, ten nodes per wavelength are obtained, while the time step is determined so that a disturbance cannot propagate through a grid spacing under one time step [6, 177]. Here, the damage has been modelled using a stiffness reduction of two orders of magnitude in a specific area that can be observed in Figure 6.3a. Thus, the modulus of elasticity drops from 73.1 *GPa, in the undamaged area, to* 731 *MPa in the bounded damaged zone, while the density and Poisson ration remain fixed at* 2780 kg/m^3 *and* 0.33*, respectively. The plate is* 1.5 *mm thick while the rest of the plate dimensions and sensor network can be observed in Figure 6.3a.*

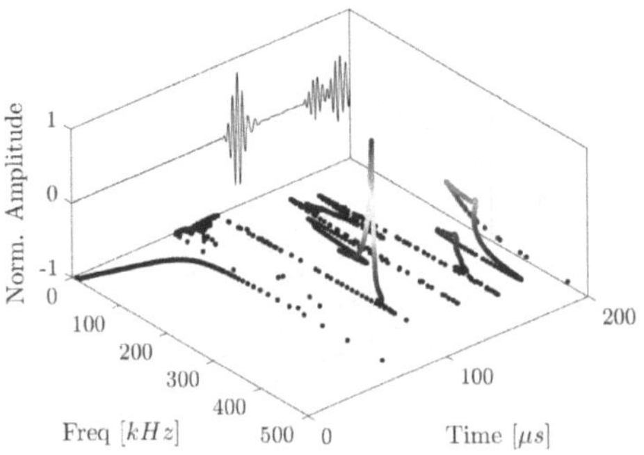

(a) HHT: 118.31 μs

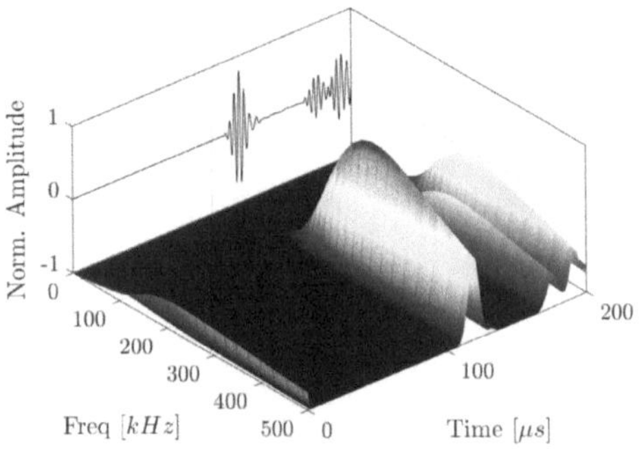

(b) CWT: 118.21 μs

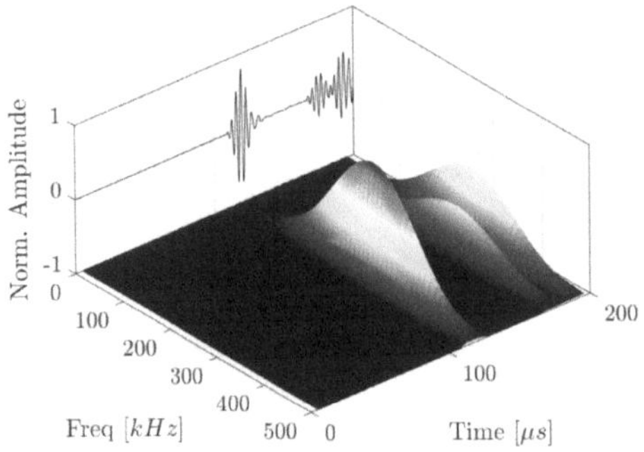

(c) STFT: 118.25 μs

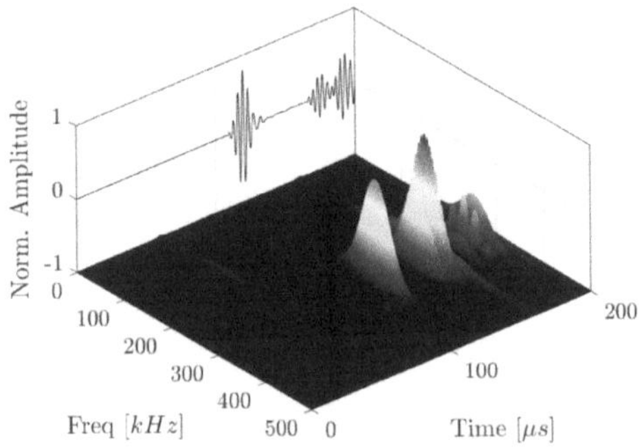

(d) WVD: 118.11 μs

Figure 6.4: Example of the outputs for the different TF models considered in this example. The time represented in each caption corresponds to the first energy peak (time of flight), which is used later for damage localisation purposes.

After obtaining the baseline signals, by carrying out an ultrasonic test in undamaged conditions of the structure, and the signals when the damage was introduced, both set of signals are subtracted in order to obtain the scattered wave generated at the damage position, as can be observed in Figure 6.3b. Then, the scattered signals are post-processed by making use of four different TF models: (1) Hilbert-Huang transform (HHT), (2) continuous wavelet transform (CWT), (3) short time Fourier transform (STFT), and (4) Wigner-Ville distribution (WVD) [51, 93, 50, 20]. Due to the different model assumptions and basic hypotheses, slightly different results are expected to be obtained, even when the same input signal is used for all of them. Figure 6.4 illustrates the output of these models applied to an arbitrary signal, different that the ones obtained in this case study. As can be observed, slightly different times of flight are obtained. This example emphasises the motivation behind the use of the Bayesian model class assessment, explained in Section 6.2.1.3, that is when several models are potentially used for the same application.

Next, a BIP is run for every model in each of the four sensors. The well-known Metropolis-Hastings (MH) [131, 90] algorithm (refer to Algorithm 1, Chapter 1) is used here to obtain samples from the posterior PDF of the model parameter, bypassing the computation of the normalising constant, or evidence. First, the posterior PDF of the dispersion parameter σ_ϵ is estimated ac-

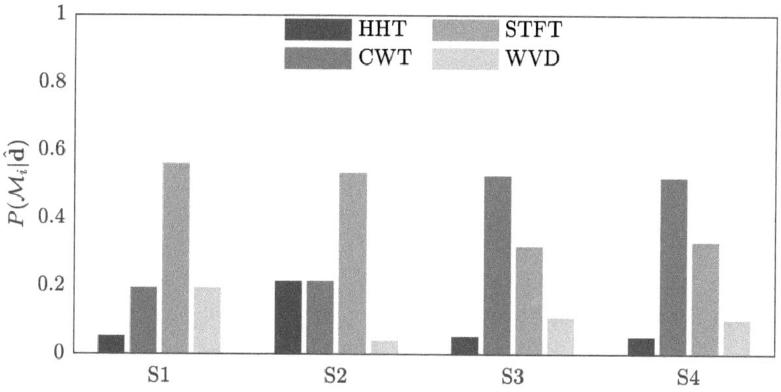

Figure 6.5: Posterior probability of each TF model for every sensor.

cording to Equation (6.6)*, and the MAP value stored to compute the TF model plausibility as per Equation* (6.8)*.* $T_s = 40,000$ *samples are used in the MH algorithm; this number is defined in order to ensure the convergence and that the burn-in period has been exceeded. The standard deviation of the proposal distribution is established so that the acceptance rate lies within the interval* $[0.2, 0.4]$ *[43, 73, 152]. The standard deviation parameter* σ_ϵ *is defined as* $\sigma_\epsilon = \rho \cdot \hat{d}_j^{(k)}$*, where* $\hat{d}_j^{(k)}$ *is the time of arrival at the* k*th sensor using the* j*th TF model, and* ρ *is a factor defined within a sufficiently large interval, which in this example is taken as* $(0, 0.5]$*. Therefore, the prior PDF of* σ_ϵ *can be expressed as a uniform distribution over the referred interval, that is* $p(\sigma_\epsilon) = \mathcal{U}(0, 0.5 \cdot \hat{d}_j^{(k)})$*. The maximum a posteriori (MAP) parameter is then computed and introduced as an input for the model class selection problem. Figure 6.5 depicts the posterior plausibilities of each TF candidate model. As can be observed, there is no clear TF model selected among the set of candidates. The STFT model shows higher plausibility in sensors S1 and S2, while CWT model is the most plausible for sensors S3 and S4. The selection of the TF model, by which the signal feature is obtained, is sensor dependent and based on purely probabilistic and mathematically rigorous principles. Finally, the values of the times of flight corresponding to the most probable TF models at each sensor are shown in Table 6.1. These values will be later used in the damage localisation problem that is described hereinafter.*

Table 6.1: Times of flight corresponding to the most probable model for sensors 1 through 4 with their corresponding model selection.

Description	Sensor 1 (S1)	Sensor 2 (S2)	Sensor 3 (S3)	Sensor 4 (S4)
TF model	STFT	STFT	CWT	CWT
Time of flight [μs]	81.50	157.30	176.55	121.45

6.2.2 Bayesian damage localisation

Once the TF model is selected, the damage localisation step may be addressed by making use of a model-based inverse problem. The model should provide an analogous output to the available data, in this case the time of flight. The nature of the model depends on the modeler choice about the information needed from the data. For instance, if the mechanical properties of the material and/or the type of damage itself are required to be known [41], the modeller may choose a complex and informative model, for example wave finite element models [124, 204]. However, if the information required by the modeller needs a lower level of detail, for example, the position of the damage, simpler models efficiently provide sufficient information, such as by using a ellipse-based model [68, 97], hereinafter referred to as time of flight (ToF) model. In this section, the use of the efficient, yet effective, ToF model is developed to address the damage localisation step in the guided-waves-based SHM. To this end, a model-based BIP is proposed so that the uncertainty information related to the damage position is captured [34]. One of the reasons for using such a Bayesian approach in monitoring aeronautical structures is to provide further support to the decision-making process in the maintenance area, while accounting for uncertainties of different sources that may produce biased decisions. The rigorous and accurate information obtained from the BIP may lead to potential savings in reducing other redundant inspections of large plate-like structures, such as the wings of an aircraft, while enabling the transition towards condition-based maintenance.

6.2.2.1 Probabilistic description of ToF model

Let's assume there are N_p actuator-sensor paths in a plate-like structure that are arbitrarily placed to excite and receive Lamb waves for damage localisation, by screening changes of their ToF. To this end, the ToF information of the scattered signals can be theoretically obtained by applying a ToF model, also known as a ellipse-based model, as follows [97]:

$$\mathrm{ToF}^{(a-s)} = \frac{\sqrt{(X_d - X_a)^2 + (Y_d - Y_a)^2}}{V_{a-d}} + \frac{\sqrt{(X_d - X_s)^2 + (Y_d - Y_s)^2}}{V_{d-s}} \tag{6.9}$$

where (X_d, Y_d) are the coordinates of the damage, (X_a, Y_a) are the actuator transducer coordinates, (X_s, Y_s) are the coordinates of one arbitrary sensor transducer. Here, a local coordinate system is strategically placed at an arbitrary location of the plate-like structure, whose centre is typically the actuator position so that $(X_a, Y_a) = (0, 0)$ m. Additionally, V_{a-d} and V_{d-s} are the wave propagation velocities of the actuator-damage and damage-sensor paths respectively. These velocities are the same, under the assumption of isotropic materials and concentrated damage within a bounded region, that is, $V = V_{a-d} = V_{d-s}$. Instead, if the damage affects a distributed region of the structure, the wave propagation velocity would change accordingly to the material's properties degradation.

The ToF model provided in Equation (6.9) is not completely defined without accounting for uncertainties coming from the data, its material properties, and from the model itself. Thus, in order to quantify those uncertainties, a set of uncertain model parameters $\boldsymbol{\theta} = \{X_d, Y_d, V\}$ is considered to describe the uncertainty about the damage coordinates as well as the wave propagation velocity. This set of parameters is further updated through Bayes' theorem, as will be explained below. The referred model error term $e \in \mathbb{R}$ is considered to account for the non-existence of a theoretical ToF model that fully represents the reality, so that

$$\mathrm{ToF}_{\mathcal{D}}^{(a-s)} = \mathrm{ToF}_{M}^{(a-s)}(\boldsymbol{\theta}) + e \tag{6.10}$$

where subscripts M and $\mathcal{D}$ from $\text{ToF}_M^{(a-s)}$ and $\text{ToF}_\mathcal{D}^{(a-s)}$ refer to modelled and measured ToF, respectively. Note in Equation (6.10) that e provides the discrepancy between $\text{ToF}_M^{(a-s)}$ and $\text{ToF}_\mathcal{D}^{(a-s)}$ values. By the PMIE, this error term is described as a zero-mean Gaussian distribution with covariance σ_e as $\mathcal{N}(0, \sigma_e)$. Thus, a probabilistic description of the ToF model from Equation (6.10) can be obtained as

$$p\left(\text{ToF}_\mathcal{D}^{(a-s)}|\text{ToF}_M^{(a-s)}(\boldsymbol{\theta})\right) = \left(2\pi\sigma_e^2\right)^{-\frac{1}{2}} \exp\left(-\frac{1}{2}\left(\frac{\text{ToF}_\mathcal{D}^{(a-s)} - \text{ToF}_M^{(a-s)}(\boldsymbol{\theta})}{\sigma_e}\right)^2\right) \tag{6.11}$$

Observe that Equation (6.11) provides a measurement of the similarity of the modelled and measured ToF. Also, note that Equation (6.11) provides a likelihood function for the $\text{ToF}_\mathcal{D}^{(a-s)}$ data under the assumption of $\text{ToF}_M^{(a-s)}(\boldsymbol{\theta})$ model.

6.2.2.2 Model parameter estimation

Given the likelihood function in Equation (6.11), one can obtain the posterior PDF of the model parameters given the ToF data $\mathbf{D} = \{D^{(1)}, \ldots, D^{(N)}\}$, where N is the total number of sensors and $D^{(j)}$ is the mean value of the most plausible probabilistic TF model in the jth sensor, as depicted in Figure 6.6, by applying the well-known Bayes' theorem as

$$p(\boldsymbol{\theta}|\mathbf{D}) = \frac{p(\mathbf{D}|\boldsymbol{\theta})p(\boldsymbol{\theta})}{p(\mathbf{D})} \tag{6.12}$$

where $p(\boldsymbol{\theta})$ is the prior PDF of the model parameters, and $p(\mathbf{D}|\boldsymbol{\theta})$ is the likelihood function for the set of data $\mathbf{D}$. Given the stochastic independence of the data, $p(\mathbf{D}|\boldsymbol{\theta})$ can be expressed as $p(\mathbf{D}|\boldsymbol{\theta}) = \prod_{k=1}^{N} p(D^{(k)}|\boldsymbol{\theta})$, where each factor $p(D^{(k)}|\boldsymbol{\theta})$ is given by Equation (6.11). Finally, $p(\mathbf{D})$ is the evidence of the data under the model specified by $\boldsymbol{\theta}$. This term, which acts as a normalising factor within the Bayes' theorem, can be bypassed through sampling, for example, using Markov Chain Monte Carlo (MCMC) methods [121]. Thus, Equation (6.12) can be rewritten as

$$p(\boldsymbol{\theta}|\mathbf{D}) \propto \left\{\prod_{k=1}^{N} p\left(D^{(k)}|\boldsymbol{\theta}\right)\right\} p(\boldsymbol{\theta}) \tag{6.13}$$

Example 16 *Considering the previous example, the ToF of the most plausible TF model class per sensor, which corresponds to* $\mathbf{D}$ *in Table 6.1, has already been obtained. These values are then used for the damage localisation reconstruction problem by using the BIP previously described. The prior information of the unknown model parameters* $\boldsymbol{\theta}$ *is defined for the damage position, and wave propagation velocity under the following PDFs:* $X_d \sim \mathcal{U}(-0.25, 0.25)$, $Y_d \sim \mathcal{U}(-0.15, 0.15)$, *and* $V \sim \mathcal{U}(1000, 4000)$, *respectively with position units in [m] and velocity in [m/s]. In this example, the asymptotic independence Markov sampling (AIMS) [25, 206] algorithm is used to draw samples from the posterior PDF,* $p(\boldsymbol{\theta}|\mathbf{D})$, *which offers a remarkable performance in multimodality scenarios. This choice is particularly relevant since the possibility of having several damage positions is feasible in real scenarios. A threshold value of* $\gamma = 1/2$ *and* $100,000$ *samples per annealing level are selected so that that the convergence of the posterior PDF is obtained. A Gaussian PDF is selected as proposal distribution, that is,* $q(\theta|\theta') = \mathcal{N}(\theta, \sigma)$, *where* σ *is the standard deviation of the M-H random walk which is selected so that the acceptance rate lies within the interval* $[0.2, 0.4]$.

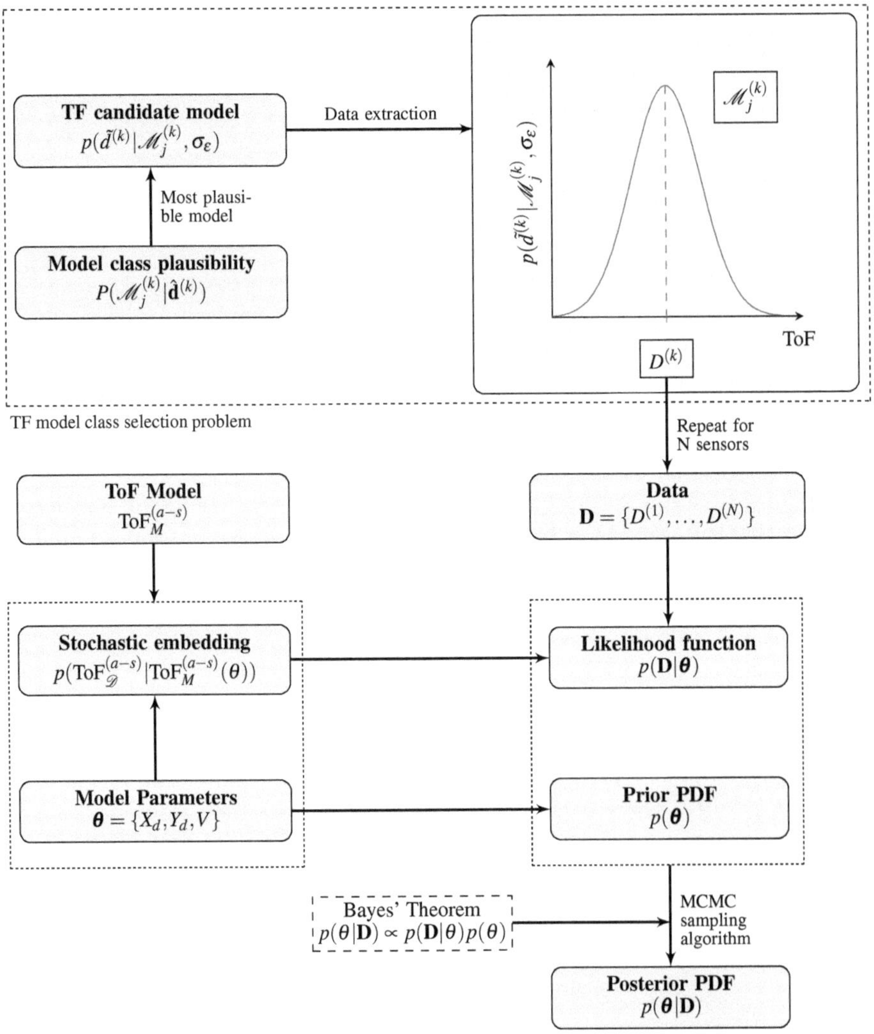

Figure 6.6: Flowchart describing the ultrasonic guided-waves based Bayesian damage localisation and its connection with the TF model class selection problem.

Figure 6.7a depicts the posterior PDF of the damage position coordinates (X_d, Y_d) *in which it can be observed that a remarkably accurate inference is obtained by applying the BIP of the damage localisation area described above. Note that one of the two areas of higher probability of the posterior PDF corresponds to the vicinity of one of the corners of the damage, which physically corresponds to the closer area of the damage to the actuator transducer, placed at* $(0,0)$ *m. The guided-wave that is excited at the center of the plate is radially transmitted towards the damage, where the wave interacts with it and the scattered signal is generated. In addition, the marginal posterior distribution of the last uncertain parameter, that is, the wave propagation velocity, can be observed in Figure 6.7b. The*

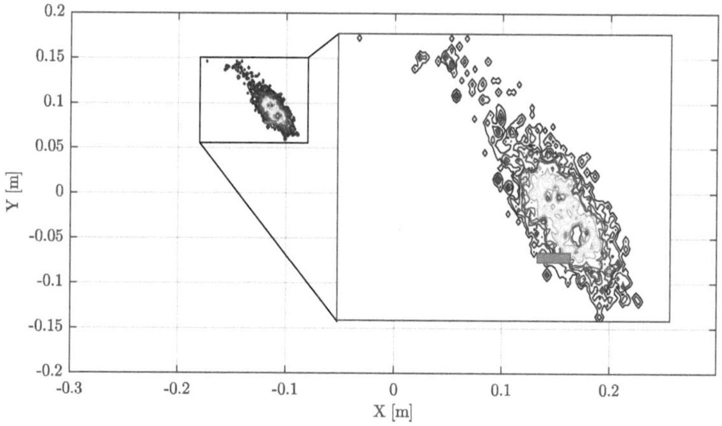

(a) Damage localisation

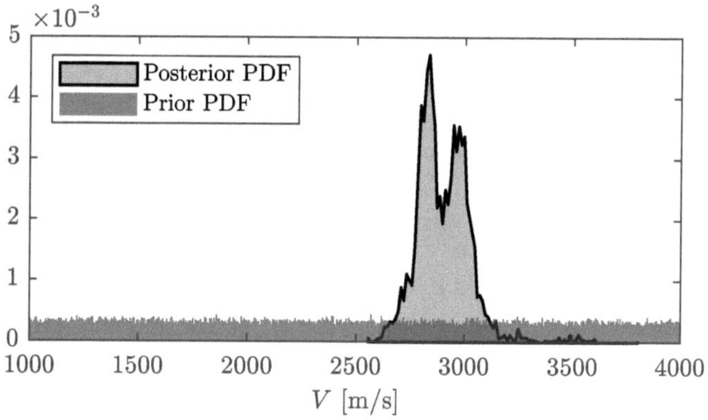

(b) Wave velocity reconstruction

Figure 6.7: Panel (a): Posterior PDF of the damage localisation variable and comparison with the ground truth about the enclosed damaged area. Panel (b): Comparison of prior and posterior PDFs of the velocity parameter.

informativeness of such a posterior PDF of the velocity parameter is depicted in the comparison of both prior and posterior PDFs. Both areas of maximum probability density of the damage position correspond to the higher probability zones observed in Figure 6.7a, although the most plausible peak corresponds to the area of probability closer to the actual damage.

6.3 Optimal sensor configuration

The optimisation approach based on value of information [35], which trades off information and cost, is presented in this section. In general terms, the value of information quantifies the increment of benefit as a consequence of the information gain about a set of uncertain model parameters $\boldsymbol{\theta}$ (e.g. damage location parameters) when ultrasound data $\mathbf{D}$ are taken into account. Note that the optimisation is addressed by using the ultrasonic guided-wave-based Bayesian damage localisation approach [34] presented in the sections above.

6.3.1 Value of information for optimal design

In order for the benefit of measuring data to be quantified, a benefit function $b(n, \boldsymbol{\theta})$ is defined dependent on the sensor configuration n and the model parameters $\boldsymbol{\theta}$. Note that the sensor configuration n entails the definition of the optimal sensor layout of the n sensors. In addition, such benefit function can be defined to be proportional to a normalised inverse of cost $f(n) : \mathbb{N} \rightarrow [0, 1]$ of each sensor configuration n and the information gained by the system, such that [35]:

$$b(n, \boldsymbol{\theta}) = f(n)\left(\alpha + \log\left[\frac{\pi(\boldsymbol{\theta}, \mathbf{D})}{p(\boldsymbol{\theta})}\right]\right) \tag{6.14}$$

where $p(\boldsymbol{\theta})$ is the prior PDF of model parameters $\boldsymbol{\theta}$, $\pi(\boldsymbol{\theta}, \mathbf{D})$ is used to denote the current PDF of $\boldsymbol{\theta}$, which could be either the prior or the posterior PDFs depending on the availability of the dataset $\mathbf{D}$, and $\log[\pi(\boldsymbol{\theta}|\mathbf{D})/p(\boldsymbol{\theta})]$ is the information gain between the aforementioned PDFs. The constant $\alpha > 0$ makes it possible to represent the basic state of information assumed in the system such that it makes $b(n, \boldsymbol{\theta}) = \alpha f(n)$ when $\pi(\boldsymbol{\theta}, \mathbf{D}) = p(\boldsymbol{\theta})$, so no relative learning has taken place. Note also that $f(n)$ is case specific so it is defined according to the manufacturing or maintenance cost law of both the structure and the SHM system.

Next, the concept of maximum prior expected benefit B', which is based on the prior information of the model parameters $p(\boldsymbol{\theta})$ [154], is defined and used to obtain the initial optimal sensor configuration n'_{opt}, as follows [113]:

$$B' = \mathbb{E}_{p(\boldsymbol{\theta})}\left[b(n'_{opt}, \boldsymbol{\theta})\right] \qquad n'_{opt} = \arg\max_{n} \int b(n, \boldsymbol{\theta}) p(\boldsymbol{\theta}) d\boldsymbol{\theta} \tag{6.15}$$

Similarly, the maximum posterior expected benefit (PEB) $B''(\mathbf{D})$ [154], which is based on the posterior distribution of the parameters given the data $p(\boldsymbol{\theta}|\mathbf{D})$, is obtained as follows [113]:

$$B''(\mathbf{D}) = \mathbb{E}_{p(\boldsymbol{\theta}|\mathbf{D})}\left[b(n''_{opt}, \boldsymbol{\theta})\right] \qquad n''_{opt} = \arg\max_{n} \int b(n, \boldsymbol{\theta}) p(\boldsymbol{\theta}|\mathbf{D}) d\boldsymbol{\theta} \tag{6.16}$$

where the conditioning on $\mathbf{D}$ is to denote that B'' depends on the data obtained by the sensors. Note that $\mathbf{D}$ can be obtained either from preliminary tests or simulations at the design stage, since real data cannot generally be used at this stage. The Bayesian inverse problem (BIP) of damage localisation used to obtain both the prior $p(\boldsymbol{\theta})$ and posterior $p(\boldsymbol{\theta}|\mathbf{D})$ distributions, with a particular sensor configuration, is based on a robust hierarchical framework, whose details of implementation can be found in [34]. Then, by subtracting both mathematical expectations evaluated at their optimal sensor configurations n''_{opt} and n'_{opt} and substituting the proposed benefit function (Eq. (6.14)), the conditional value of information (CVI) on $\mathbf{D}$ is given by [35]:

$$\mathrm{CVI}(\mathbf{D}) = f(n''_{opt})\mathrm{KL}(p(\boldsymbol{\theta}|\mathbf{D})\|p(\boldsymbol{\theta})) - \alpha\left[f(n'_{opt}) - f(n''_{opt})\right] \tag{6.17}$$

where $\mathrm{KL}(p(\boldsymbol{\theta}|\mathbf{D})\|p(\boldsymbol{\theta}))$ denotes the Kullback-Leibler divergence between the posterior and the prior PDFs of the uncertain model parameters $\boldsymbol{\theta}$; the constant α, which is also present in Equation (6.14), acts here as a trade-off variable between the weighed information gain (left term of Equation (6.17)) and the relative cost of implementation [35] (right term). Note that Equation (6.17) is defined for only one damage scenario, to which the sensors acquire the data $\mathbf{D}$.

6.3.2 Expected value of information

The optimisation process of a sensor configuration needs to consider a number of varied damage scenarios rather than a single one. The CVI criterion, which is formulated for only one dataset, needs then to be generalised so that a greater set of different damage scenarios $\mathfrak{D} \ni \mathbf{D}$ are assessed. To this end, a mathematical expectation of the CVI criterion in Equation (6.17), known as the expected value of information (EVI) [35], is computed as follows:

$$\mathrm{EVI} = \int_{\mathfrak{D}} \left[f(n''_{opt})\mathrm{KL}(p(\boldsymbol{\theta}|\mathbf{D})\|p(\boldsymbol{\theta}))\right] p(\mathbf{D})d\mathbf{D} - \int_{\mathfrak{D}} \alpha[f(n'_{opt}) - f(n''_{opt})]p(\mathbf{D})d\mathbf{D} \tag{6.18}$$

Observe that the expected KL divergence in Equation (6.18) involves a double multidimensional integral that cannot be solved analytically in most cases, except for linear relations between the quantity of interest and the model parameters. Conversely, a more general method that is applicable to both linear and nonlinear models, for example, through sampling, is preferred as long as the computational burden of the model evaluations is acceptable. Therefore, Equation (6.18) is numerically approximated using the MC method as follows [155, 8, 92]:

$$\begin{aligned}\mathrm{EVI} \approx\ & f(n''_{opt})\frac{1}{N_{out}}\sum_{m=1}^{N_{out}}\left[\log_2 p(\mathbf{D}_{(m)}|\boldsymbol{\theta}_{(m)}) - \log_2\left(\frac{1}{N_{in}}\sum_{k=1}^{N_{in}} p(\mathbf{D}_{(m)}|\boldsymbol{\theta}_{(k)})\right)\right] \\ & - \alpha[f(n'_{opt}) - f(n''_{opt})]\end{aligned} \tag{6.19}$$

where $\boldsymbol{\theta}_{(m)}$ is a sample drawn from the prior distribution $p(\boldsymbol{\theta})$; $\mathbf{D}_{(m)}$ is a sample dataset drawn from the likelihood distribution $p(\mathbf{D}|\boldsymbol{\theta} = \boldsymbol{\theta}_{(m)})$ (refer to Equation (6.11)); and N_{out} and N_{in} are the number of Monte Carlo samples used in the outer and inner sums, respectively.

Note that, at the design stage, experimental data are not typically available, and therefore the data are obtained by using a model and prior knowledge about the model parameters. In other words, a pre-posterior analysis is carried out by maximising the value of information, that is, minimising the pre-posterior parameter uncertainty. Thus, Equation (6.19) is adopted here as optimality criterion to obtain the optimal sensor configuration considering an area of possible damage locations. It is also worth highlighting that the EVI criterion establishes a trade-off between the information gained by a candidate sensor configuration and its relative cost of implementation. Therefore, by using this criterion, sensor configurations with an excessive number of sensors will be penalised based on a given cost law. Similarly, sensor layouts providing low information will be avoided by this criterion. Instead, a balance between these two characteristics (i.e. information gain and cost) will be obtained.

6.3.2.1 Algorithmic implementation

The optimisation problem based on the maximisation of the EVI requires an appropriate algorithmic search strategy to deal with a potentially large space of possible sensor configurations.

Potential candidates include evolutionary algorithms such as genetic algorithms [77] or the covariance matrix adaptation evolutionary strategy [89]. Nonetheless, a practical approach is the forward sequential sensor search algorithm [141, 144], which optimises the location of the nth sensor based on the optimal locations of the remaining $i-1$ sensors and has been adopted in this chapter (see Algorithm 10). In particular, the optimisation of the nth sensor location is obtained here by carrying out an exhaustive search over a discrete grid of possible sensor locations, given that a relatively small search space is considered. The objective function $U(\mathcal{C}^n) : \mathbb{N} \to \mathbb{R}$ for a candidate sensor configuration $\mathcal{C}^n$, defined by a spatial configuration of n sensors, is given by the average of the EVI in Equation (6.19) as follows

$$
\begin{aligned}
U(\mathcal{C}^n) = f(n)\frac{1}{N_{set}}\sum_{h=1}^{N_{set}}\Bigg\{\frac{1}{N_{out}}\sum_{m=1}^{N_{out}}\Bigg[&\log p(\mathbf{D}_{(m,h)}|\boldsymbol{\theta}_{(m,h)},\mathcal{C}^n) \\
&- \log\left(\frac{1}{N_{in}}\sum_{k=1}^{N_{in}} p(\mathbf{D}_{(m,h)}|\boldsymbol{\theta}_{(k,h)},\mathcal{C}^n)\right)\Bigg]\Bigg\} - \alpha[f(n'_{opt}) - f(n)]
\end{aligned}
\tag{6.20}
$$

The expectation of the KL divergence (recall Eq. (6.19)) is addressed efficiently using $N_{out} = N_{in}$ prior samples, which are reused in both the inner and outer sums of Equation (6.20) at the cost of a small increment in the bias of the estimator [92, 8]. Nevertheless, the Monte Carlo approximation method implies a numerical error that is inversely proportional to the number of samples, leading to stochastic uncertainty of the objective function $U(\mathcal{C}^n)$. This numerical error is inevitably present due to the approximative nature of the method; however it can be reduced if $U(\mathcal{C}^n)$ is averaged over N_{set} sets of of $N_{out} = N_{in}$ samples for each configuration $\mathcal{C}^n$, as shown in Equation (6.20).

Furthermore, note that n'_{opt} in Equation (6.20) is obtained by maximising the prior expected benefit (recall Equation (6.15)). Given the independence of the prior expected benefit on the sensor configuration, n'_{opt} is only dependent on the inverse cost function $f(n)$ only. Thus, n'_{opt} provides the most economically beneficial amount of sensors given that $f(n)$ is decreasing, regardless of their position.

Algorithm 10: Pseudo-code implementation of forward sequential search algorithm for geometrically unconstrained sensor configurations.

/* Preamble */
Define N_s; // number of possible sensor locations
Define $n_{s,max}$; // maximum number of sensors locations considered in the optimization
/* Algorithm */
$n'_{opt} \leftarrow \arg\max_n f(n)$; // Optimal number of sensors under prior information
$\{\{\boldsymbol{\theta}_{(m,h)}\}_{m=1}^{N}\}_{h=1}^{N_{set}} \sim p(\boldsymbol{\theta})$; // N_{set} sets of $N = N_{out} = N_{in}$ samples drawn from the prior PDF
$\{\{\mathbf{D}_{(m,h)}\}_{m=1}^{N}\}_{h=1}^{N_{set}} \sim p(\mathbf{D}|\boldsymbol{\theta}_{(\cdot,h)})$; // N_{set} sets of $N = N_{out} = N_{in}$ samples drawn from the likelihood PDF
$\mathfrak{C} \leftarrow \emptyset$; // Initialize an empty vector of optimal sensors placement
/* Forward sequential sensor for loop */
for $n = 1$ **to** N_s **do**

```
    /* Exhaustive search for loop                                          */
    for i = 1 to n_{s,max} do
        C^n ← C^{n-1} ∪ C^i;          // Define C^n by concatenating the previous
                                       configuration to the ith sensor
        Obtain J(C^n);                 // Evaluate the objective function according
                                       to Eq. (6.20)
    end
    Return: C^n_opt and J(C^n_opt);    // Optimal sensor layout and its
                                       corresponding objective function value
    ℭ ← C^n_opt;                       // Store the nth optimal sensor location for
                                       the next iteration
end
n''_opt ← arg max_n J(ℭ);              // Optimal number of sensors under posterior
                                       information
```

The optimisation of the sensor configuration is solved here by the forward sequential sensor search algorithm [141] as presented above in Algorithm 10. This approach will provide an approximate solution of the global optimum, given that the search space is drastically reduced with respect to the fully combinatorial search. To illustrate such computational savings, let us assume a grid of $N_s = 40$ possible sensor locations and $n = 5$ sensors to be optimised. The exhaustive combinatorial search would involve $N_s!/n!(N_s - n)! = 40!/5!35! = 658,008$ possible sensor configurations and their corresponding objective function evaluations. By adopting the forward sequential search approach, the search space reduces up to $n(2N_s - n + 1)/2 = 5(80 - 5 + 1)/2 = 190$. Nonetheless, the solution provided by the sequential approach is in general suboptimal since it does not necessarily explore the optimal candidate solution. However, such a suboptimal solution is assumed to be near-global and sufficiently robust [116, 197].

Example 17 *The objective of this example is to illustrate how the optimal sensor configuration is achieved for a plate-like structure with a central region of possible damage occurrence. To this end, a 600mm×400mm aluminium plate with* 1.5*mm thickness is considered, where the uncertainty distribution of the wave velocity is quantified by a Gaussian PDF, so* $V \sim \mathcal{N}(2800, 40^2)$ *with units in [m/s]. Notice that the mean value of the velocity corresponds to the antisymmetric 0 guided-wave mode at* 300*kHz for an aluminium alloy 2024. Further, the concentrated damage may occur anywhere within the enclosed designated area assuming two probability density distributions: (1) a log-normal distribution for the damage coordinate* $X_d \sim \mathcal{LN}(-2.05, 0.56) - 0.2[m]$ *where the MAP value lies in* $X_d = -0.1[m]$*, while the* Y_d *coordinate is uniformly distributed in* $\mathcal{U}(-0.1, 0.1)$ *with units in [m]; and (2) a uniform distribution for both* X_d *and* Y_d *coordinates, so* $X_d \sim \mathcal{U}(-0.2, 0.2)$ *and* $Y_d \sim \mathcal{U}(-0.1, 0.1)$ *with units in [m]. The spatial representations of the prior probability densities of the damage coordinates for both scenarios are depicted in Figures 6.8a and 6.8c by means of a grayscale colour map. In this map, the darker the colour, the higher the probability density. The inverse cost function* $f(n)$ *is defined mathematically with a decreasing monotonic function, so* $f(n) = 200/(n^2 + 200)$ *for illustration purposes.*

The optimisation problem is addressed using $N_{set} = 10$ *sets of* $N_{in} = N_{out} = 1000$ *samples from the prior PDF of the model parameters* $\boldsymbol{\theta}$*. A grid of sixty-four candidate sensor locations situated between the area of possible damage occurrence and the boundaries of the plate is considered. Additionally, one actuator is considered at the centre of plate. The optimal sensor results in Figures 6.8a and 6.8c show different layouts of three sensors for both types of prior information*

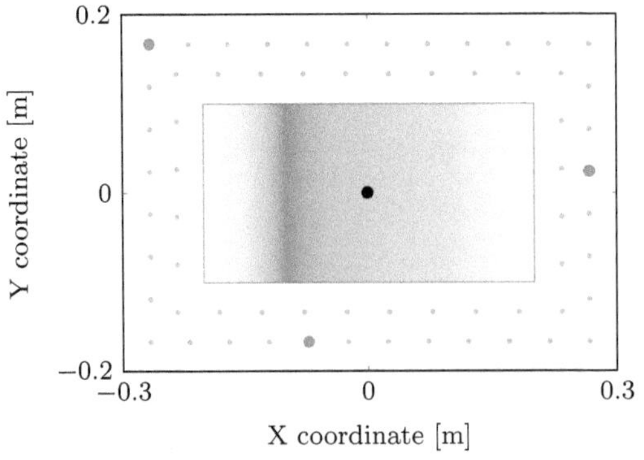

(a) Optimal layout – log-normal PDF

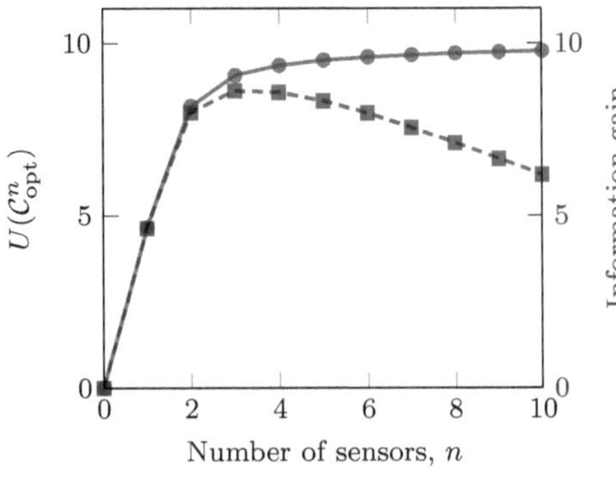

(b) EVI vs information – log-normal PDF

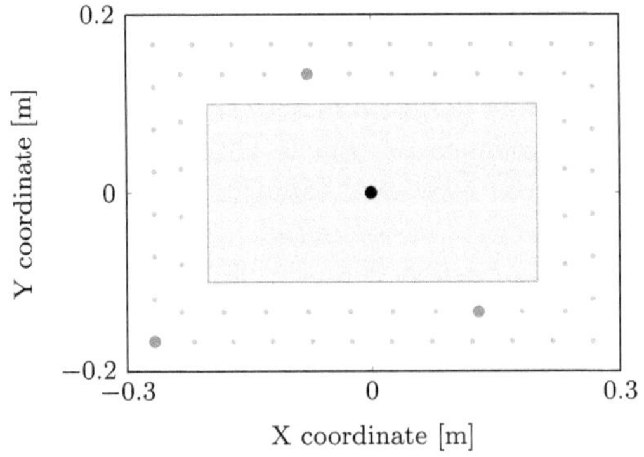

(c) Optimal layout – uniform PDF

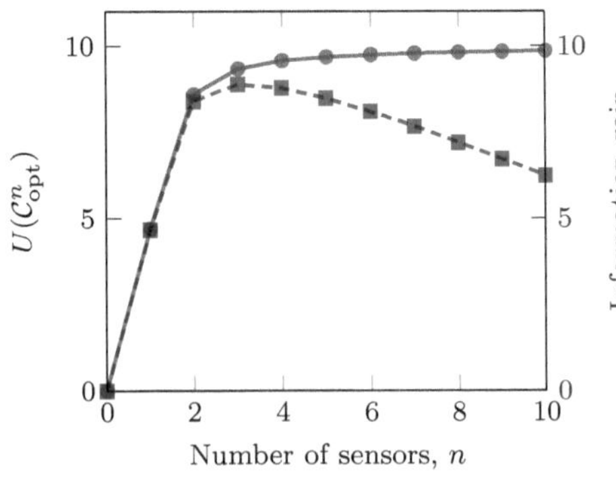

(d) EVI vs information – uniform PDF

Figure 6.8: Optimal sensor layouts for both prior PDFs, namely, log-normal (a) and uniform (c). Legend: ● – actuators and ● – sensor. Comparison of value (squares) against amount (circles) of information for each number of sensors in (b) and (d).

about the damage distribution. Sensors are placed in a triangular shape around the damage area, with slight differences. While the case of a log-normal PDF (Figure 6.8a) provides sensors in the farthest locations from the damage, the case of a uniform PDF (Figure 6.8c) has the sensors placed alternatively between the farther and closer grids of possible locations. Additionally, a comparison between the classical optimisation criterion of the amount of information gain [24, 142] and the newest one based on the value of information [35] is provided in Figures 6.8b and 6.8d. Observe that the EVI criterion (in black) provides a clear global optimum at three sensors, while the amount of information (in blue) fails to do so, hence showing the superiority of the EVI in providing an optimal number of sensors in addition to their optimal locations.

To demonstrate the effectiveness of the optimal sensor layouts in localising damage, the reconstruction of one sample of damage within the possible area of occurrence is carried out using the ultrasonic guided-wave based Bayesian damage localisation approach described in Section 6.2. Different damage samples are assumed for the different optimal sensor configurations, that is, located at (-0.10m, -0.06m) for the case of a log-normal prior PDF, and at (0.15m, 0.05m) for the case

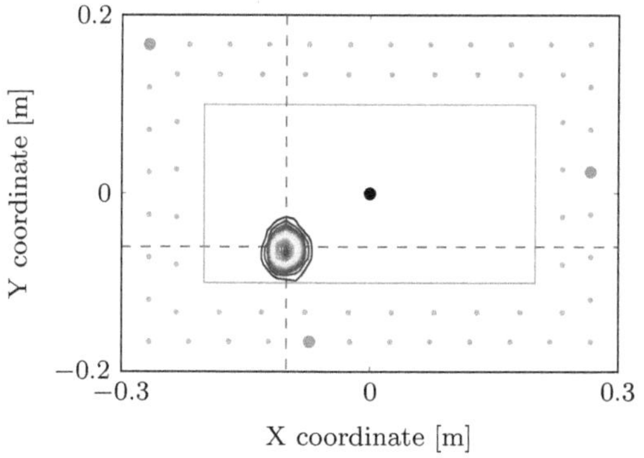

(a) Log-normal prior PDF

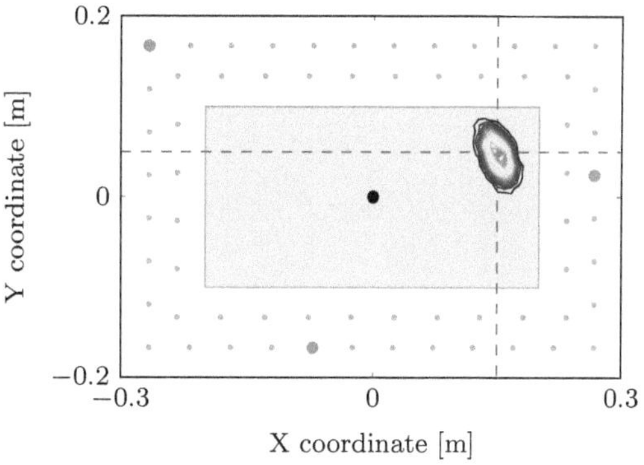

(b) Uniform prior PDF

Figure 6.9: Damage location reconstruction (contour lines) using the optimal sensor configurations for both cases of prior PDFs, that is, log-normal PDF in (a) and uniform PDF in (b). The actual damage coordinates are represented by -+- at (-0.1m, -0.06m) for (a) and (0.15m, 0.05m) for (b).

of a uniform prior PDF. The reconstruction results are shown in Figure 6.9. Note that both optimal sensor configurations reconstruct the damage in a very accurate and precise manner, matching the actual damage location with the area of maximum probability of the posterior PDF. Therefore, the importance of having an optimal sensor configuration for maximising the identifiability of a structural system has been revealed.

6.4 Summary

This chapter has illustrated the benefits of Bayesian methods in ultrasonic guided-wave based SHM in order to obtain the optimal sensor configuration that maximises the damage identification performance. First, several signal post-processing techniques were rigorously ranked through Bayesian model class assessment. Then, a Bayesian inverse problem for damage localisation was formulated using the most plausible model class obtained in the first step as input data. Finally, the optimal sensor configuration problem is presented using the expected value of information as optimality criterion, which minimises the pre-posterior parameter uncertainty of the damage localisation problem.

Note that different sources of uncertainties (e.g. material properties or measurement noise) are quantified using the proposed methodology. These uncertainties stem from the original stage of signal post-processing to the damage localisation itself. Thus, the optimal sensor configuration can be expected to be sufficiently robust to error-related sources of uncertainty.

7

Fast Bayesian Approach for Stochastic Model Updating using Modal Information from Multiple Setups

Wang-Ji Yan,[1,*] *Lambros Katafygiotis*[2] and *Costas Papadimitriou*[3]

This chapter illustrates the suitability of advanced Bayesian techniques to extract the modal properties of vibrating structures and stochastic model updating. A technique for variable separation is presented so that the interaction between spectrum variables (e.g. frequency, damping ratio as well as the amplitude of modal excitation and prediction error) and spatial variables (e.g. mode shape components) can be decoupled completely. Based on the uncertainties of the modal properties, a novel Bayesian methodology is developed for structural stochastic model updating by incorporating the local mode shape components identified from different clusters automatically without prior assembling. The suitability and potential of the presented methodology is illustrated using a numerical example and an experimental case study.

7.1 Introduction

Structural health monitoring (SHM) based on ambient vibration responses has become a hot research topic in the field of civil engineering. The primary concern of SHM usually involves data acquisition, damage feature extraction, and structural condition assessment. Nowadays, the deployment of dense sensor arrays has enabled the generation of a large bulk of measurements, which calls for advanced computational methods to translate the measurements into effective system representations [39]. The task of transferring these raw data into damage-sensitive features usually necessitates adoption of system identification (SI) approaches. As a general term that refers to the extraction of information about structural behaviour directly from experimental data, SI can be implemented either by building dynamic models from measured data or without the use of a model [111]. The results produced by SI are expected to be further utilised to assess structural health status, which allows asset owners to improve the safety of critical structures.

In the context of SHM, SI mainly includes two branches: modal-parameter identification and physical-parameter identification. Modal-parameter identification aims to identify structural modal

[1] University of Macau (China).
[2] Hong Kong University of Science and Technology (China).
[3] University of Thessaly (Greece).
[*] Corresponding author: wangjiyan@um.edu.mo

properties, while physical-parameter identification focuses on the extraction of stiffness, mass, and damping. Conventionally, SI results are usually restricted to deterministic optimal values while the parametric uncertainties cannot be quantified. However, as discussed in Chapter 1, there are many kinds of uncertainties that deserve considering in engineering systems, including uncertainties due to incomplete information, errors due to imperfect modelling of physical phenomena as well as measurement noise [199]. To ensure the robustness of the identified results, SHM is best tackled as a probabilistic inference problem, which not only gives the optimal values for the various identified parameters but also provides a quantitative assessment of their accuracy [26, 108, 145, 143]. For example, the associated uncertainty information during damage detection is important for engineers when they make judgements about repairing works [126]. Therefore, the primary focus of this chapter is to provide rigorous probabilistic algorithms for processing SHM data using Bayesian statistics. The issue of probabilistic modelling for frequency-domain ambient vibration responses is first investigated. Based on the analytical probabilistic model of frequency-domain responses, a two-stage fast Bayesian approach for operational modal analysis is proposed to address the variability in the identification of modal characteristics. The Bayesian structural model updating problem is then formulated by incorporating statistical modal properties corresponding to different setups.

This chapter is organised as follows: the probability density functions (PDFs) of different kinds of frequency-domain ambient vibration responses such as raw fast Fourier transform (FFT) coefficients, power spectral density (PSD) matrix, and the trace of PSD matrix are revisited in Section 7.2. Based on the analytical probabilistic model of frequency-domain responses, a two-stage fast Bayesian spectral density approach is proposed in Section 7.3 to identify the most probable modal properties as well as their uncertainties. In Section 7.4, the model updating problem is then formulated as one minimising an objective function, which can incorporate statistical local mode shape components corresponding to different setups automatically without prior assembling or processing. The efficiency and accuracy of all these methodologies are verified by numerical examples and experimental studies in Section 7.5 and 7.6.

7.2 Probabilistic consideration of frequency-domain responses

7.2.1 PDF of multivariate FFT coefficients

For a linear system with n_d degree-of-freedoms (DOFs), it is assumed that discrete acceleration responses are available for n_o ($\leq n_d$) measured DOFs and the sampling time interval is assumed to be Δt. Assume that there are n_s sets of independent and identically distributed time histories for n_o measured DOFs, and $\boldsymbol{y}_j(n\Delta t) \in R^{n_o}$ $(n = 1, 2, \cdots N)$ denotes the discrete-time stochastic vector process corresponding to the j-th set of sampled data. The Fast Fourier transform (FFT) of $\boldsymbol{y}_j(n)$ at f_k is defined as

$$\mathbf{Y}_j(k) = \boldsymbol{Y}_{\boldsymbol{R},j}(k) + \boldsymbol{i}\boldsymbol{Y}_{\boldsymbol{I},j}(k) = \sqrt{\frac{\Delta t}{2\pi N}} \sum_{n=0}^{N-1} \boldsymbol{y}_j(n) \exp\left(-\boldsymbol{i}2\pi f_k n \Delta t\right) \tag{7.1}$$

where $\boldsymbol{i}^2 = -1$, $f_k = k\Delta f$, $k = 1, 2, \cdots, \mathrm{Int}\left(\frac{N}{2}\right)$, and $\Delta f = \frac{1}{(N\Delta t)}$. In Equation (7.1), $\boldsymbol{Y}_{\boldsymbol{R},j}(k)$ and $\boldsymbol{Y}_{\boldsymbol{I},j}(k)$ denote the real and imaginary part of $\boldsymbol{Y}_j(k)$, respectively. In this work, 'k' shown in the bracket or in the subscript denotes the frequency point f_k.

It has been proved by [202] that the random vector $\boldsymbol{Y}_j(k)$ follows a Gaussian distribution as $N \to \infty$, with mean and covariance matrix given by

$$E\left(\boldsymbol{Y}_j(k)\right) = 0; \quad \boldsymbol{\Sigma}_k = E\left[\boldsymbol{S}_{y,j}(k)\right] \tag{7.2}$$

where $E\left[\boldsymbol{S}_{y,j}(k)\right]$ denotes the mathematical expectation of the power spectral density (PSD) matrix of $\boldsymbol{y}_j(n)$, which is defined as

$$\boldsymbol{S}_{y,j}(k) = \boldsymbol{Y}_j(k)\boldsymbol{Y}_j(k)^* \tag{7.3}$$

where $*$ denotes the conjugate transpose of a complex vector. As a result, the analytical PDF of $\boldsymbol{Y}_j(k)$ is given by

$$p\left(\boldsymbol{Y}_j(k)\right) = \frac{1}{\pi^{n_o}|\det \boldsymbol{\Sigma}_k|} \exp\left(-\boldsymbol{Y}_j(k)^*\boldsymbol{\Sigma}_k^{-1}\boldsymbol{Y}_j(k)\right) \tag{7.4}$$

7.2.2 PDF of PSD matrix

If n_s sets of discrete FFT coefficients $\boldsymbol{Y}_j(k)$ follows independent and identical complex Gaussian distribution, then one can prove that the sum of the PSD matrix

$$\boldsymbol{S}_k^{sum} = \sum_{j=1}^{n_s} \boldsymbol{Y}_j(k)\left(\boldsymbol{Y}_j(k)\right)^* \tag{7.5}$$

follows the central complex Wishart distribution of dimension n_o with n_s degrees of freedom [201]. The PDF of central complex Wishart distribution is given by [78]

$$p\left(\boldsymbol{S}_k^{sum}\right) = \frac{1}{\Gamma_{n_o}(n_s)} \frac{1}{|\boldsymbol{\Sigma}_k|^{n_s}} |\boldsymbol{S}_k^{sum}|^{(n_s - n_o)} \exp\left\{-\left(\boldsymbol{\Sigma}_k^{-1}\boldsymbol{S}_k^{sum}\right)\right\} \tag{7.6}$$

where $\Gamma_{n_o}(n_s) = \pi^{\frac{n_o(n_o-1)}{2}} \prod_{i=1}^{n_o}(n_s - i)!$, and $(\cdot)$ and $|\cdot|$ denote the trace and determinant of a matrix, respectively.

7.2.3 PDF of the trace of the PSD matrix

For an arbitrary Wishart matrix $\mathbf{A}$ with n_s degrees of freedom and covariance matrix $\boldsymbol{\Sigma}$, it can be proved that $((\boldsymbol{A}) - n_s(\boldsymbol{\Sigma}))/\sqrt{2n_s(\boldsymbol{\Sigma}^2)}$ asymptotically follows standard Gaussian distribution as n_s is large [129]. In Section 7.2.2, it has been illustrated that the superposition of n_s sets of the PSD matrix (i.e. $\boldsymbol{S}_k^{sum}$) follows Wishart distribution, then the trace of $\boldsymbol{S}_k^{sum}$ should asymptotically follow Gaussian distribution with mean $n_s(\boldsymbol{\Sigma}_k)$ and covariance $2n_s\left(\boldsymbol{\Sigma}_k^2\right)$ when n_s is large. Thus, the PDF of the trace of the PSD matrix denoted by $(\boldsymbol{S}_k^{sum})$ is given by

$$p\left((\boldsymbol{S}_k^{sum})\right) = \frac{1}{\sqrt{4\pi n_s\left(\boldsymbol{\Sigma}_k^2\right)}} \exp\left(-\frac{\left((\boldsymbol{S}_k^{sum}) - n_s(\boldsymbol{\Sigma}_k)\right)^2}{4n_s\left(\boldsymbol{\Sigma}_k^2\right)}\right) \tag{7.7}$$

When the number of discrete data points $N \to \infty$, it has been proved that $\boldsymbol{Y}_j(k)$ and $\boldsymbol{Y}_j(k')$ are uncorrelated as $k \neq k'$ [201]. Correspondingly, $\boldsymbol{S}_k^{sum}$ and $\boldsymbol{S}_{k'}^{sum}$ as well as $(\boldsymbol{S}_k^{sum})$ and $\left(\boldsymbol{S}_{k'}^{sum}\right)$ for $k \neq k'$ are also independently distributed:

$$p(\boldsymbol{y}_k, \boldsymbol{y}_{k'}) = p(\boldsymbol{y}_k)\, p(\boldsymbol{y}_{k'}) \tag{7.8a}$$

$$p\left(\boldsymbol{S}_k^{sum}, \boldsymbol{S}_{k'}^{sum}\right) = p\left(\boldsymbol{S}_k^{sum}\right) p\left(\boldsymbol{S}_{k'}^{sum}\right) \tag{7.8b}$$

$$p\left((\boldsymbol{S}_k^{sum}), (\boldsymbol{S}_{k'}^{sum})\right) = p\left((\boldsymbol{S}_k^{sum})\right) p\left((\boldsymbol{S}_{k'}^{sum})\right) \tag{7.8c}$$

7.3 A two-stage fast Bayesian operational modal analysis

The Bayesian spectral density approach (BSDA) formulated previously [109] is a promising candidate for ambient modal analysis since it presents a strict way for obtaining optimal modal properties and their associated uncertainties. However, difficulties such as computational inefficiency and ill-conditioning have severely hindered the realistic application of BSDA. Motivated by Bayesian spectral density approach (BSDA) and the fast Bayesian FFT framework proposed recently by [11, 12, 13], a two-stage fast Bayesian spectral density approach was proposed for ambient modal analysis by [194, 195].

In the two-stage fast Bayesian operational modal analysis, the interaction between spectrum variables (e.g. frequency, damping ratio, as well as the magnitude of modal excitation and prediction error) and spatial variables (e.g. mode shape components) can be decoupled completely. The spectrum variables and their associated uncertainties can be identified by the 'fast Bayesian spectral trace approach' (FBSTA) in the first stage based on the statistical properties of the trace of the spectral density matrix illustrated in Section 7.2.3. Then, the spatial variables and their uncertainties can be extracted instantaneously through the 'fast Bayesian spectral density approach' (FBSDA) in the second stage by using the statistical properties of the PSD matrix introduced in Section 7.2.2. The original formulation can be found in [194, 195] and only the main concept is reviewed in this section.

7.3.1 Prediction error model connecting modal responses and measurements

Assume that there are n_s sets of independent and identically distributed time histories for n_o measured DOFs. The j-th measured response denoted by $\boldsymbol{y}_j(n)\,(j = 1, 2, \cdots, n_s)$ at the n-th time step $n\Delta t$ $(n = 1, 2, \cdots, N)$ is modeled as

$$\boldsymbol{y}_j(n) = \boldsymbol{x}_j(n) + \boldsymbol{\mu}_j(n) \tag{7.9}$$

where $\boldsymbol{x}_j(n)$ is the j-th model response, a function of the model parameters ϑ to be identified; $\boldsymbol{\mu}_j(n)$ is the j-th prediction error process, which can be represented by a discrete zero-mean Gaussian white noise vector process. As a result, the covariance matrix of $\boldsymbol{Y}_k$ is given by

$$\boldsymbol{\Sigma}_k = E\left[\boldsymbol{S}_{y,j}(k)\right] = E\left(\boldsymbol{S}_x(k)\right) + E\left(\boldsymbol{S}_\mu(k)\right) \tag{7.10}$$

For the case of separated modes primarily considered in this chapter, there is only one mode to be identified over the specified frequency band $R \in [k_1\Delta f, k_2\Delta f]$. For clarity, the natural frequency, damping ratio, the amplitude of the modal excitation, and the prediction error are denoted by $f_s, \varsigma_s, s_{\mu s}$, and S_{fs}, respectively, while the mode shape is denoted by $\boldsymbol{\psi}_s$. The subscript 's' denotes 'separated modes'. The covariance matrix $\boldsymbol{\Sigma}_k$ in the vicinity of the resonant frequency shown in Equation (7.10) reduces to [11]

$$\boldsymbol{\Sigma}_k(\vartheta) = \Lambda_k \boldsymbol{\psi}_s \boldsymbol{\psi}_s^T + s_{\mu s}\boldsymbol{I}_{n_o} \tag{7.11}$$

It is worth mentioning that Λ_k in Equation (7.11) is given by $\Lambda_k = S_{fs}P_k^{-1}$ where $P_k = \left[\left(\beta_{sk}^2 - 1\right)^2 + \left(2\beta_{sk}\varsigma_s\right)^2\right]$ with $\beta_{sk} = \frac{f_s}{f_k}$. As a result, the parameters ϑ to be identified involve

$$\vartheta = \{f_s, \varsigma_s, s_{\mu s}, S_{fs}, \boldsymbol{\psi}_s\} \tag{7.12}$$

7.3.2 Spectrum variables identification using FBSTA

In the first stage, the spectrum variables can be separated from the full set of parameters through FBSTA by employing the statistical properties of trace of the PSD matrix. Conditioned on the set of trace $(\boldsymbol{S}_R^{sum}) = \{(\boldsymbol{S}_k^{sum})\,|k = k_1, \cdots k_2\}$ formed over $R \in [k_1\Delta f, k_2\Delta f]$, the updated PDF of ϑ is given, according to the Bayes' Theorem as

$$p\left(\vartheta|\left(\boldsymbol{S}_R^{sum}\right)\right) = c_o p(\vartheta)\, p\left(\left(\boldsymbol{S}_R^{sum}\right)|\vartheta\right) \tag{7.13}$$

where c_o is a normalising constant and the likelihood function $p\left(\left(\boldsymbol{S}_R^{sum}\right)|\vartheta\right)$ is equal to

$$p\left(\left(\boldsymbol{S}_R^{sum}\right)|\vartheta\right) = \prod_{k=k_1}^{k_2} p\left(\left(\boldsymbol{S}_k^{sum}\right)|\vartheta\right) \tag{7.14}$$

where $p\left(\left(\boldsymbol{S}_k^{sum}\right)|\vartheta\right)$ is given by Equation (7.7) with $\boldsymbol{\Sigma}_k(\vartheta)$ given in Equation (7.11). As a result, the optimal spectrum variables can be identified by manipulating the following 'negative log-likelihood function' (NLLF)

$$L_{FBSDA}(\vartheta) = \frac{1}{2}\sum_{k=k_1}^{k_2} ln\left(4\pi n_s\left(\boldsymbol{\Sigma}_k^2(\vartheta)\right)\right) + \sum_{k=k_1}^{k_2}\frac{\left[\left(\boldsymbol{S}_k^{sum}\right) - n_s\left(\boldsymbol{\Sigma}_k(\vartheta)\right)\right]^2}{4n_s\left(\boldsymbol{\Sigma}_k^2(\vartheta)\right)} \tag{7.15}$$

Based on Equation (7.11), $(\boldsymbol{\Sigma}_k(\vartheta))$ and $\left(\boldsymbol{\Sigma}_k^2(\vartheta)\right)$ are given by

$$(\boldsymbol{\Sigma}_k(\vartheta)) = \Lambda_k\left(\ \ {}_s\boldsymbol{\psi}_s^T\right) + \left(s_{\mu s}\boldsymbol{I}_{n_o}\right) \tag{7.16}$$

$$\left(\boldsymbol{\Sigma}_k^2(\vartheta)\right) = \Lambda_k^2 + 2s_{\mu s}\Lambda_k + n_o s_{\mu s}^2 \tag{7.17}$$

Substituting Equation (7.16) and (7.17) into Equation (7.15) lead to

$$L_{FBSTA}(\vartheta_{spe}) = \frac{1}{2}\sum_{k=k_1}^{k_2} \ln 4\pi n_s\left(\Lambda_k^2 + 2s_{\mu s}\Lambda_k + n_o s_{\mu s}^2\right) + \sum_{k=k_1}^{k_2}\frac{\left[\left(S_k^{sum}\right) - n_s\left(\Lambda_k + n_o s_{\mu s}\right)\right]^2}{4n_s\left[\Lambda_k^2 + 2s_{\mu s}\Lambda_k + n_o s_{\mu s}^2\right]} \tag{7.18}$$

From Equation (7.18), one can observe that the NLLF depends only on four spectrum variables, that is, $\vartheta_{spe} = \{f_s, \varsigma_s, S_{fs}, s_{\mu s}\}$. Moreover, there is no complicated matrix operation (e.g. determinant, inverse and singular value decomposition, etc.) in Equation (7.18). Thus, the optimal value can be obtained by unconstrained numerical optimisation without much computational effort. The updated PDF of the identified parameters can be well-approximated by a Gaussian PDF centred at the optimal parameters with covariance matrix $\boldsymbol{C}_{FBSTA}$ equal to the inverse of the Hessian matrix of the NLLF $L_{FBSTA}(\vartheta_{spe})$. The Hessian matrix $\boldsymbol{\Gamma}_{FBSTA}(\vartheta_{spe})$ contains the second derivatives of $L_{FBSTA}(\vartheta)$ with respect to all the spectrum parameters $\vartheta_{spe} = \{f_s, \varsigma_s, S_{fs}, s_{\mu s}\}$ arranged into an one-dimensional array. The derivatives of $\boldsymbol{\Gamma}_{FBSTA}(\vartheta_{spe})$ can be calculated through analytical derivation shown in [194].

7.3.3 Mode shape identification using FBSDA

Once the spectrum variables are extracted, the mode shapes as well as their uncertainties can be identified by FBSDA using the statistical information of the PSD matrix in the second stage. It is assumed that the spectral density set $S_R^{sum} = \{S_k^{sum}, k = k_1, \cdots, k_2\}$ formed over the frequency band $R \in [k_1\Delta f, k_2\Delta f]$ is employed for ambient modal analysis. Based on Bayes' Theorem, the updated PDF of ϑ is given, as

$$p(\vartheta|\boldsymbol{S}_R^{sum}) = c_o p(\vartheta)\, p(\boldsymbol{S}_R^{sum}|\vartheta) \tag{7.19}$$

Using Equations (7.11) and (7.6), one can obtain the NLLF corresponding to FBSDA

$$L_{FBSDA}(\vartheta) = n_s n_f (n_o - 1) \ln s_{\mu s} + n_s \sum_{k=k_1}^{k_2} \left[\ln(s_{\mu s} + \Lambda_k) + s_{\mu s}^{-1}\left(d - \boldsymbol{\psi}_s^T \Xi \boldsymbol{\psi}_s\right)\right] \tag{7.20}$$

where $d = \sum_{k=k_1}^{k_2} (\boldsymbol{S}_k^{sum})$ and $\boldsymbol{\Xi} = \sum_{k=k_1}^{k_2} \left[1 + \frac{s_{\mu s}}{\Lambda_k}\right]^{-1} \boldsymbol{S}_k^{sum}$. In Equation (7.20), one can observe that the mode shape has been expressed in an explicit manner. Since $L_{FBSDA}(\lambda)$ depends on $\boldsymbol{\psi}_s$ solely through the last quadratic term, the most probable values of $\boldsymbol{\psi}_s$ can be obtained just by minimising the last quadratic term

$$\min \quad L_{FBSDA}(\boldsymbol{\psi}_s) = -s_{\mu s}^{-1} \boldsymbol{\psi}_s^T \boldsymbol{\Xi} \boldsymbol{\psi}_s \tag{7.21a}$$

$$\text{s.t.} \quad \boldsymbol{\psi}_s^T \boldsymbol{\psi}_s = 1 \tag{7.21b}$$

Equation (7.21) falls into classical Rayleigh Quotient minimisation problem, and the optimal $\boldsymbol{\psi}_s$ is just the eigenvector of $\boldsymbol{\Xi}$ corresponding to the largest eigenvalue [11]. The posterior covariance matrix of the mode shape can be approximated by the inverse of the Hessian matrix of the $L_{FBSDA}(\boldsymbol{\psi}_s)$. The general formulation of the derivatives of the Rayleigh quotient with respect to $\boldsymbol{\psi}_s$ can be found in [194]. Let $\{\varrho_1, \varrho_2, \cdots, \varrho_{n_o}\}$ arranged in ascending order be the eigenvalues of the Hessian matrix $\Gamma_{FBSDA}(\boldsymbol{\psi}_s)$, while their corresponding eigenvectors are assumed to be $\{\boldsymbol{v}_1, \boldsymbol{v}_2, \cdots, \boldsymbol{v}_{n_o}\}$. The covariance matrix $C_{FBSDA}(\boldsymbol{\psi}_s)$ can be evaluated properly via its eigen-basis representation with the first zero eigenvalue ignored [11]

$$\boldsymbol{C}_{FBSDA}(\boldsymbol{\psi}_s) = \sum_{i=2}^{n_o} \varrho_i^{-1} \boldsymbol{v}_i \boldsymbol{v}_i^T \tag{7.22}$$

7.3.4 Statistical modal information for model updating

In real application, the DOFs of interest are usually measured in different setups with common 'reference' DOFs present across different setups. As is illustrated in Figure 7.1, the two-stage Bayesian operational modal analysis can be implemented by using a multiple-level hierarchical architecture. In the first stage, the auto-spectral density of all measured setups can be collected and summed together, and then the spectrum variables such as natural frequencies and damping ratios, etc. can be identified by making full use of the relevant information contained in all response measurements. In the second stage, a group of local mode shape components are identified from different sensor setups sharing some reference sensors. It is worth mentioning that multiple-level hierarchical architecture in Figure 7.1 fuses the information of all sensors together to identify the spectral variables, indicating that, instead of achieving different spectral variables for different setups, only a single set of spectral parameters $\vartheta_{spe} = \{f_s, \varsigma_s, S_{fs}, s_{\mu s}\}$ is available.

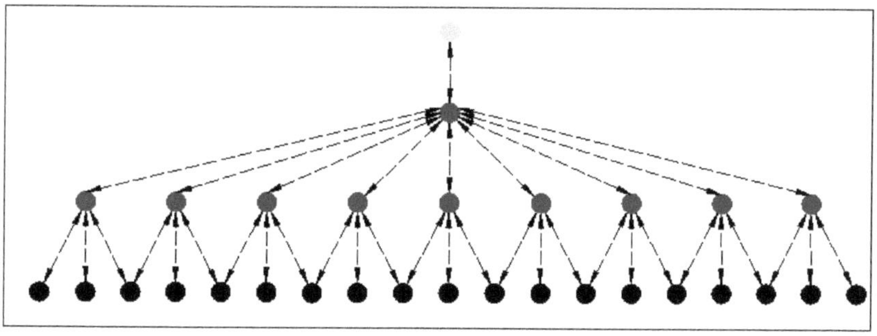

Figure 7.1: Multiple-level hierarchical architecture of modal analysis.

In this chapter, the modal data to be utilised in the structural model updating problem introduced in Section 7.4 consist of $\wp = \{\omega_r, \boldsymbol{\psi}_{r,i}, \boldsymbol{C}_{\psi_{r,i}}\}$ with $r = 1, 2, \cdots, n_m$ and $i = 1, 2, \cdots, n_t$. Here n_m and n_t denote the number of modes and the number of setups, respectively. ω_r denotes the r-th optimal natural frequency (in rad/s). $\boldsymbol{\psi}_{r,i} \in R^{n_i}$ and $\boldsymbol{C}_{\psi_{r,i}} \in R^{n_i \times n_i}$ denote the optimal values and covariance matrix of the local mode shape confined to the measured DOFs of the i-th setup for the r-th given mode; n_i is the number of DOFs measured in the i-th setup. The total number of measured DOFs denoted by n_l should be equal to the number of distinct measured DOFs from all setups.

7.4 Bayesian model updating with modal data from multiple setups

7.4.1 Structural model class

For a linear structural model with n_d DOFs, the stiffness matrix $\boldsymbol{K} \in R^{n_d \times n_d}$ and mass matrix $\boldsymbol{M} \in R^{n_d \times n_d}$ are parameterised by $\boldsymbol{\theta} = \{\theta_1, \theta_2, \cdots, \theta_{n_\theta}\}$ and $\boldsymbol{\rho} = \{\rho_1, \rho_2, \cdots, \rho_{n_\rho}\}$ respectively as follows,

$$\boldsymbol{K}(\theta) = \boldsymbol{K}_0 + \sum_{i=1}^{n_\theta} \theta_i \boldsymbol{K}_i \tag{7.23a}$$

$$\boldsymbol{M}(\rho) = \boldsymbol{M}_0 + \sum_{i=1}^{n_\rho} \rho_i \boldsymbol{M}_i \tag{7.23b}$$

where $\boldsymbol{K}_i$ and $\boldsymbol{M}_i$ are nominal substructures contributing to the global stiffness matrix and global mass matrix, which can be obtained from the conventional finite-element model of the structure; $\theta_i\,(i = 1, 2, \cdots, n_\theta)$ and $\rho_i\,(i = 1, 2, \cdots, n_\rho)$ are scaling factors which enable the modification of the nominal substructure stiffness and mass matrix so as to be more consistent with the real structure.

The prior PDF of $\boldsymbol{\theta}$ and $\boldsymbol{\rho}$ are taken to be independent Gaussian with mean $\boldsymbol{\theta}_0$ and $\boldsymbol{\rho}_0$, that is,

$$p(\theta) = \exp\left[-\frac{1}{2}(\boldsymbol{\theta} - \boldsymbol{\theta}_0)^T \boldsymbol{C}_{\boldsymbol{\theta}}^{-1} (\boldsymbol{\theta} - \boldsymbol{\theta}_0)\right] \tag{7.24a}$$

$$p(\rho) = \exp\left[-\frac{1}{2}(\boldsymbol{\rho} - \boldsymbol{\rho}_0)^T \boldsymbol{C}_{\boldsymbol{\rho}}^{-1} (\boldsymbol{\rho} - \boldsymbol{\rho}_0)\right] \tag{7.24b}$$

where $\boldsymbol{C}_\theta$ and $\boldsymbol{C}_\rho$ denote the covariance matrix of $\boldsymbol{\theta}$ and $\boldsymbol{\rho}$, and they are taken to be diagonal matrices. It is worth noting that the variances of the prior covariance of $\boldsymbol{\rho}$ are taken to be small so as to avoid making the identification problem ill-posed when treating $\boldsymbol{\theta}$ and $\boldsymbol{\rho}$ as uncertain variables simultaneously [48, 49].

7.4.2 Formulation of Bayesian model updating

7.4.2.1 The introduction of instrumental variables system mode shapes

Instrumental variables system mode shapes $\boldsymbol{\Phi}_r \in R^{n_d}, (r = 1, 2, \cdots, n_m)$ proposed previously [189] are introduced in this study as extra parameters to be updated. 'System mode shapes' represent the actual mode shapes of the structure without being constrained to be eigenvectors of the structural model. Introducing the instrumental variables will bring some advantages [48, 199]. For example, there is no need to perform eigenvalue decomposition and mode-matching. Also, it can skip the step of modal expansion from measured DOFs to the unmeasured DOFs, which is able to connect the full mode shape information with partial mode shape information. As will be found in the remaining part of this chapter, one appealing advantage not mentioned before is that the instrumental variables help to incorporate the local mode shape components identified from multiple setups automatically without prior assembling or processing.

7.4.2.2 Probability model connecting 'system mode shapes' and measured local mode shape

Let $\boldsymbol{\Phi}_r' \in R^{n_l}$ be the components of $\boldsymbol{\Phi}_r$ only confined to the measured DOFs, and let $\boldsymbol{\Phi}_{r,i} \in R^{n_i}$ be the components $\boldsymbol{\Phi}_r$ only confined to the measured DOFs in the i-th setup. Then $\boldsymbol{\Phi}_{r,i}$ can be mathematically related to $\boldsymbol{\Phi}_r$ by

$$\Phi_{r,i} = \boldsymbol{L}_i \boldsymbol{\Phi}_r' = L_i \boldsymbol{\Phi}_r \tag{7.25}$$

where $\boldsymbol{L}_i = \boldsymbol{L}_i\boldsymbol{L}_o$. Here $\boldsymbol{L}_o \in R^{n_l \times n_d}$ is a selection matrix which picks the observed DOFs from the 'system mode shape' $\boldsymbol{\Phi}_r$; $\boldsymbol{L}_i \in R^{n_i \times n_l}$ is another selection matrix with $\boldsymbol{L}_i(j,k) = 1$ if the j-th data channel in the i-th setup gives the k-th DOF of $\boldsymbol{\Phi}_r'$ and zero otherwise.

Similar to the mode shape assembly problem [10], the discrepancy between $\frac{\boldsymbol{\Phi}_{r,i}}{\|\boldsymbol{\Phi}_{r,i}\|}$ and the identified local mode shape $\psi_{r,i}$ both of unit Euclidean norms, instead of the discrepancy between $\boldsymbol{\Phi}_{r,i}$ and $\psi_{r,i}$, ought to be bridged since the measured local mode shapes are normalised to unit norm (i.e. $\|\psi_{r,i}\| = 1$). As illustrated in Section 7.3.3, the identified local mode shape can be well-approximated by a Gaussian PDF with mean $\{\psi_{r,i} : r = 1, \cdots, n_m; i = 1, \cdots, n_t\}$ and covariance matrix $\{\boldsymbol{C}_{\psi_{r,i}} : r = 1, \cdots, n_m; i = 1, \cdots, n_t\}$. Under the Bayesian framework, $\frac{\boldsymbol{\Phi}_{r,i}}{\|\boldsymbol{\Phi}_{r,i}\|}$ should best fit the identified counterparts by assigning a weight according to the data quality. Therefore, the likelihood function reflecting the contribution of the i-th local mode shape can be written explicitly in terms of $\boldsymbol{\Phi}_{r,i}$ as

$$p(\hat{\psi}_{r,i}, \mathbf{C}_{\psi_{r,i}} | \boldsymbol{\Phi}_{r,i}) = \frac{1}{\sqrt{(2\pi)^{n_d} |\mathbf{C}_{\psi_{r,i}}|}} \exp\left(-\frac{1}{2} (\tilde{\mathbf{L}}_i \boldsymbol{\Phi}_r / \|\tilde{\mathbf{L}}_i \boldsymbol{\Phi}_r\| - \hat{\psi}_{r,i})^T (\mathbf{C}_{\psi_{r,i}}^{-1}) (\tilde{\mathbf{L}}_i \boldsymbol{\Phi}_r / \|\tilde{\mathbf{L}}_i \boldsymbol{\Phi}_r\| - \hat{\psi}_{r,i}) \right) \tag{7.26}$$

It is assumed that local mode shapes identified from different setups are statistically independent. Therefore, the likelihood function reflecting the contribution from multiple setups $\{\psi_{r,i}, \boldsymbol{C}_{\psi_{r,i}} : i = 1, \cdots, n_t\}$ is given by [193]

$$p\left(\{\psi_{r,i}, \boldsymbol{C}_{\psi_{r,i}} : i = 1, \cdots, n_t\} | \boldsymbol{\Phi}_r\right) = \prod_{i=1}^{n_t} p\left(\psi_{r,i}, \boldsymbol{C}_{\psi_{r,i}} | \boldsymbol{\Phi}_{r,i}\right) \tag{7.27}$$

So far, the discrepancy between the system mode shape and the measured local mode shapes has been well-bridged. The probability model shown in the above equation is reasonable since the local mode shapes not well-identified in particular setups should have less influence on determining the final 'system mode shape'. As a result, less weight should be assigned for the unreliable setups because their data quality is relatively poor. This is explicitly accounted for through the inverse of $\boldsymbol{C}_{\psi_{r,i}}$ in Equation (7.26) and Equation (7.27).

7.4.2.3 Probability model for the eigenvalue equation errors

For the purpose of model updating, structural model parameters should also be connected with the measured frequencies and 'system mode shape' through a mathematical model. In this study, the eigenvalue equation errors will be employed in that it can measure how well the identified modal properties are matched with the counterparts from the updated structural model [200]:

$$\boldsymbol{\varepsilon}_r = \boldsymbol{K}(\boldsymbol{\theta})\boldsymbol{\Phi}_r - \omega_r^2 \boldsymbol{M}(\boldsymbol{\rho})\boldsymbol{\Phi}_r \tag{7.28}$$

The prediction-error vector $\boldsymbol{\varepsilon}_r$ $(r = 1, 2, \cdots, n_m)$ can be modeled as independent Gaussian variables

$$\boldsymbol{\varepsilon}_r \sim N(0, \delta_r \boldsymbol{I}_{n_d}) \tag{7.29}$$

where $\boldsymbol{I}_{n_d}$ is a n_d by n_d identity matrix and δ_r denotes the variance of the prediction error which is unknown. Equation (7.29) can be justified by the maximum entropy principle stating that the Gaussian PDF gives the largest uncertainty for any statistical distribution with specified means and variances [100]. Therefore, the likelihood function expressing the contribution from the measured frequencies ω_r is given by

$$\begin{aligned} p\left(\omega_r|\boldsymbol{\rho},\boldsymbol{\theta},\boldsymbol{\Phi}_r,\delta_r\right) &= \left(\sqrt{(2\pi)^{n_d}\left|\delta_r \boldsymbol{I}_{n_d}\right|}\right)^{-1} \exp\left(-\frac{1}{2}\left(\eta_r\Phi_r\right)^T\left(\delta_r \boldsymbol{I}_{n_d}\right)^{-1}\left(\eta_r\boldsymbol{\Phi}_r\right)\right) \\ &= \left(2\pi\delta_r\right)^{-\frac{n_d}{2}} \exp\left(-\frac{1}{2}\delta_r^{-1}\|\boldsymbol{\eta}_r\boldsymbol{\Phi}_r\|^2\right) \end{aligned} \tag{7.30}$$

where $\|\cdot\|$ denotes the Euclidean norm and $\boldsymbol{\eta}_r = \boldsymbol{K}(\boldsymbol{\theta}) - \omega_r^2\boldsymbol{M}(\boldsymbol{\rho})$. As a result, a relationship is established between the 'system mode shape' ($\boldsymbol{\Phi}_r$), the measured natural frequencies (ω_r), and the model parameters ($\boldsymbol{\rho},\boldsymbol{\theta}$) connected through the Gaussian probability model for the eigenvalue equation errors.

7.4.2.4 Negative log-likelihood function for model updating

Under the assumption that modal parameters of different modes are statistically independent, thus the likelihood function $p(\wp|\lambda, M)$ is given by

$$\begin{aligned} p\left(\wp|\lambda, M\right) &= \prod_{r=1}^{n_m} p\left(\{\omega_r, \boldsymbol{\psi}_{r,i}, \boldsymbol{C}_{\boldsymbol{\psi}_{r,i}} : i=1,\cdots,n_t\}|\boldsymbol{\rho},\boldsymbol{\theta},\boldsymbol{\Phi}_r,\delta_r\right) \\ &= \prod_{r=1}^{n_m} p\left(\{\boldsymbol{\psi}_{r,i}, \boldsymbol{C}_{\boldsymbol{\psi}_{r,i}} : i=1,\cdots,n_t\}|\boldsymbol{\Phi}_r\right) p\left(\omega_r|\boldsymbol{\rho},\boldsymbol{\theta},\boldsymbol{\Phi}_r,\delta_r\right) \end{aligned} \tag{7.31}$$

After substituting Equations (7.27) and (7.30) into Equation (7.31), the likelihood function $p(\wp|\lambda, M)$ can be obtained. Then substituting Equation (7.31) and the prior information from Equation (7.24) into the Bayes' Theorem give the updated probabilities $p(\lambda|\wp, M)$. The most probable values (MPV) of the unknown parameters based on the modal data are given by maximising $p(\lambda|\wp, M)$. For this optimisation problem, it is more convenient to work with the *negative logarithm likelihood function* (NLLF) given by [193]:

$$\begin{aligned} L_{upd}(\lambda) = &\frac{1}{2}\left(\boldsymbol{\theta}-\boldsymbol{\theta}_0\right)^T \boldsymbol{C}_{\boldsymbol{\theta}}^{-1}\left(\boldsymbol{\theta}-\boldsymbol{\theta}_0\right) + \frac{1}{2}\left(\boldsymbol{\rho}-\boldsymbol{\rho}_0\right)^T \boldsymbol{C}_{\boldsymbol{\rho}}^{-1}\left(\boldsymbol{\rho}-\boldsymbol{\rho}_0\right) + \frac{1}{2}\sum_{r=1}^{n_m}\{n_d ln\delta_r + \delta_r^{-1}\|\eta_r\boldsymbol{\Phi}_r\|^2\} \\ &+ \frac{1}{2}\sum_{r=1}^{n_m}\sum_{i=1}^{n_t}\left(\frac{\boldsymbol{L}_i\boldsymbol{\Phi}_r}{\|\boldsymbol{L}_i'\boldsymbol{\Phi}_r\|} - \boldsymbol{\psi}_{r,i}\right)^T \boldsymbol{C}_{\boldsymbol{\psi}_{r,i}}^{-1}\left(\frac{L_i\boldsymbol{\Phi}_r}{\|L_i'\boldsymbol{\Phi}_r\|} - \boldsymbol{\psi}_{r,i}\right) \end{aligned} \tag{7.32}$$

It is worth noting that the constant terms independent of the parameters to be identified are not included in Equation (7.32). As can be seen from Equation (7.32), there is no need to implement eigenvalue decomposition, mode-matching, and modal expansion in the procedure of formulating the objective function. Moreover, the local mode shape components identified from different setups have been incorporated together in an automated manner.

7.4.3 Solution strategy

In Equation (7.32), the full set of parameters to be identified is composed of

$$\lambda = \{\boldsymbol{\rho},\boldsymbol{\theta},\boldsymbol{\Phi}_r,\delta_r : r=1,2,\cdots,n_m\} \tag{7.33}$$

The most probable parameter vector λ can be obtained by minimising $L_{upd}(\lambda)$, while the covariance matrix of λ can be obtained by taking the inverse of the Hessian matrix calculated at $\lambda = \hat{\lambda}$. Note that the function $L_{upd}(\lambda)$ is not quadratic about the unknown parameters λ, and a numerical optimisation algorithm needs to be employed for solving the desired optimisation problem.

However, the use of numerical optimisation methods usually requires evaluation of gradients or Hessians with respect to λ, which is computationally challenging due to the high dimensional and nonlinear features of the problem at hand. One effective way to solve Equation (7.32) is to decompose the complicated optimisation problem into several coupled simple optimisation problems with analytical solutions. Then the objective function can be optimised iteratively through a sequence of linear optimisation problems until satisfactory convergence is achieved. As can be seen from Equation (7.32), the objective function is quadratic about $\boldsymbol{\rho}$ and $\boldsymbol{\theta}$. Moreover, the objective function in terms of δ_r has the form of $a \log x + \frac{b}{x}$. Therefore, it is easy to get the optimal analytical solutions for $\boldsymbol{\rho}$, $\boldsymbol{\theta}$, and $\boldsymbol{\delta}_r$. However, the objective function is not quadratic about $\boldsymbol{\Phi}_r$, and does not fall into certain functional forms that are easy to optimise. To bypass this difficulty, a Lagrange multiplier approach is employed in this study. The auxiliary variables $\chi_{r,i}$ are introduced, defined as [193]:

$$\chi_{r,i}^2 = \frac{1}{\|\boldsymbol{L}_i \boldsymbol{\Phi}_r\|^2} \tag{7.34}$$

This means that the objective function in (7.32) can be reformulated as

$$\begin{aligned} L_{upd}(\lambda) = {} & \frac{1}{2} (\boldsymbol{\theta} - \boldsymbol{\theta}_0)^T \boldsymbol{C}_{\boldsymbol{\theta}}^{-1} (\boldsymbol{\theta} - \boldsymbol{\theta}_0) + \frac{1}{2} (\boldsymbol{\rho} - \boldsymbol{\rho}_0)^T \boldsymbol{C}_{\boldsymbol{\rho}}^{-1} (\boldsymbol{\rho} - \boldsymbol{\rho}_0) + \frac{1}{2} \sum_{r=1}^{n_m} \left\{ n_d ln\delta_r + \delta_r^{-1} \|\eta_r \boldsymbol{\Phi}_r\|^2 \right\} \\ & + \frac{1}{2} \sum_{r=1}^{n_m} \sum_{i=1}^{n_t} (\chi_{r,i} L_i \boldsymbol{\Phi}_r - \boldsymbol{\psi}_{r,i})^T \boldsymbol{C}_{\boldsymbol{\psi}_{r,i}}^{-1} (\chi_{r,i} L_i \boldsymbol{\Phi}_r - \boldsymbol{\psi}_{r,i}) + \sum_{r=1}^{n_m} \sum_{i=1}^{n_t} \beta_{r,i} \left(\chi_{r,i}^2 \|\boldsymbol{L}_i \boldsymbol{\Phi}_r\|^2 - 1 \right) \end{aligned} \tag{7.35}$$

where $\beta_{r,i}$ are Lagrange multipliers that enforce the definition of Equation (7.34). As a result, after the objective function is rearranged, the full set of model parameters to be identified includes:

$$\lambda = \{\chi_{r,i}, \beta_{r,i}, \boldsymbol{\Phi}_r, \delta_r, \boldsymbol{\rho}, \boldsymbol{\theta};\ r = 1, \cdots, n_m; i = 1, \cdots, n_t\} \tag{7.36}$$

The dimension of λ is $n_{upd} = n_m n_d + 2n_m n_t + n_m + n_\theta + n_\rho$. By making use of the special properties of the objective function in Equation (7.35), a sequence of iterations comprised of a number of linear optimisation problems can be finished until the prescribed convergence criteria is satisfied once the nominal values of the parameters λ are assigned in advance. More details about solving Equation (7.35) can be found in [193].

The posterior PDF of λ can be well approximated by a Gaussian distribution centered at the optimal parameters and with covariance matrix $\boldsymbol{C}_{upd}$ equal to the inverse of the Hessian matrix $\Gamma_{upd}(\lambda)$ calculated at the optimal parameters λ. The Hessian matrix is given by

$$\boldsymbol{\Gamma}_{upd}(\lambda) = \begin{pmatrix} \boldsymbol{L}^{(\boldsymbol{\Phi\Phi})} & \boldsymbol{L}^{(\boldsymbol{\Phi}\chi)} & \boldsymbol{L}^{(\boldsymbol{\Phi}\beta)} & \boldsymbol{L}^{(\boldsymbol{\Phi}\delta)} & \boldsymbol{L}^{(\boldsymbol{\Phi}\rho)} & \boldsymbol{L}^{(\boldsymbol{\Phi}\theta)} \\ \boldsymbol{L}^{(\chi\boldsymbol{\Phi})} & \boldsymbol{L}^{(\chi\chi)} & \boldsymbol{L}^{(\chi\beta)} & \boldsymbol{L}^{(\chi\delta)} & \boldsymbol{L}^{(\chi\rho)} & \boldsymbol{L}^{(\chi\theta)} \\ \boldsymbol{L}^{(\beta\boldsymbol{\Phi})} & \boldsymbol{L}^{(\beta\chi)} & \boldsymbol{L}^{(\beta\beta)} & \boldsymbol{L}^{(\beta\delta)} & \boldsymbol{L}^{(\beta\rho)} & \boldsymbol{L}^{(\beta\theta)} \\ \boldsymbol{L}^{(\delta\boldsymbol{\Phi})} & \boldsymbol{L}^{(\delta\chi)} & \boldsymbol{L}^{(\delta\beta)} & \boldsymbol{L}^{(\delta\delta)} & \boldsymbol{L}^{(\delta\rho)} & \boldsymbol{L}^{(\delta\theta)} \\ \boldsymbol{L}^{(\rho\boldsymbol{\Phi})} & \boldsymbol{L}^{(\rho\chi)} & \boldsymbol{L}^{(\rho\beta)} & \boldsymbol{L}^{(\rho\delta)} & \boldsymbol{L}^{(\rho\rho)} & \boldsymbol{L}^{(\rho\theta)} \\ \boldsymbol{L}^{(\theta\boldsymbol{\Phi})} & \boldsymbol{L}^{(\theta\chi)} & \boldsymbol{L}^{(\theta\beta)} & \boldsymbol{L}^{(\theta\delta)} & \boldsymbol{L}^{(\theta\rho)} & \boldsymbol{L}^{(\theta\theta)} \end{pmatrix} \tag{7.37}$$

In general, $L^{(xy)}$ denotes the derivatives of L_{upd} with respect to any two vectors x and y. Here, x or y represents an one-dimensional array arranging from the following parameters: (1) $\boldsymbol{\Phi} = \left[\boldsymbol{\Phi}_1^T, \cdots, \boldsymbol{\Phi}_r^T, \cdots, \boldsymbol{\Phi}_{nm}^T\right]^T$; (2) $\boldsymbol{\chi} = \left[\chi_1^T, \cdots, \chi_r^T, \cdots, \chi_{nm}^T\right]^T$ with the r-th block $\boldsymbol{\chi}_r^T = \left[\chi_{r,1}, \cdots, \chi_{r,i}, \cdots, \chi_{r,n_t}\right]^T$; (3) $\boldsymbol{\beta} = \left[\beta_1^T, \cdots, \boldsymbol{\beta}_r^T, \cdots, \beta_{nm}^T\right]^T$ with the r-th block $\boldsymbol{\beta}_r^T = \left[\beta_{r,1}, \cdots, \beta_{r,i}, \cdots, \beta_{r,n_t}\right]^T$; (4) $\boldsymbol{\delta} = \left[\delta_1, \cdots, \delta_r, \cdots, \delta_{nm}\right]^T$; (5) $\boldsymbol{\rho} = \left[\rho_1, \cdots, \rho_i, \cdots, \rho_{n_\rho}\right]^T$ (6) $\boldsymbol{\theta} = \left[\theta_1, \cdots, \theta_i, \cdots, \theta_{n\theta}\right]^T$. $\boldsymbol{\Gamma}_{upd}$ is a symmetrical matrix which can be computed analytically and one can refer to [193] for more details.

7.5 Numerical example

Simulated data of a fifteen-story shear building are processed firstly to illustrate the accuracy of the proposed model updating algorithms. The stiffness to mass ratio at each floor is assumed to be $2500s^{-2}$. The substructure mass matrix and the substructure stiffness matrix are given by

$$\boldsymbol{M}_1 = \rho_1 \times \begin{bmatrix} 1 & 0_{1\times14} \\ 0_{14\times1} & 0_{14\times14} \end{bmatrix}; \boldsymbol{M}_i = \rho_i \times 2500 \times \begin{bmatrix} & 0_{(i-2)\times15} & & \\ 0_{1\times(i-2)} & 1 & 0 & 0_{1\times(15-i)} \\ 0_{1\times(i-2)} & 0 & 1 & 0_{1\times(15-i)} \\ & 0_{(15-i)\times15} & & \end{bmatrix} \tag{7.38a}$$

$$\boldsymbol{K}_1 = \theta_1 \times 2500 \times \begin{bmatrix} 1 & 0_{1\times14} \\ 0_{14\times1} & 0_{14\times14} \end{bmatrix}; \boldsymbol{K}_i = \theta_i \times 2500 \times \begin{bmatrix} & 0_{(i-2)\times15} & & \\ 0_{1\times(i-2)} & 1 & -1 & 0_{1\times(15-i)} \\ 0_{1\times(i-2)} & -1 & 1 & 0_{1\times(15-i)} \\ & 0_{(15-i)\times15} & & \end{bmatrix} \tag{7.38b}$$

Here θ_i $(i = 1, 2, 3, \cdots, 15)$ represents the scaling factor (in percentage) of the substructural stiffness. Classical Rayleigh damping with the damping ratios for the first two modes set to be 1% is assumed here for the simulation. Two scenarios have been considered for the structure: (i) no damage occurs in the structure; (ii) damages occurs in 1st, 3rd, and 10th stories. The stiffness scaling factors θ_i for the first scenario should be unit while θ_i for the second scenario should be unit except for $\theta_1 = 0.70$, $\theta_3 = 0.80$, and $\theta_{10} = 0.90$. The structure is excited by Gaussian white noise and its auto-spectral intensity is 0.25 m^2s^{-3}. The prediction error is also assumed to be Gaussian white noise.

7.5.1 Robustness test of the probabilistic model of trace of PSD matrix

The nonparametric Kolmogorov-Smirnov test is employed to verify the accuracy of the Gaussian approximation for $(\boldsymbol{S}_k^{sum})$ when the number of data sets n_s varies, that is, $n_s = 2, 5, 10, 20, 50, 100$. For each n_s considered, five hundred independent samples of (S_k^{sum}) are generated to calculate the empirical cumulative distribution function (CDF). Figure 7.2 shows the comparisons between the CDF of standard normal distribution (denoted by 'Normal' in Figure 7.2) and CDFs of (S_k^{sum}) with different n_s in the vicinity of the first and the third resonant frequencies, respectively. It can be seen from Figure 7.2 that the curves fit the standard normal distribution well. This indicates that the trace of the PSD matrix can be well approximated by Gaussian distribution. In this study, twenty

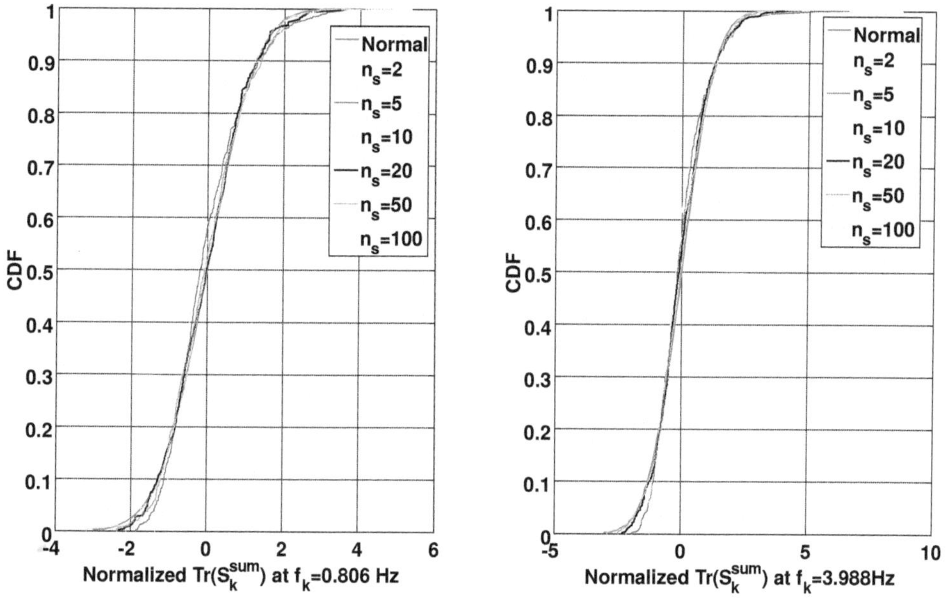

Figure 7.2: CDFs of $(\boldsymbol{S}_k^{sum})$ at $f_k = 0.806$ Hz (left) and $f_k = 3.988$ Hz (right) with different n_s.

sets of data (i.e. $n_s = 20$) are employed for ambient modal analysis with each data set lasting for 500 seconds.

7.5.2 Bayesian operational modal analysis

It is assumed that no sensors are installed on the 8th, 12th, and 14th floors. The measured twelve DOFs can be divided into three setups, and the setup information is shown in Table 7.1. Twenty sets of acceleration response are generated for both the healthy state and damaged state. The two-stage fast Bayesian spectral density approach is then employed to identify the modal properties as well as their uncertainties. To identify the spectrum variables, the auto-spectral densities of all measured DOFs are collected together to form (S_k^{sum}). The resonant peaks of the sum of the auto-spectral density are used for locating the initial guess of the natural frequencies. The frequency bands for modal identification can also be selected in the vicinity of these peaks, which are shown in the second column of Table 7.2. Then the spectrum variables as well as their uncertainties can be estimated based on the procedures summarised in Section 7.3. Each mode is identified separately. The actual natural frequencies, the identified natural frequencies and their coefficients of variance (c.o.v.) in healthy state and damaged state are shown in Table 7.2. Once the spectrum variables are identified in the first stage, the mode shape can then be identified with spectrum variables fixed at their optimal values.

As can be seen from Table 7.2, the most probable values identified by the proposed method and the exact values corresponding to the structural model that generates the ambient data agree very well with each other. Moreover, the results show that the c.o.v. values of the frequencies are much smaller than those of damping ratios, indicating that frequencies can be identified with less

Table 7.1: Setup information of the measured DOFs.

Setup	Measured DOFs
1	1, 2, 3, 4, 5, 6
2	5, 6, 7, 9, 10
3	9, 10, 11, 13, 15

Table 7.2: Identified modal properties for the 2D shear building.

Mode	Parameter	Undamaged State			Damaged State		
		Actual	MPV	c.o.v.(%)	Actual	MPV	c.o.v.(%)
1	f_1	0.806	0.806	0.068	0.772	0.772	0.070
	ξ_1	0.010	0.011	7.109	0.010	0.011	7.021
2	f_2	2.410	2.410	0.046	2.317	2.317	0.050
	ξ_2	0.010	0.010	7.270	0.010	0.011	7.445
3	f_3	3.989	3.993	0.095	3.892	3.895	0.073
	ξ_3	0.014	0.016	15.751	0.014	0.014	13.357

uncertainty than damping, which is consistent with the intuition. Figure 7.3 shows the conditional PDFs of the first natural frequency and the first damping ratio (keeping the remaining parameters at their optimal values) calculated using the Bayesian approach and their corresponding Gaussian approximation. The conditional PDFs obtained from the Bayesian approach in solid lines and the Gaussian approximation in dashed lines are extremely close to each other, indicating that the Gaussian approximation for the identified parameters is effective.

7.5.3 Bayesian model updating

The modal properties identified in Section 7.5.2 are employed for structural model updating. The initial guess for the stiffness scaling factors are taken to be $\theta_0 = \{10, 10, \cdots, 10\}$ so that the nominal values are significantly overestimated. The exact stiffness scaling factors, most probable values, as well as the c.o.v. values of ten stiffness parameters are shown in Table 7.3. It can be clearly seen from the values that the 1st floor, the 3rd floor, and the 10th floor have damage extents close to the exact values. These evidences indicate that the proposed method is able to update the structural model parameters even though the modal information employed is not complete. Figure 7.4 shows the iterative histories for the most probable values of stiffness parameters corresponding to the structure in healthy state and damaged state, respectively. This figure indicates that convergence occurs within a few iterations for both scenarios.

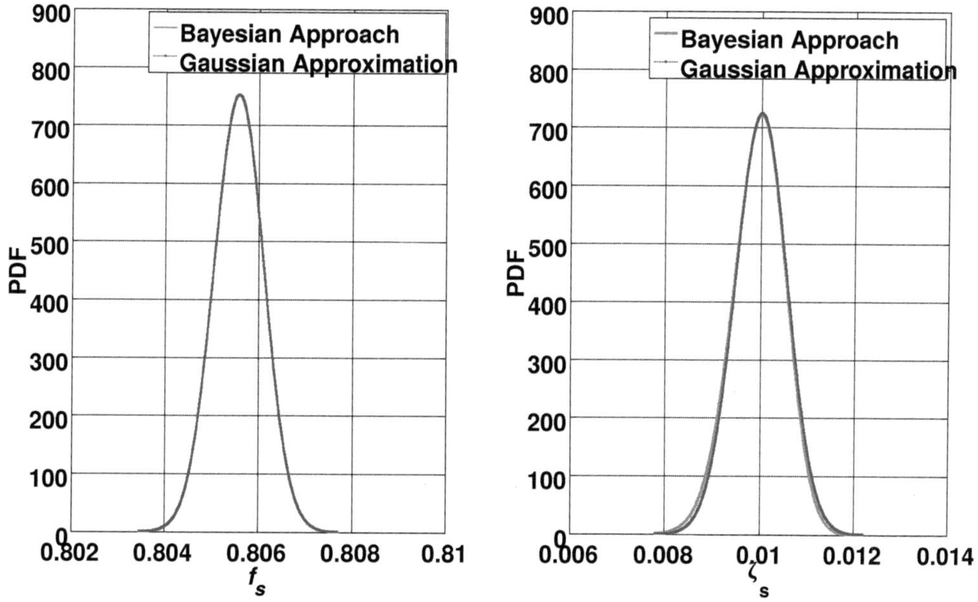

Figure 7.3: Conditional PDFs of f_s (left) and ς_s (right) for the shear building.

Table 7.3: Identified stiffness scaling factors for the 2D shear building.

Parameter	Healthy state			Damaged state		
	Actual	MPV	c.o.v.(%)	Actual	MPV	c.o.v.(%)
θ_1	1.000	0.997	0.278	0.700	0.699	0.280
θ_2	1.000	0.998	0.281	1.000	0.998	0.283
θ_3	1.000	0.998	0.286	0.800	0.799	0.289
θ_6	1.000	0.998	0.318	1.000	0.998	0.320
θ_9	1.000	0.997	0.378	1.000	0.977	1.038
θ_{10}	1.000	0.996	0.408	0.900	0.898	0.411
θ_{13}	1.000	0.995	0.576	1.000	0.996	4.264
θ_{14}	1.000	1.002	0.706	1.000	1.005	6.412
θ_{15}	1.000	0.982	1.000	1.000	0.980	12.517

7.6 Experimental study

The performance of the methods presented in this chapter is experimentally investigated with a three-story shear building. The model shown in Figure 7.5 is constructed with aluminium with dimensions of 401 (*width*) × 314 (*depth*) × 1158 (*height*) in mm. The building was fixed on a shake table which can generate ground motion horizontally. The experimental setups are composed of a

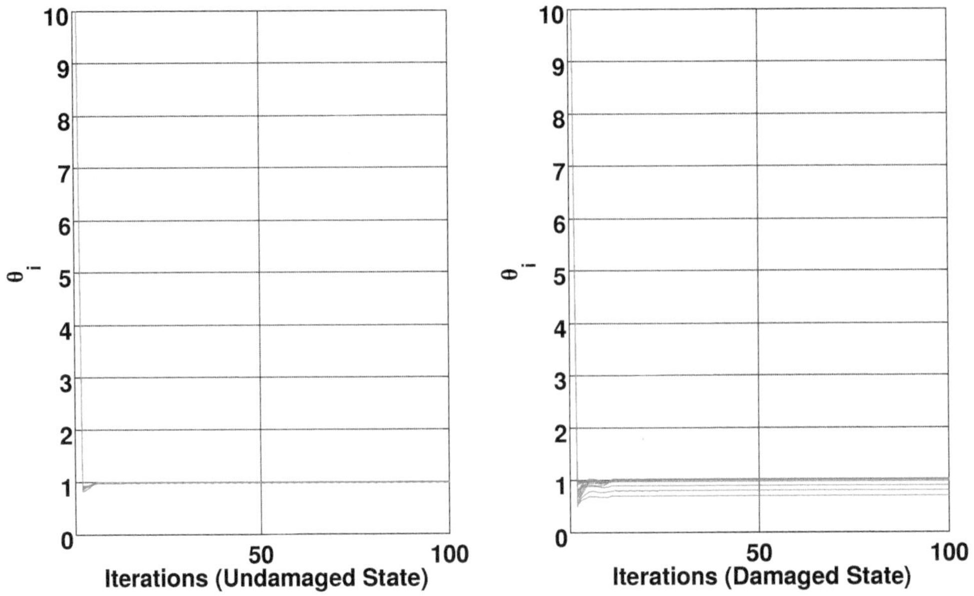

Figure 7.4: Iteration histories of θ_i for the shear building in the healthy state and the damaged state.

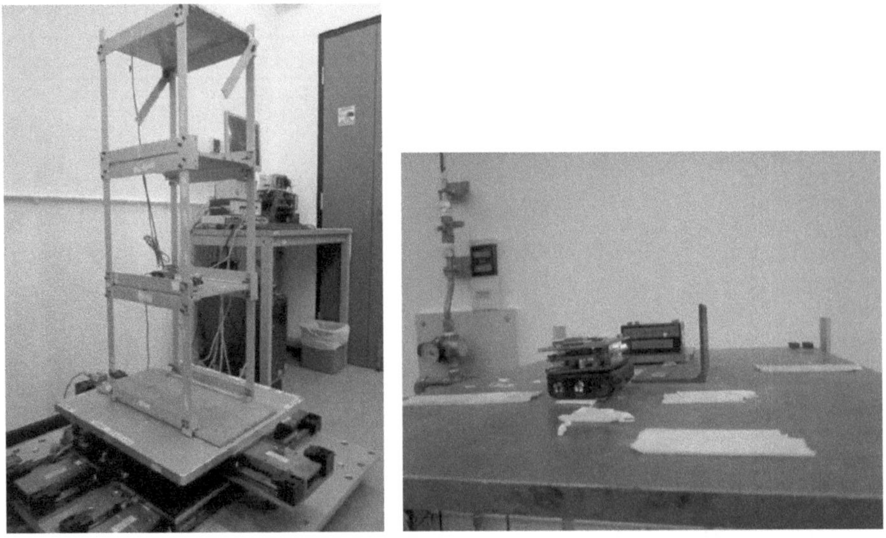

(a) Tested building

(b) Sensor placement on the top floor

Figure 7.5: Shear building used for laboratory testing.

shaker table and wireless sensor network system composed of a laptop and the Crossbow Imote2 platforms. A side view of the sensor placement on the top floor is shown in Figure 7.5b. The gateway node connected to the laptop directly works as a bridge communicating between the laptop and leaf nodes. The main hardware components of a leaf node include an Imote2 board, a high-sensitivity SHM-H sensor board [104], and a battery board. The engineering analysis software employed in this study is the ISHMP Services Toolsuite provided by the Illinois SHM Project (http://shm.cs.uiuc.edu). A RemoteSensing component available in ISHMP Services Toolsuite was employed to collect precisely synchronised data from the leaf nodes, and transfer them back to the gateway node, and write the output to the file. More details on the experimental issues are referred to [196, 195].

The structure can be simplified as a 3-DOF shear building, with its substructure stiffness given by

$$\boldsymbol{K}_1 = \theta_1 \times \begin{bmatrix} 1 & 0_{1\times 2} \\ 0_{2\times 1} & 0_{2\times 2} \end{bmatrix}; \quad \mathbf{K}_i = \theta_i \times \begin{bmatrix} & 0_{(i-2)\times 3} & & \\ 0_{1\times(i-2)} & 1 & -1 & 0_{1\times(3-i)} \\ 0_{1\times(i-2)} & -1 & 1 & 0_{1\times(3-i)} \\ & 0_{(3-i)\times 3} & & \end{bmatrix} \tag{7.39}$$

Here, $\theta_i\,(i = 1, 2, 3)$ represents the stiffness factors of the substructure stiffness matrix. The linear stiffness coefficients obtained from a static test were $k_1 = 20.88kN/m$, $k_2 = 22.37kN/m$, and $k_3 = 24.21kN/m$ [38], which could be used as the initial guess for structural model updating. The mass coefficients were measured as $m_1 = 5.63kg$, $m_2 = 6.03kg$, and $m_3 = 4.66kg$. The substructure mass matrix are given by

$$\boldsymbol{M}_1 = \rho_1 \times \begin{bmatrix} 1 & 0_{1\times 2} \\ 0_{2\times 1} & 0_{2\times 2} \end{bmatrix}; \quad \boldsymbol{M}_i = \rho_i \times \begin{bmatrix} & 0_{(i-2)\times 3} & & \\ 0_{1\times(i-2)} & 1 & 0 & 0_{1\times(3-i)} \\ 0_{1\times(i-2)} & 0 & 1 & 0_{1\times(3-i)} \\ & 0_{(3-i)\times 3} & & \end{bmatrix} \tag{7.40}$$

where $\rho_1 = 5.63kg$, $\rho_2 = 6.03kg$, and $\rho_3 = 4.66kg$.

7.6.1 Bayesian operational modal analysis

Horizontal accelerations were measured by each wireless sensor node with a sampling rate of 100 Hz and a 40 Hz cut-off frequency. The measured acceleration time histories lasted 15 minutes with 90,000 points, and twenty data sets were recorded for ambient modal identification. Figure 7.6 shows sample acceleration time history corresponding to the top story as well as the trace of the PSD matrix, from which the initial guess of natural frequencies and frequency bands adopted for modal identification can be determined. The identified most probable values (MPV) of spectrum variables as well as their c.o.v. values are shown in Table 7.4.

The frequencies identified using two different methods agree well with each other. However, the damping ratios identified using the proposed method are much smaller than those obtained from a free vibration response. This coincides with the conclusion by intuition that, under the ambient vibration condition, the accuracy of the damping ratio estimation is dependent on the length of the collected time series. Therefore, the time duration should be increased for the measurement to get more precise damping ratios. As seen from Table 7.4, the c.o.v. values for the parameters except for the prediction errors increased with the modes, indicating that the lower modes can be identified with higher accuracy.

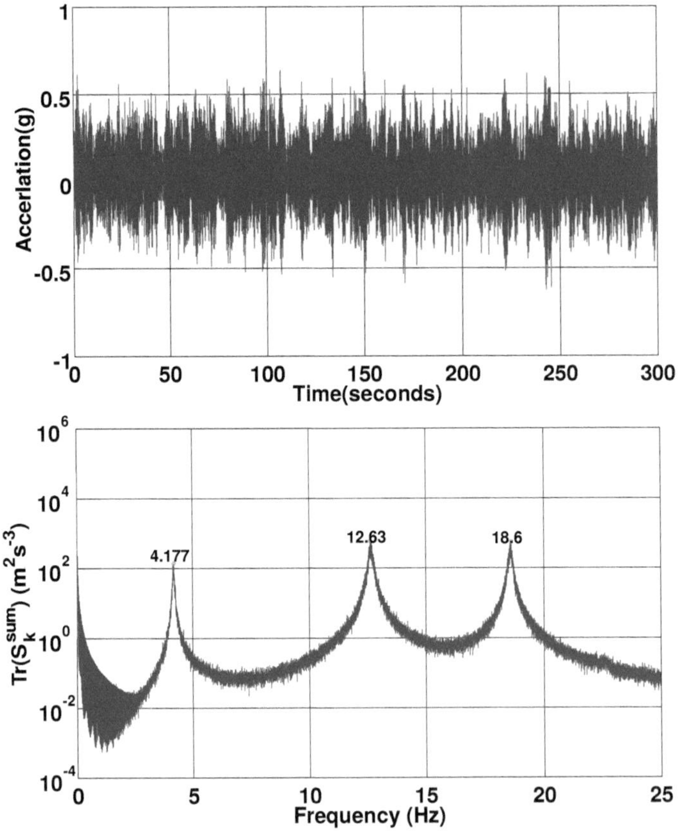

Figure 7.6: Acceleration of the top story and the trace of PSD matrix.

7.6.2 Bayesian model updating

Four scenarios are considered for structural model updating. In scenario 1, the mode shape components of three measured DOFs are identified simultaneously and all modal data are involved in model updating. In scenario 2, the sensors are divided into two local groups. Each group is composed of two sensor nodes, with the one in the middle story being overlapping (reference) nodes. For scenario 3, the mode shape components of the top floor are not used for model updating. For scenario 4, the third mode shape is not employed for model updating. Figure 7.7 presents the convergence curves of the stiffness factors for four different scenarios, which indicate that the stiffness factors converge quickly by using the proposed method.

The identified most probable stiffness factors as well their associated c.o.v. for four scenarios are shown in Table 7.5. The results from the static test by [38] are also presented for comparison in Table 7.5. As seen from Table 7.5, the results given by the static test and the proposed technique seem to be consistent. Figure 7.8 shows the most probable 'system mode shape' for different scenarios.

Table 7.4: Identified spectrum variables for the laboratory shear building model.

Mode	Variables	Bayesian MPV	Bayesian c.o.v(%)
1	f_s	4.1677	0.027
	ξ_s	0.0086	0.698
	$\sqrt{S_{fs}}$	0.0103	5.087
	$\sqrt{s_{\mu s}}$	0.0462	87.936
2	f_s	12.6450	0.032
	ξ_s	0.0068	10.430
	$\sqrt{S_{fs}}$	0.0158	8.486
	$\sqrt{s_{\mu s}}$	0.0966	79.643
3	f_s	18.5740	0.032
	ξ_s	0.0047	14.784
	$\sqrt{S_{fs}}$	0.0093	16.600
	$\sqrt{s_{\mu s}}$	0.2470	26.163

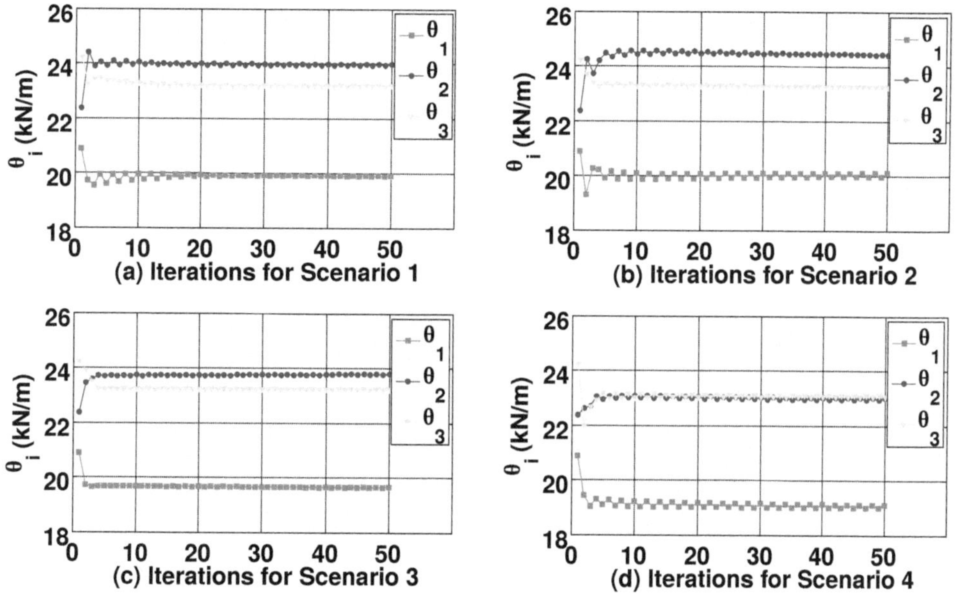

Figure 7.7: Iteration histories of model updating for four scenarios.

The 'system mode shapes' identified from different scenarios also agree well with each other, indicating that the proposed method is able to treat the case of missing information and the local mode shape components can be incorporated automatically. We make no final statement as to whether the experimental studies can give accurate uncertainty information or not since it depends

Table 7.5: Identified stiffness parameters for four different scenarios.

Param.	Static	Scenario 1		Scenario 2		Scenario 3		Scenario 4	
		MPV (kN/m)	c.o.v.(%)	MPV (kN/m)	c.o.v.(%)	MPV (kN/m)	c.o.v.(%)	MPV (kN/m)	c.o.v.(%)
θ_1	20.88	19.926	6.28	20.014	7.42	19.602	6.63	18.951	6.72
θ_2	22.37	23.937	5.79	24.446	5.97	23.807	5.77	22.902	6.23
θ_3	24.21	23.153	5.8	23.209	5.67	23.21	5.05	23.062	5.66

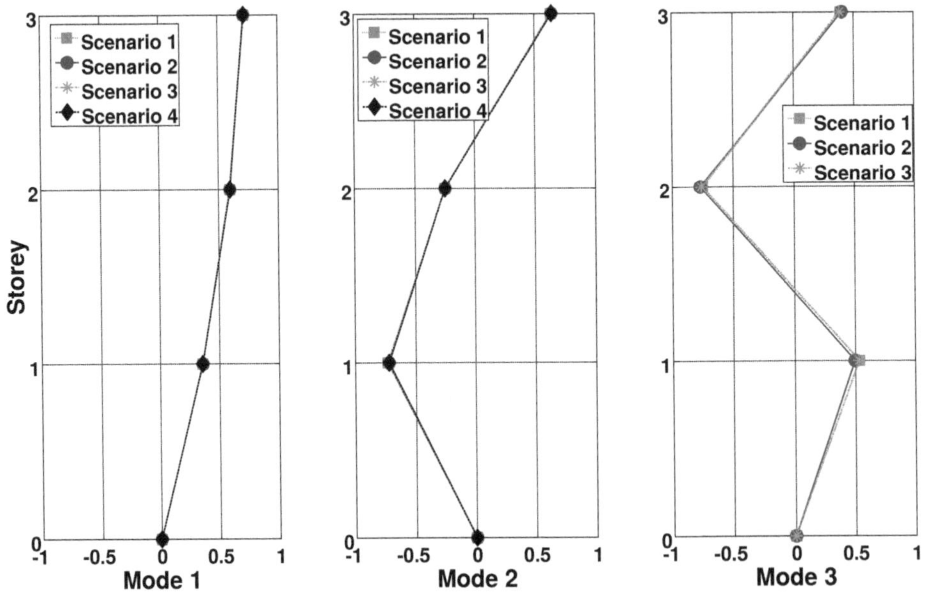

Figure 7.8: Identified optimal 'system mode shapes' for different scenarios.

on the identification model, the environmental conditions, or even the nature of the parameter under question. Rather, the experimental observations serve to give an idea of the order of uncertainty magnitude in the test conditions stated.

7.7 Concluding remarks

SHM based on ambient vibration responses are inevitably affected by multiple uncertainties, which will distort intrinsic information reflecting the real state of structures. Therefore, it is of great significance to accommodate these uncertainties and improve the robustness and accuracy of SHM results. In this chapter, analytical probabilistic models of the FFT coefficients, power spectral density (PSD), as well the trace of PSD are examined. Based on the probabilistic models of frequency-domain responses, a two-stage fast Bayesian operational modal analysis approach is proposed to extract modal properties and their associated uncertainties. A novel technique for variable separation is developed so that the interaction between spectrum variables (e.g. frequency, damping ratio, as well as the amplitude of modal excitation and prediction error) and spatial variables (e.g. mode shape components) can be decoupled completely.

Based on the uncertainties of the identified modal properties (e.g. natural frequencies and partial mode shapes of some modes), a novel Bayesian methodology is developed for structural model updating. The model updating problem is formulated as one minimising an objective function, which can incorporate the local mode shape components identified from different clusters automatically without prior assembling or processing. A fast-iterative scheme is proposed to efficiently compute the optimal parameters so as to resolve the computational burden required for optimising the objective function numerically. The posterior uncertainty of the model parameters can also be derived analytically. The efficiency and accuracy of all these methodologies are verified by a numerical example. Experimental studies are also conducted by employing laboratory shear building models installed with advanced wireless sensors. Successful validation of the proposed methods using measured acceleration demonstrate the potential of Bayesian approaches to accommodate multiple uncertainties in SHM.

8

A Worked-out Example of Surrogate-based Bayesian Parameter and Field Identification Methods

Noémi Friedman,[1,*] *Claudia Zoccarato,*[2] *Elmar Zander*[3] and *Hermann G. Matthies*[3]

The aim of this chapter is to compare and present available surrogate-based methods suitable for the assimilation of measurement data of some model response in order to characterize the model's input parameters or input fields. The comparison is done through a worked-out example of assimilating seabed subsidence data for the compressibility field identification of a reservoir. Throughout the example we provide a general framework for any engineering problem where we wish to identify input parameters or fields—and by that to reduce the uncertainties of the predicted response—by assimilating observations of the output of the model. We here provide algorithms of the compared methods and link the examples to on-line accessible MATLAB® codes (see https://ezander.github.io/ParameterAndFieldIdentification/).

8.1 Introduction

The production of fluids from deep hydrocarbon reservoirs causes a variation of the pore pressure within the subsurface rock formation. The consequent compaction of the reservoir can be appreciated as deformation on the land or seabed surface. In case of offshore reservoirs, the vertical displacements of the seafloor can cause induced earthquakes and severe damages, such as in the well-known literature cases of the Goose Creek field south of Huston or the Groningen field in the Netherlands [208]. Thus the target is to continuously improve the accuracy of estimation of the compaction caused by the reservoir development as new information (e.g. measurements) are available.

Here, a three-dimensional (3D) finite element method (FEM) model [70] is used to describe the geomechanical behaviour of an offshore reservoir exploited for gas production. The main focus is at estimating the parameters characterising the model response and the associated uncertainty. In particular, the uniaxial vertical compressibility is considered for calibration, as it represents the most influencing parameter controlling the amount of deep compaction.

Different approaches are available to deal with the estimation of compressibility. The simplest way is by some laboratory tests for characterising the geological formations using well-logging tech-

[1] Institute for Computer Science and Control (SZTAKI), Budapest, Hungary.
[2] University of Padova, Italy.
[3] Technische Universität Braunschweig, Germany.
* Corresponding author: n.friedman@ilab.sztaki.hu

niques. Here a different approach is considered, based on the inversion of seabed surface data. The idea here is to use indirect observations of seafloor subsidence to infer a heterogeneous compressibility field. In particular, a time-lapse bathymetric map of the seabed collected over the reservoir domain is used. Such method allows a relatively cheap non-destructive monitoring of the reservoir by updating the compressibility field when new measurements of the subsurface displacements are available. One way to update our initial uncertainty of the compressibility field represented by some prior distribution is to use the Bayesian posterior, an updated distribution of the compressibility field given the measured seabed displacements. This posterior is proportional to the product of the likelihood and the prior distribution of the compressibility field. The likelihood is the probability of how likely is it to measure the given values of displacements for a given realisation of the compressibility field. Due to the fact that the likelihood can not be written here with a closed form one rather uses sampling based procedures. The Metropolis-Hastings random walk can be used as our example of a sampling technique, which results in a Markov-chain Monte Carlo (MCMC) procedure. One problem is the slow convergence of Monte Carlo (MC) based methods, necessitating a large number of MC simulations; for an sufficiently accurate statistic, one is often required to run many hundred thousands of samples. This means for the evaluation of the likelihood that one needs to compute for every sample of the compressibility field the displacement field with the help of the FEM model. This can be highly demanding from a computational point of view in large-scale systems [209].

Another way to update the compressibility field is through an optimised estimator based on the conditional expectation (see Chapter 4). The task here is to find an estimator—a map from the measured data to the updated input—that minimises in a mean squared sense the difference between the prior input parameter and the estimated one computed from the measured data. The Kalman smoother or its frequently used version the Ensemble Kalman smoother (EnKF) [210] are equivalent to such approach if we restrict the estimators to be some linear function of the measured data, and the prior to be Gaussian. As it is shown in the context of the compressibility field identification, limiting the estimator to linear maps may be a too strong restriction for non-linear problems, and the assumption of Gaussian prior also limit the method. To improve the performance, non-linear estimators are also tested here for comparison.

In this study, the computational cost of all the above mentioned methods is kept here in a manageable scale by approximating the dependence of the FEM model solution, i.e., the seabed surface vertical displacements, on the compressibility field via a purely mathematical surrogate or proxy model using the so-called polynomial chaos expansion (PCE). In contrast to the FEM solver, the PCE can be evaluated with little computational cost. We generate a surrogate model for two scenarios: (I) for the case when we restrict the compressibility to be constant throughout the domain, and (II) for a stochastic spatially varying compressibility field input. The dependence of the seabed displacement on the value of the compressibility is smooth, and thus the PCE approximation can be computed by a few runs of the deterministic FEM model. However, when soil conditions are such that, a spatially varying compressibility field has to be assumed, the whole random field of the compressibility has to be updated. To enable the use of the same framework of the surrogate modelling when the random input is not a random variable but a random field, we discuss here separated representations of random fields. In this chapter this discretisation of the compressibility random field is accomplished via a truncated Karhunen-Loève Expansion (KLE). Using the KLE, the compressibility field can be represented as a function of a finite number of independent random variables. With such a representation, the surrogate modelling is done in the same way as in the scalar case, but the dimension of the stochastic problem is higher, which necessitates a higher number of runs of the deterministic FEM solver for the determination of the proxy model. With the help of the PCE surrogate model, classical stochastic inversion techniques can be applied in an efficient and straight forward manner. Using the PCE framework, the Kalman filter posterior can also be given in a PCE form, resulting in a completely sample-free inversion method [153]. Further improvements can be achieved by identifying non-linear estimators from the minimisation problem but still limiting these maps to some finite dimensional space spanned by some given basis functions. This procedure is described in detail in Chapter 4.

This chapter is organised as follows: Section 8.2 briefly describes the FEM geomechanical model, that is, the deterministic solver and its reformulation using a probabilistic framework. The PCE spectral representation of the model response, that is, the surrogate or proxy model, is explained in Section 8.3. Here, an orthogonal projection is used for the computation of a surrogate model. In the following Section 8.4, the separated representation of a random field is explained through the KLE and the POD theories. Finally, the different data assimilation methods, that is the MCMC, the PCE-based Kalman filter, and the nonlinear filters are explained and compared in Section 8.5 for the reservoir compressibility field identification. We conclude the work in Section 8.6, and also discuss an actual ongoing research outlook, and possibilities of further improvements.

8.2 Numerical modelling of seabed displacement

Modelling the geomechanical behaviour of a producing reservoir aims at computing the stress and displacements fields generated by pore-pressure changes in space and time due to reservoir activities such as fluid extraction from the subsurface. In this section, the mathematical formulation of the problem is provided, along with its extension to the stochastic dimension. This extension enables us to put forward the uncertainties of the model due to our lack of knowledge of the exact input.

8.2.1 The deterministic computation of seabed displacements

The geomechanical analysis of a producing gas reservoir is carried out here by solving the governing partial differential equations of poro-elasticity with the aid of a FEM model. A one-way coupling approach is implemented, assuming that the mechanics-to-fluid coupling is negligibly weak. The pore pressure increment in space and time is first solved for the fluid flow dynamics, and then used as an external source to solve the mechanical equilibrium equations. Here, we just briefly recall the basic formulation and the most important model features. Interested readers are directed to Appendix A for some more details, and to [70] for the whole model description. The equilibrium equations governing the phenomenon of consolidations and the boundary conditions read

$$\begin{aligned} \nabla \cdot \boldsymbol{\sigma}' - \alpha \nabla p &= \rho \mathbf{g} + \mathbf{b} && \text{in} \quad \mathcal{D}, \\ \mathbf{u} &= \mathbf{u}_0 && \text{on} \quad \Gamma_D, \\ \boldsymbol{\sigma} \cdot \mathbf{n} &= \mathbf{t} && \text{on} \quad \Gamma_N, \quad \Gamma_N \cup \Gamma_D = \Gamma, \end{aligned} \tag{8.1}$$

with ∇ the Nabla operator, $\boldsymbol{\sigma}'$ and $\boldsymbol{\sigma}$ the effective and the total stress tensors, α the Biot's coefficient, p the pore pressure, ρ the fluid density, $\mathbf{g}$ the gravity acceleration, $\mathbf{b}$ the external body forces, and $\mathbf{t}$ the total force per unit surface on the Neumann boundary Γ_N. By deriving the weak formulation of the problem and applying a spatial discretisation of the displacement and pressure fields (see derivation and the equation in a more detailed form in Appendix A), we get a nonlinear system of equations

$$\mathcal{A}(\mathbf{a}, \mathbf{p}) = \Pi(\mathbf{a}) - \mathbf{f}(\mathbf{p}) = \mathbf{0}, \tag{8.2}$$

with the nodal displacements $\mathbf{a}$ and the nodal pressure values $\mathbf{p}$. This equation can be solved for the unknown $\mathbf{a}$ with the help of a Newton solver by initiating a solution vector $\mathbf{a}_0$ and solving in a sequential manner the linear system of equations

$$-\mathbf{K}_T \Delta \mathbf{a}_n = \mathcal{A}(\mathbf{a}_n, \mathbf{p}) \tag{8.3}$$

for the increment of nodal displacements $\Delta\mathbf{a}$, and computing the new solution vector $\mathbf{a}_{n+1} = \mathbf{a}_n + \Delta\mathbf{a}_\mathbf{n}$, until Equation (8.2) is satisfied with a residual error whose norm is within a certain threshold. In Equation (8.3) $\mathbf{K}_T$ is the tangent stiffness matrix, the Jacobian of Π. The important thing here is that $\mathbf{K}_T$ is scaled with the inverse of the oedometric compressibility c_M (see Equation A11 in Appendix A). According to previous experiments, the compressibility can be described by a power law relation

$$c_M(\sigma_z) = c_{M0} \left(\frac{\sigma'_z}{f_0} \right)^{-\lambda}, \tag{8.4}$$

where σ'_z is the vertical effective stress, f_0 is a fixed value of stress enabling a dimensionless form, and λ and c_{M0} are material coefficients estimated via measurements, for example, by a radioactive marker technique (RMT). The problem of the actual analysis is that no in-situ measurements for λ and c_{M0} were available. However, measurements were available for similar case studies, which allowed a pre-calibration of the c_M model parameters, that is $c_{M0} = 1.0044 \cdot 10^{-2}$ MPa^{-1} and $\lambda = 1.1347$ with $f_0 = 1$ MPa. Due to the strong compartmentalisation of the analysed fault formation, the compressibility may vary from block to block of the reservoir. In due course, a horizontally varying function $f_{c_M}(\hat{x}, \hat{y})$[1] is introduced to scale the compressibility c_M[2], so Equation (8.4) modifies to

$$c_M(\hat{x}, \hat{y}, \sigma_z) = f_{c_M}(\hat{x}, \hat{y}) c_{M0} \left(\frac{\sigma'_z}{f_0} \right)^{-\lambda}. \tag{8.5}$$

To conclude, for a prediction of the subsidence, our task is to solve for the nodal displacements $\mathbf{a}$ the nonlinear system of equations (8.2), where the Jacobian depends on the actual value of the stress state through the compressibility parameter c_M. Figure 8.2 shows a visualisation of this task, in the case of knowing the exact value of the scaling parameter f_{c_M}. The figure shows that given a specific pressure change, the FEM code solves for the nodal displacements $\mathbf{a}$, which can be mapped to the displacement field $\mathbf{u} = [u_{\hat{x}}, u_{\hat{y}}, u_{\hat{z}}]^T$. This can be further mapped to the seabed subsidence evaluated at all n_y points where measurements are available. To be consistent with Chapter 4, let us note that this forward model is a concrete example of the abstract solution operator $\mathcal{G}$, that is,

$$\mathbf{y} = \mathcal{G}(f_{cM}(\hat{x}, \hat{y})) \quad [\mathbf{y}]_i := u_{\hat{z}}(\hat{x}_i, \hat{y}_i) \quad i = 1...n_y. \tag{8.6}$$

In our numerical examples, we used the mesh shown in panel (A) of Figure 8.1 with $n = 320,901$ mesh nodes. The $n_y = 60$ assimilation points — where measurements are available — are shown in panels (B) and (D) of the figure. The coordinates of assimilation points were chosen such, that

[1] We use $\hat{x}$, $\hat{y}$ and $\hat{z}$ letters for the spatial variables to distinguish it from the measurable response $\mathbf{y}$ and the measurement $\mathbf{z}$.

[2] Please note, it would make sense to introduce another scaling factor, scaling the exponent λ, but for simplicity, for now we stick to the problem of one uncertain field.

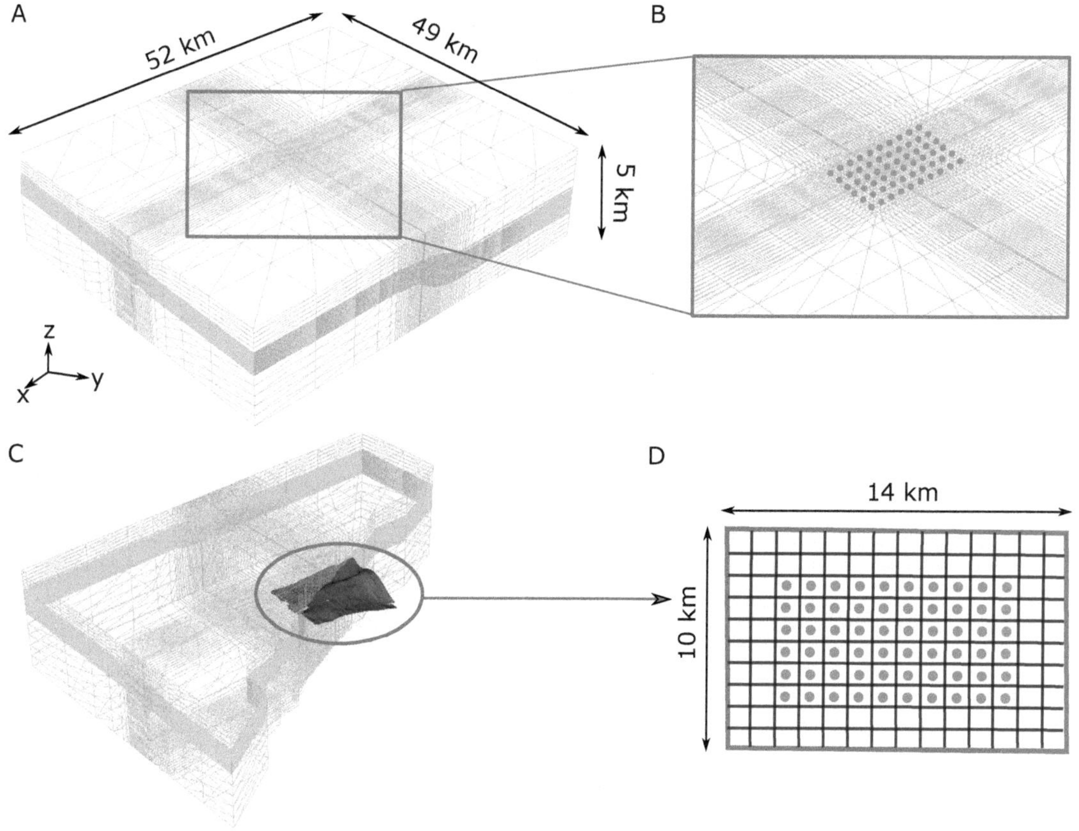

Figure 8.1: (A) Axonometric view of the FEM grid used for the geomechanical simulations, (B) zoom of the central portion of the domain with location of the $n_y = 60$ data assimilation points on the seabed, (C) 3D section of (A) where the producing layers are highlighted, and (D) discretisation of the compressibility field by a 14x10 square cells with location of the 60-data assimilation points.

they are located at the center points of the FEM elements. The subsurface displacements of the reservoir are mainly influenced by the central portion of the spatial domain shown in panel (D). Our goal was to identify the compressibility field in this central portion $\mathcal{D}_s$ of the domain.

8.2.2 Modified probabilistic formulation

Due to a lack of accurate information on the scaling factor f_{c_M}, we use the available seabed data of vertical displacements $\mathbf{y}$ for calibrating this scaling factor in order to have a more accurate prognosis for the seabed subsidence in the future. In other words, the parameter f_{c_M} should be inferred from measurement data on the history of seabed subsidence $\mathbf{y}$. Unfortunately, from a given measured subsidence $\mathbf{z}_m$ we can not give an explicit form of the parameter f_{c_M}. That would mean to invert the deterministic solver shown in Figure 8.2, or to be more exact, the operator $\mathcal{G}$ in Equation (8.6).

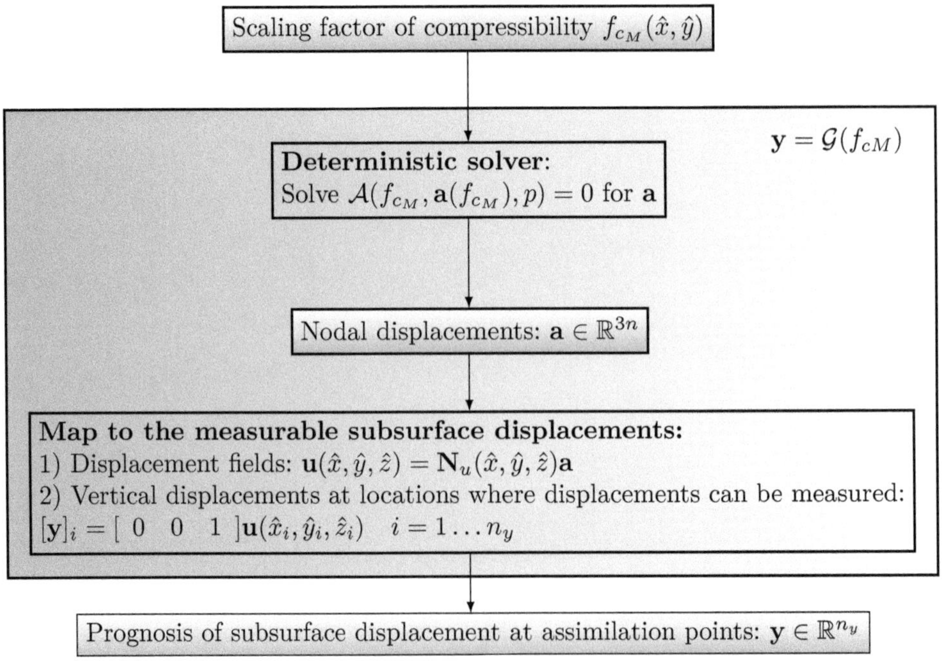

Figure 8.2: Schematic flowchart of the deterministic solver, the computation of subsidence, and the measurable expression.

$\mathcal{G}$ is typically not invertible, and hence the inverse problem is unfortunately an ill-posed one, for example, because there is no solution or no unique solution. Instead of releasing this problem by some rather ad hoc regularisation method, we follow a different path, namely, we put the whole problem in a probabilistic setting, by handling the unknown parameter f_{c_M} as a random field, that is

$$F_{c_M}(\hat{x}, \hat{y}, \omega) : \mathcal{D}_s \times \Omega \to \mathbb{R}, \tag{8.7}$$

where ω[3] is an event or a realisation from the space of all possible outcomes Ω, and $\mathcal{D}_s$ is the crucial part of the spatial domain shown in the (D) panel of Figure 8.1[4]. To represent our knowledge about the possible values of F_{cM}, we assign to it some *a priori* distribution function $\pi_F(f_{c_M})$, and use the Bayesian inversion to update this *a priori* distribution using the measurement data $\mathbf{z}_m$.

First, let us choose the *a priori* distribution function of the scaling factor. This should be based on some professional geotechnical expertise. Careful attention has to be paid so that we do not modify the problem in such a way that we ruin the essential properties of the physics. In our case, it is important that the matrix $\mathbf{C}$ stays positive definite, as naturally a negative compressibility would not make any sense. To keep this important property, we use the semi-bounded Lognormal distribution with support $I_F = (0, +\infty)$. The fact that λ and c_{M0} were measured from similar case

[3]One can think of ω as an abstract notation showing random nature (see more in Chapter 4).

[4]Furthermore we assign capital letters for the random variables and random fields, and small letters for a realisation of them. For example F_{cM} is a random variable and one realisation of it is $F_{cM}(\omega_i) = f_{cM,i}$.

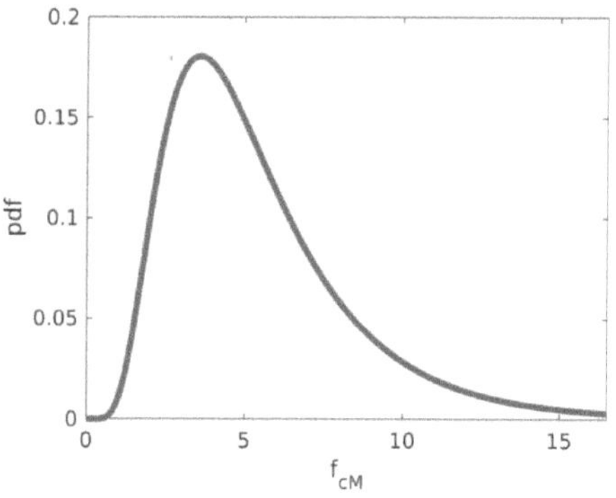

Figure 8.3: Lognormal prior probability distribution of f_{cm}.

studies would suggest a constant expected value of the scaling factor equal to one. Nevertheless, due to the experienced subsidence being much higher then what the field experts expected, we assumed a higher expected value of 5.5, that is

$$\mathbb{E}[F_{cM}] = \int_{I_F} f_{cM} \pi_F(f_{cM})\, \mathrm{d}f_{cM} = 5.5, \tag{8.8}$$

and a variance chosen in such a way that with 99.5% probability $f_{cM} < 10$. Accordingly, we define the prior distribution of the scaling factor to be

$$F_{cM} \sim \ln\mathcal{N}(\mu, \sigma^2\Sigma), \quad \mu(\hat{x}, \hat{y}) = 1.562, \quad \sigma(\hat{x}, \hat{y}) = 0.534. \tag{8.9}$$

We assumed a constant variance and mean over the spatial domain. Σ is the correlation function of the random field. The marginal distribution at all spatial points is shown in Figure 8.3. We consider two different scenarios:

(I) The first, and simpler, scenario assumes that the compressibility field is spatially constant throughout the spatial domain. In such a case, the random field is fully correlated, which means that for any two points with coordinates $(\hat{x}, \hat{y})$ and $(\hat{x}', \hat{y}')$, the correlation is one

$$\Sigma(\hat{x}, \hat{y}, \hat{x}', \hat{y}') = 1; \tag{8.10}$$

and thus the scaling factor can be described by one scalar random variable.

(II) The second scenario allows the update to identify a spatially varying compressibility field. Although in this way the complexity of the model is higher, the strong compartmentalisation suggests that the compressibility may not be constant throughout the spatial domain. We assume that at the analysed height the compressibility field is smoothed out and thus its values at different spatial points are strongly correlated, and that this correlation depends only on the distance d in between the points. According to geotechnical expertise, the dependence can be described by the Matérn covariance function[5] $C_{\nu_c}(d/l_c)$

$$\Sigma(\hat{x},\hat{y},\hat{x}',\hat{y}') = C_{\nu_c}(d/l_c)^5, \quad \nu_c = 2, \quad l_c = 4000m, \tag{8.12}$$

where ν_c is a non-negative parameter of the covariance function—influencing the smoothness of the field realisations, d is the distance $\sqrt{(\hat{x}-\hat{x}')^2+(\hat{y}-\hat{y}')^2}$, and l_c is the correlation length. The smaller the correlation length is, the faster the covariance values decay with the distance and the wilder the realisations of the field can get.

From a mathematical point of view, it is much nicer to work with variables, which are in a vector space (so that any linear combination of the variable is also in the same space). For this purpose, we rather represent the field F_{cM} as a function of some underlying Gaussian random field. The map from the Gaussian to the Lognormal random field is straightforward:

$$F_{cM}\left(\Theta(\hat{x},\hat{y},\omega)\right) = e^{\mu+\sigma\Theta} \quad \Theta \sim \mathcal{N}(0,\Sigma), \tag{8.13}$$

where $\Theta : \mathcal{D}_s \times \Omega \to \mathbb{R}$ is the underlying Gaussian field. To further simplify the description of the stochastic input, we write the Gaussian field as a function of some random variables $\mathbf{Q}$

$$\Theta(\hat{x},\hat{y},\omega) = \mathcal{F}(\mathbf{Q}(\omega)). \tag{8.14}$$

This task is trivial for a spatially constant field, that is for the problem description (I). In this scenario, the field Θ is fully correlated and thus independent of the spatial variables $\hat{x}$ and $\hat{y}$, and accordingly can be written as a standard Gaussian random variable, so $\mathbf{Q}$ is just one standard Gaussian random variable Q, and $\mathcal{F}$ is simply the identity map

$$\Theta(\hat{x},\hat{y},\omega) = \mathcal{F}(Q) = Q(\omega) \quad Q \sim \mathcal{N}(0,1). \tag{8.15}$$

In the case of a spatially varying field, the task is a bit more complicated. Fortunately, with the help of the Karhunen-Loève Expansion (KLE, see Section 8.4) we can represent any 'decent' stochastic field as a linear combination of the product of some square integrable spatial eigenfunctions $r_i(\hat{x},\hat{y})$ and Gaussian independent random variables X_i in the form

$$\Theta(\hat{x},\hat{y},\omega) \approx \mu_\Theta(\hat{x},\hat{y}) + \sum_{i=1}^{L} \sigma_i r_i(\hat{x},\hat{y}) X_i(\omega), \tag{8.16}$$

where μ_Θ is the mean of the field Θ, which is due to Equation (8.13) is constant zero. For the properties and the numerical computation of the KLE, see Section 8.4. Setting these X_i standard Gaussian random variables to be Q, we have the desired set of input random variables and the corresponding $\mathcal{F}$ map

$$\mathcal{F}(\mathbf{Q}) = \mu_\Theta(\hat{x},\hat{y}) + \sum_{i=1}^{L} \sigma_i r_i(\hat{x},\hat{y}) Q_i(\omega), \quad \mathbf{Q} \sim \mathcal{N}(\mathbf{0},\mathbf{I}_L), \tag{8.17}$$

[5]The Matérn covariance function reads

$$C_{\nu_c}(d/l_c) = \sigma^2 \frac{2^{1-\nu_c}}{\Gamma(\nu_c)} \left(\sqrt{2\nu_c}\frac{d}{\ell_c}\right)^{\nu_c} K_{\nu_c}\left(\sqrt{2\nu_c}\frac{d}{\ell_c}\right) \tag{8.11}$$

where Γ is the gamma function, and K_{ν_c} is the modified Bessel function of the second kind.

with $\mathbf{I}_L$ being the identity matrix of size $L \times L$. Later, in the update process, instead of directly updating the F_{cM} field, we update first these random variables $\mathbf{Q}$, and then use the above defined maps to compute the updated scaling factor. Such a general framework enables the algorithms to be used not only for the task to update a field input, but also for the identification of a set of uncertain parameters. In such a case, the $\mathbf{Q}$ parameters can directly be the uncertain parameters if they are Gaussian.

With the new probabilistic description of the scaling factor

$$F_{c_M}(\hat{x}, \hat{y}, \omega) = e^{\mu + \sigma \mathcal{F}(\mathbf{Q}(\omega))} \tag{8.18}$$

the modified formulation of the compressibility in Equation (8.5) reads

$$C_M(\sigma'_z, \hat{x}, \hat{y}, \omega) = F_{c_M}(\hat{x}, \hat{y}, \omega)\, c_{M0} \left(\frac{\sigma'_z}{f_0}\right)^{-\lambda}. \tag{8.19}$$

Within the probabilistic framework, the nonlinear operator given in Equation (8.2) becomes a stochastic operator, as it depends on the actual realisation of the random field F_{cM},

$$\mathcal{A}_s(F_{c_M}(\hat{x}, \hat{y}, \omega), \mathbf{a}(F_{c_M}(\hat{x}, \hat{y}, \omega)), \mathbf{p}) = 0, \tag{8.20}$$

or in a shorter form

$$\mathcal{A}_s(\omega, \mathbf{a}(\omega), \mathbf{p}) = 0. \tag{8.21}$$

Naturally, in this way the nodal displacements $\mathbf{a}$ and accordingly, the displacement field $U_z = U_z(\hat{x}, \hat{y}, \omega)$ are also random expressions, depending on the actual realisation ω of the random vector $\mathbf{Q}$. For the assimilation of the compressibility field, we use measurements of the subsurface displacements $\mathbf{Y} = \mathbf{Y}(\mathbf{Q}(\omega))$, which is also a random vector. This prediction of the displacements computed by the FEM solver has to be compared with the measured data $\mathbf{z}_m$. However, the mathematical model is just a simplification of reality, which means that the true displacement field can differ from the predicted value, even if we know the exact compressibility field. Furthermore, our measurements are usually poisoned by some measurement errors. Supposing an additive measurement noise and modelling error, the measurement can be written as

$$\mathbf{Z}(\omega) = \mathbf{Y}(\omega) + \mathbf{E}_{FEM}(\omega) + \mathbf{E}(\omega), \tag{8.22}$$

where $\mathbf{E}_{FEM}$ and $\mathbf{E}$ are the random vectors of the modelling error of the FEM code and of the measurement noise, respectively. As for this chapter, we do not use real measurements, but virtual ones that we generate with the FEM model, we can ignore the modelling error, so the measurement model reads

$$\mathbf{Z} = \mathbf{Y} + \mathbf{E}. \tag{8.23}$$

In our examples, we suppose that the measurement error is a mean-free Gaussian random vector

$$\mathbf{E} \sim \mathcal{N}(\mathbf{0}, \mathbf{C}_E), \quad [\mathbf{C}_E]_{jj} = \sigma^2_{\epsilon,j} = (0.15 \cdot [\mathbf{z}_m]_j)^2, \quad \pi_E(\boldsymbol{\epsilon}) = \prod_{j=1}^{60} \frac{1}{\sqrt{2\pi\sigma^2_{\epsilon,j}}} e^{-\frac{\epsilon_j^2}{2\sigma^2_{\epsilon,j}}}, \tag{8.24}$$

with $\mathbf{C}_E \in \mathbb{R}^{60\times 60}$ the measurement error covariance matrix, a diagonal matrix so that the measurement errors at the different locations are independent of each other. $\sigma_{\epsilon,j}$ is the standard deviation of the measurement error at the jth assimilation point and is modelled proportional to the measured value of the displacements $\mathbf{z}_m$. π_E is the probability density function of the measurement noise model, and $[\boldsymbol{\epsilon}]_j = \epsilon_j$ is a realisation of the measurement error at the jth assimilation node.

For real-life applications, a modelling error is advisable to be also included in the measurement model. This error is in general a correlated random field, and its properties are not known in general. For such problems, a different error model can be included. Then the assimilation will include the identification of these parameters of the modelling error. If we rather suppose that the modelling error is just a white noise, then it can be handled together with the measurement noise model. The danger of ignoring a systematic modelling error can result in an update process that compensates a missed-out term by an improper shift of the uncertain input parameter(s).

8.3 Surrogate modelling

To mitigate the high computational cost of the different update procedures, it is advantageous to build a surrogate model which mimics the expensive FEM solver with an approximate, but computationally cheap, mathematical model. The surrogate model allows a fast evaluation of the displacements $\mathbf{y}$ for a specific realisation $\mathbf{q}$ of $\mathbf{Q}$, or indirectly for a specific realisation of the scaling factor F_{cM}. This is schematically shown in Figure 8.4. Depending on the representation of the surrogate, an evaluation of statistics and sensitivities is often also very cheap. In the following, we describe the so-called Polynomial Chaos Expansion (PCE) for the surrogate modelling.

For mathematical convenience, we represent all random expressions with some fixed 'reference random variables'. We choose these variables to be a set of uncorrelated Gaussian random variables $\{\Xi_i\}_{i=1}^S$ collected in a random vector $\boldsymbol{\Xi} \sim \mathcal{N}(\mathbf{0}, \mathbf{I})$ with distribution π_Ξ given by

$$\pi_\Xi(\boldsymbol{\xi}) = \prod_{j=1}^{S} \frac{1}{\sqrt{2\pi}} e^{-\frac{\xi_j^2}{2}} \tag{8.25}$$

As the variables Ξ_i are Gaussian and uncorrelated, they are independent as well. One can think of $\boldsymbol{\Xi}$ as an orthogonal coordinate system, such as the local, reference Cartesian coordinate system in the FEM procedure, only that this one is in the parametric and not in the spatial domain. For different types of distributions of the uncertain parameters or fields, we can define functions that map the reference random variables $\boldsymbol{\Xi}$—also called the *germ*—to the uncertain parameters or fields of interest. In our example, the parameter $\mathbf{Q}$ is a set of independent random variables and thus it could directly be the germ. The reason to introduce the parameter $\mathbf{Q}$ besides the germ (see Figure 8.4) is because we will need to modify this parameter in the calibration process, by obtaining an updated map from the germ to the parameter $\mathbf{Q}$. A practical and efficient approach is to make a surrogate of $\mathbf{y} = G\left(f_{cM}\left(\mathbf{Q}\left(\boldsymbol{\xi}\right)\right)\right)$ in the form of a linear combination of some multivariate stochastic polynomials $\{\Phi_i(\xi_1, \xi_2, \cdots \xi_s)\}_{i=0}^{M-1}$. Also for mathematical convenience, we choose orthogonal polynomials with respect to the Gaussian density, satisfying the orthogonality condition

$$\mathbb{E}[\Phi_i(\boldsymbol{\xi})\Phi_j(\boldsymbol{\xi})] = \int_{\mathbb{R}^n} \Phi_i(\boldsymbol{\xi})\Phi_j(\boldsymbol{\xi})\pi_\Xi(\boldsymbol{\xi})\mathrm{d}\boldsymbol{\xi} = \gamma_i\delta_{ij}, \tag{8.26}$$

where δ_{ij} is the Kronecker delta, and γ_i is the squared norm of the polynomials, that is

$$\gamma_i = \mathbb{E}[\Phi_i(\boldsymbol{\xi})\Phi_i(\boldsymbol{\xi})]. \tag{8.27}$$

The polynomials orthogonal with respect to the Gaussian density π_Ξ are the so-called Hermite polynomials[6]. Then the measurable displacement $\mathbf{y}$ can be written as a linear combination of these polynomials Φ_i with coefficients $\boldsymbol{v}_i$

$$\mathbf{y}_h = \sum_i \boldsymbol{v}_i \Phi_i = \boldsymbol{\Upsilon}\boldsymbol{\Phi}_M, \tag{8.28}$$

where in the last expression we have collected the polynomials in a vector $\boldsymbol{\Phi}_M = [\Phi_0, \Phi_1, \cdots \Phi_{M-1}]^T$ and the PCE coefficients of the vertical displacement $\boldsymbol{v}_i \in \mathbb{R}^{n_y}$ in a matrix $\boldsymbol{\Upsilon} \in \mathbb{R}^{n_y \times M}$, whose j^{th} row corresponds to the PCE coefficients of the displacement at the jth assimilation point and its ith column corresponds to the ith stochastic basis function.

Setting up the surrogate model consists of two steps:

1. Represent the input random field f_{CM} as a function of the reference random variable $\boldsymbol{\xi}$ (see Figure (8.4)).

2. Determine the surrogate model by computing the coefficients $\boldsymbol{v}_i$ of the expansion of $\mathbf{y}_h$ in Equation (8.28).

For the first task, we have already represented (Equation (8.18)) the field F_{cM} with a set of L (in the first scenario $L = 1$) Gaussian random variables $\mathbf{Q}$. Now we only need to express $\mathbf{Q}$ in the reference coordinate system, a set of independent standard Gaussian random variables. As in our example, $\mathbf{Q}$ is already a set of independent random variables, for its description we need $S = L$ reference random variables and because the elements of $\mathbf{Q}$ have already standard Gaussian distribution, the relationship can be given now by simply providing the identity map

$$\mathbf{Q} = \mathcal{I}(\Xi) = \Xi. \tag{8.29}$$

The second task, the computation of the coefficients $\boldsymbol{v}_i$ can be done by different approaches, which are usually classified into so-called intrusive and non-intrusive methods. Intrusive methods can sometimes lead to more stable approximations, but may need modifications of the deterministic solver and are thus harder to implement. Non-intrusive methods use the deterministic solver as it is, doing only pointwise evaluations. Interested readers are directed to [179] for a detailed description of these methods. Here, we choose a relatively straightforward non-intrusive approach, the orthogonal projection, which will be described in the next section.

8.3.1 Computation of the surrogate by orthogonal projection

The target here is to find a best approximation of $\mathbf{y}(\boldsymbol{\xi})$ in the form of a linear combination of the stochastic polynomials $\Phi_i(\boldsymbol{\xi})$, where 'best' is meant in the least squares sense. Therefore, we want

[6]For non-Gaussian random vector $\mathbf{Q}$ it is also possible to use a set of non-Gaussian reference random variables, and thus the orthogonal polynomials are not Hermite polynomials but ones that are orthogonal with respect to the probability density of these reference variables [192], in this case the approximation is termed 'Generalized Polynomial Chaos Expansion' (gPCE). Because some of the later described update methods are restrained to the update of Gaussian random variables we require here the input parameter $\mathbf{Q}$ to be Gaussian.

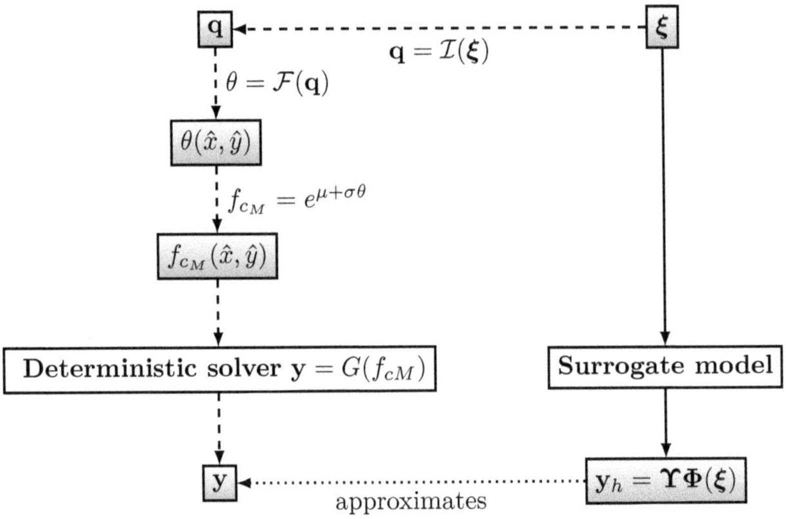

Figure 8.4: Replacing the deterministic solver by a PCE surrogate model: the dashed line shows the path without a surrogate model and the continuous line shows the one when a surrogate model is used. f_{cM} is the random field input, θ is the underlying Gaussian field, $\mathbf{q}$ is a Gaussian random vector representing the field θ, $\boldsymbol{\xi}$ is the germ of the surrogate, a set of independent standard Gaussian random variables serving as the reference coordinate system of the stochastic dimension, $\mathbf{y}$ is the measurable response, and $\mathbf{y}_h$ is its surrogate approximation.

to choose the coefficients such that the expression

$$\mathbb{E}\left[\boldsymbol{\epsilon}_h^2\right] = \mathbb{E}\left[(\mathbf{y} - \mathbf{y}_h)^2\right] = \int_{\mathbb{R}^s} (\mathbf{y}(\boldsymbol{\xi}) - \mathbf{y}_h(\boldsymbol{\xi}))^2 \, \pi_\Xi(\boldsymbol{\xi}) \, \mathrm{d}\boldsymbol{\xi} \tag{8.30}$$

is minimised. It is derived in Appendix D that the minimisation problem leads to the following equation for the coefficients

$$\boldsymbol{v}_i = \frac{1}{\gamma_i}\mathbb{E}\left[\mathbf{y}\Phi_i\right] = \frac{1}{\gamma_i}\int_{\mathbb{R}^s} \mathbf{y}(\boldsymbol{\xi})\Phi_i(\boldsymbol{\xi})\pi_\Xi(\boldsymbol{\xi}) \, \mathrm{d}\boldsymbol{\xi}. \tag{8.31}$$

Unfortunately, this integral expression can not be computed in closed form, as the dependence of $\mathbf{y}$ on $\boldsymbol{\xi}$ is through the simulation and cannot be explicitly given. However, since we can compute the displacement $\mathbf{y}$ at specific values of $\boldsymbol{\xi}$ by the FEM solver, the integral expression can be evaluated numerically by the quadrature rule

$$\boldsymbol{v}_i \approx \frac{1}{\gamma_i}\sum_{j=1}^{N} w_j \mathbf{y}(\boldsymbol{\xi}_j)\Phi_i(\boldsymbol{\xi}_j) \tag{8.32}$$

where N is the number of integration points and $\boldsymbol{\xi}_j$ and w_j are the points and weights of the integration rule. For an overview of the integration rules for stochastic and high-dimensional problems, see example [179]. For the reservoir problem here, we used mostly sparse Smolyak grids with Gauss-Hermite quadrature rules. To sum up, the algorithm to compute the coefficients of the surrogate model of $\mathbf{y}$ is as follows:

Algorithm 11: Computation of the gPCE coefficients by orthogonal projection

1. Define the prior distribution of the input field $F_{cM}(\hat{x}, \hat{y})$ and determine the map from the set of independent reference random variables $\boldsymbol{\xi} = [\xi_1, \xi_2, \cdots, \xi_L]^T$ to f_{cM}.
2. Specify the orthogonal basis polynomials $\{\Phi_i\}_{i=0}^{M-1}$ used for the expansion (see Appendix B.1).
3. Compute the squared norms γ_i of the basis polynomials (see Appendix B.2 for the norms of univariate polynomials and Appendix B.2 for multivariate polynomials).
4. Get the integration points $\boldsymbol{\xi}_j$ and weights w_j for all $j = 1 \ldots N$ from a suitable integration rule.
5. Map the integration points $\boldsymbol{\xi}_j$ to the uncertain field $f_{cM,j}(\hat{x}, \hat{y})$ by the map defined in point 1.
6. Compute the measurable response $\mathbf{y}$ by the deterministic solver with the field $f_{cM,j}(\hat{x}, \hat{y})$ for all $j = 1 \ldots N$.
7. Evaluate all basis functions $\{\Phi_i\}_{i=0}^{(M-1)}$ at all the integration points $\{\boldsymbol{\xi}_j\}_{j=1}^{N}$.
8. Compute the gPCE coefficients from $\boldsymbol{\upsilon}_i \approx \frac{1}{\gamma_i} \sum_{j=1}^{N} w_j \mathbf{y}(\boldsymbol{\xi}_j) \Phi_i(\boldsymbol{\xi}_j)$ and collect them in the matrix $\boldsymbol{\Upsilon} = [\boldsymbol{\upsilon}_0, \ldots, \boldsymbol{\upsilon}_{M-1}]$.

The gPCE of $\mathbf{y}$ then reads $\mathbf{y}_h(\boldsymbol{\xi}) = \boldsymbol{\Upsilon}\boldsymbol{\Phi}_M(\boldsymbol{\xi})$, where $\boldsymbol{\Phi}_M$ is a vector collecting the basis functions defined in point 2, or the same in function of the $\mathbf{Q}$ input parameter using Equation (8.29):

$$\mathbf{y}_h(\boldsymbol{\xi}) = \boldsymbol{\Upsilon}\boldsymbol{\Phi}_M(\mathcal{I}^{-1}(\mathbf{q})) \tag{8.33}$$

Example 1: Surrogate model of seabed displacement by orthogonal projection. Here, we present an example for determining a PCE surrogate model for one nodal displacement computed by orthogonal projection. The example is developed for the univariate case, that is, for the problem where we assume a homogeneous scaling factor. (See Examples 3 and 4 when the input is a random field.)

1. **Define the prior distribution of the random field F_{cM} and determine the map from the germ to F_{cM}.**

   The prior distribution of $F_{cM}(\hat{x}, \hat{y}, \omega)$ is defined in Equation (8.9). The variables θ and Q and the maps in between are defined in Equations (8.15), (8.18), and (8.29). As is shown there, for the homogeneous scenario, $q = \xi$ is just one standard Gaussian random variable. The map from the germ to the input field is defined by

$$f_{c_M}(\xi) = e^{\mu+\sigma\xi} = e^{1.562+0.534\xi}. \tag{8.34}$$

2. **Specify the orthogonal basis polynomials used for the expansion of y.**

   As we have only one variable ξ, we use univariate polynomials, and as we have Gaussian germ, we use the Hermite ones. We choose here a polynomial basis of maximum degree three and generate the polynomials using the three-term recurrence relation (see Appendix B.1). The vector of PCE basis functions is then

$$\mathbf{\Phi}_4 = \begin{bmatrix} H_0(\xi) \\ H_1(\xi) \\ H_2(\xi) \\ H_3(\xi) \end{bmatrix} = \begin{bmatrix} 1 \\ \xi \\ \xi^2 - 1 \\ \xi^3 - 3\xi \end{bmatrix}. \tag{8.35}$$

3. **Compute the squared norms h_i of the polynomials.**

   The norms of the polynomials can be also computed easily from the sequences of the three-term recurrence relation (see Appendix B.2). For our Hermite polynomials the norms read

$$\gamma_i = h_i = i! \quad \gamma_0 = h_0 = 1 \quad \gamma_1 = h_1 = 1 \quad \gamma_2 = h_2 = 2 \quad \gamma_3 = h_3 = 6. \tag{8.36}$$

4. **Get the integration points and weights.**

   We approximate the integral (8.31) by numerical integration. As the variable ξ is Gaussian, we use the Gauss-Hermite quadrature rules [71]. The integration points and the weights can be computed from an eigenvector problem given in Appendix B.3. If we assume that $y_k(\xi)$ can be well approximated by polynomials of maximum degree $d = 3$, then the $y_k(\xi)\Phi_i(\xi)$ term can be approximated by a polynomial of maximum degree $2d = 6$. Polynomials of maximum degree $2N - 1$ can be integrated exactly by an N point Gauss integration rule. Accordingly, we choose the number of points such that $2N - 1$ is bigger or equal to $2d = 6$, that is, we go with a four-point rule $N = d + 1 = 4$, corresponding to the roots of the $\Phi_4(\xi)$ polynomial. The integration points and weights for the four-point Gauss-Hermite rule are

$$\begin{aligned} &\xi_1 = -2.3344 \quad &&\xi_2 = -0.7420 \quad &&\xi_3 = 0.7420 \quad &&\xi_4 = 2.3344, \\ &w_1 = 0.0459 \quad &&w_2 = 0.4541 \quad &&w_3 = 0.4541 \quad &&w_4 = 0.0459. \end{aligned} \tag{8.37}$$

5. **Map the integration points ξ_j to $f_{cM,j}$.**

   The ξ_j integration points are mapped to $f_{cM,j}(\hat{x}, \hat{y})$ by Equation (8.34)

$$f_{cM,1} = 1.3694 \qquad f_{cM,2} = 3.2073 \qquad f_{cM,3} = 7.0885 \qquad f_{cM,4} = 16.6019. \tag{8.38}$$

6. **Compute the measurable response.**

   This step is the computationally expensive one. Now we have four different homogeneous scaling factors. We have to call the deterministic solver $N = 4$ times to get the nodal displacements for the different scaling values. Here, we only show the results for one specific nodal displacement

($k = 25$, which corresponds to the node in the third row from the bottom and the fifth node from the left of the nodes shown by the blue dots in the D panel of Figure 8.1).

$$y_k(\xi_1) = -0.0015 \qquad y_k(\xi_2) = -0.0038 \qquad y_k(\xi_3) = -0.0087 \qquad y_k(\xi_4) = -0.0206. \quad (8.39)$$

The numbers are in meters and the negative sign means a downward displacement.

7. **Evaluate the basis functions at all integration points.**

   The basis functions evaluated at the integration points are

$$\Phi_4(\xi_1) = \begin{bmatrix} 1.0000 \\ -2.3344 \\ 4.4495 \\ -5.7181 \end{bmatrix} \quad \Phi_4(\xi_2) = \begin{bmatrix} 1.0000 \\ -0.7420 \\ -0.4495 \\ 1.8174 \end{bmatrix} \quad \Phi_4(\xi_3) = \begin{bmatrix} 1.0000 \\ 0.7420 \\ -0.4495 \\ -1.8174 \end{bmatrix} \quad \Phi_4(\xi_4) = \begin{bmatrix} 1.0000 \\ 2.3344 \\ 4.4495 \\ 5.7181 \end{bmatrix}. \quad (8.40)$$

8. **Compute the PCE coefficients.**

   The PCE coefficients of the 25th node which are written in the 25th row of the $\boldsymbol{\Upsilon}$ coefficient matrix

$$\boldsymbol{\Upsilon} = \begin{bmatrix} \vdots & \vdots & \vdots & \vdots \\ -0.0067 & -0.0037 & -0.0010 & -0.0002 \\ \vdots & \vdots & \vdots & \vdots \end{bmatrix}. \quad (8.41)$$

   As the response surface – the dependence of the $\mathbf{y}$ displacement on ξ – is a smooth function, the absolute values of the PCE coefficients decay fast.

8.3.2 Computation of statistics

Statistics like the mean and the variance of the displacements can be computed cheaply from the surrogate model, due to the advantageous orthogonality property of the basis. The mean of the displacements $\mathbf{Y}$, for example, can be computed directly from the PCE coefficients corresponding to the zeroth polynomial $\Phi_0 = 1$ as

$$\begin{aligned} \mathbb{E}[\mathbf{Y}] \approx \mathbb{E}[\mathbf{Y}_h] &= \mathbb{E}\left[\sum_{i=0}^{M-1} \boldsymbol{v}_i \Phi_i(\Xi)\right] \\ &= \sum_{i=0}^{M-1} \boldsymbol{v}_i \mathbb{E}[\Phi_i(\Xi)] = \boldsymbol{v}_0, \end{aligned} \quad (8.42)$$

because of $\mathbb{E}[\Phi_0] = 1$ and $\mathbb{E}[\Phi_{i\neq 0}] = 0$ due to the orthogonality condition (8.26). The autocovariance of $\mathbf{Y}$ can be computed from the rest of the coefficients by

$$\begin{aligned} \operatorname{cov}[\mathbf{Y}, \mathbf{Y}] \approx \operatorname{cov}[\mathbf{Y}_h, \mathbf{y}_h] &= \mathbb{E}\left[\left(\sum_{i=0}^{M-1} \boldsymbol{v}_i \Phi_i(\Xi) - \mathbb{E}[\mathbf{y}_h]\right)\left(\sum_{i=0}^{M-1} \boldsymbol{v}_i \Phi_i(\Xi) - \mathbb{E}[\mathbf{y}_h]\right)^T\right] \\ &= \sum_{i=1}^{M-1} \boldsymbol{v}_i \boldsymbol{v}_i^T \gamma_i, \end{aligned} \quad (8.43)$$

again because of the orthogonality condition (8.26). Suppose, the random variable $\mathbf{Q}$ is also given with a PCE form in the same PCE basis

$$\mathbf{Q}_h = \sum_{i=0}^{M-1} \hat{\mathbf{Q}}_i \Phi_i(\Xi). \tag{8.44}$$

Then the covariance $\text{cov}[\mathbf{Q}, \mathbf{Y}]$ can be approximated by

$$\text{cov}[\mathbf{Q}_h, \mathbf{Y}_h] = \sum_{i=1}^{M-1} \hat{\mathbf{q}}_i \boldsymbol{v}_i^T \gamma_i \tag{8.45}$$

Example 2: Statistics from PCE expansion. In this example, we show how to compute mean and variance of the kth nodal displacement resulting from the prior uncertainties of the scaling factor. We use the PCE coefficients derived in Example 1 for the computation. The mean is computed from Equation (8.42), which is directly the zeroth coefficient

$$\mathbb{E}(y_k) = v_{k,0} = v_{0k} = -0.0067. \tag{8.46}$$

The variance of the displacement can be computed from Equation (8.43), from the square of the rest of the coefficients

$$\text{var}[y_{h,25}, y_{h,25}] = \sum_{i=1}^{3} v_{25,i}^2 h_i = (-0.0037)^2 \cdot 1 + (-0.0010)^2 \cdot 2 + (-0.0002)^2 \cdot 6 = 1.5678e-05. \tag{8.47}$$

The displacement at the 25th assimilation point can also be computed at any value of the f_{c_M} scaling factor. One computation of the nodal displacements take hours, but evaluation of the surrogate model doesn't even take seconds. The PCE of y_{25} is

$$y_{h,25}(\xi) = \begin{bmatrix} -0.0067 & -0.0037 & -0.0010 & -0.0002 \end{bmatrix} \begin{bmatrix} 1 \\ \xi \\ \xi^2 - 1 \\ \xi^3 - 3\xi \end{bmatrix}. \tag{8.48}$$

For example, when $f_{c_M} = 4.7681$, the nodal displacement can be computed from

$$\xi = \frac{\ln(4.7681) - \mu}{\sigma} = 0, \tag{8.49}$$

$$y_{h,k}(0) = \begin{bmatrix} -0.0067 & -0.0037 & -0.0010 & -0.0002 \end{bmatrix} \begin{bmatrix} 1 \\ 0 \\ -1 \\ 0 \end{bmatrix} = -0.0057. \tag{8.50}$$

8.3.3 Validating surrogate models

Often, prior information on the smoothness of the response y as a function of ξ is not available, which makes it difficult to come up with a good idea of which PCE basis one should use, and by which method the coefficients should be computed. For this reason, it is always recommended to carry

Table 8.1: Validation of different degree surrogate models with the normed relative error (values are in percentage) computed on $N_v = 10$ quasi-Monte Carlo validation points.

PCE degree d	1	2	3	4	5
Number of solver calls N	2	3	4	5	6
Normed rel. error $\|\epsilon_{rh}\|_2$	115.48	26.32	5.21	1.47	1.16

out some kind of validation of the computed PCE approximation. Here, we use a set of validation points $\{\xi_j\}_{j=1}^{N_v}$ sampled from the distribution of Ξ using a quasi-random sampling method, in this case the Halton-sequence. First, the responses are computed by the FEM solver and then the error is evaluated at these validation points. The relative averaged squared error in the kth spatial node

$$\epsilon_{rh,k}^2 = \frac{\sum_{j=1}^{N_v} \left([\mathbf{y}(\xi_j)]_k - [\mathbf{y}_h(\xi_j)]_k\right)^2}{\sum_{j=1}^{N_v} \left([\mathbf{y}(\xi_j)]_k\right)^2}, \tag{8.51}$$

is then evaluated and compared to decide which surrogate model to use for the inverse method. Table 8.1 shows the $\|\epsilon_{rh}\|_2$ normed relative errors with proxy models of different degree polynomial bases. In the later identification process, we used polynomial degree five, which gives a sufficiently accurate surrogate model.

8.4 Efficient representation of random fields

Random fields are a collection of infinitely many (correlated) random variables; one at every point of the field. Unfortunately, it is practically infeasible to work with infinitely many variables. We can handle, however, the fields with their so-called separated representations which represent the field as a sum of products of (often uncorrelated) random variables and pure spatial functions. Two prominent examples, the so-called Karhunen-Loéve Expansion (KLE) and the proper orthogonal decomposition (POD) shall be examined in the following.

8.4.1 Karhunen-Loève Expansion (KLE)

When the soil conditions are treated as spatially varying, the scaling factor becomes a random field. As that consists of uncountably many random variables – one at each position $(\hat{x}, \hat{y})$ – we cannot use it directly in a computation. However, given certain smoothness of the random field, it can be represented as a countable series of products of spatial functions times scalar random variables with decreasing magnitude, such that this series can be truncated after finitely many terms with diminishing error. Such a representation can be given using the Karhunen-Loève Expansion (KLE), which expands the Gaussian stochastic field into a series

$$\Theta(\hat{x}, \hat{y}, \omega) = \mu_\Theta(\hat{x}, \hat{y}) + \underbrace{\sum_{i=1}^{\infty} \sigma_i r_i(\hat{x}, \hat{y}) X_i(\omega)}_{\tilde{\Theta}}, \tag{8.52}$$

where the $r_i(\hat{x},\hat{y})$ are orthogonal, square integrable spatial functions, the $X_i(\omega)$ are independent standard Gaussian random variables, and the σ_i are scaling factors. μ_Θ is the mean field and $\tilde{\Theta}$ is the fluctuating part of the stochastic field.

The KLE spatial functions r_i have a structure which is typical for the specific random field and are optimal for the representation of the field in the sense that the truncated expansion minimizes the mean squared error and maximizes the captured variance. It is shown in Appendix C that when the problem is discretised in the spatial domain, and the functions r_i are written as a linear combination

$$r_i(\hat{x},\hat{y}) \approx r_{h,i}(\hat{x},\hat{y}) = \sum_{j=1}^{n} \Psi_j(\hat{x},\hat{y}) v_{ji} = \mathbf{\Psi}\mathbf{V} \tag{8.53}$$

of some n given spatial basis functions that fulfill the Kronecker delta property $\Psi_j(\hat{x}_k,\hat{y}_l) = \delta_{kl}$ (e.g. the FEM nodal basis functions) then these optimal spatial functions can be found by solving the generalized eigenvalue problem (see Equation C21 in Appendix C)

$$\mathbf{GCG}\mathbf{v}_i = \underbrace{\lambda_i}_{\sigma_i^2} \mathbf{G}\mathbf{v}_i. \tag{8.54}$$

Here, $\mathbf{G}$ is the Gramian matrix of the basis functions (also often called the mass matrix), $\mathbf{C}$ is the covariance matrix, and λ_i and $\mathbf{v}_i$ are the generalized eigenvalues and eigenvectors (see Appendix C). Once Equation (8.54) is solved for λ_i and $\mathbf{v}_i$, the Gaussian Θ field can be approximated by

$$\begin{aligned}\Theta(\hat{x},\hat{y},\omega) \approx \mu_\Theta + \sum_{i=1}^{L} \sigma_i r_i(\hat{x},\hat{y}) X_i(\omega) &= \mu_\Theta + \sum_{i=1}^{L} \sigma_i \sum_{k=1}^{n} \Psi_k(\hat{x},\hat{y}) v_{ki} X_i(\omega) \\ &= \mu_\Theta + \mathbf{\Psi}(\hat{x},\hat{y})\mathbf{VSX}(\omega),\end{aligned} \tag{8.55}$$

where $L \le n$ is the truncated number of the eigenvectors and $\mathbf{S}$ is an L by L diagonal matrix with the $\sigma_i = \sqrt{\lambda_i}$ values. In a more detailed form, the second part of the equation can be expressed as

$$\begin{aligned}&\mathbf{\Psi}(\hat{x},\hat{y})\mathbf{VSX}(\omega) = \\ &= \begin{bmatrix}\Psi_1(\hat{x},\hat{y}) & \Psi_2(\hat{x},\hat{y}) & \cdots & \Psi_n(\hat{x},\hat{y})\end{bmatrix} \begin{bmatrix} v_{11} & v_{12} & \cdots & v_{1L} \\ v_{21} & v_{22} & \cdots & v_{2L} \\ \vdots & \vdots & \ddots & \vdots \\ v_{n1} & v_{n2} & \cdots & v_{nL}\end{bmatrix} \begin{bmatrix}\sigma_1 & & & \\ & \sigma_2 & & \\ & & \ddots & \\ & & & \sigma_L\end{bmatrix} \begin{bmatrix} X_1(\omega) \\ X_2(\omega) \\ \vdots \\ X_L(\omega)\end{bmatrix}.\end{aligned} \tag{8.56}$$

Algorithm 12: Computation of the truncated KLE of a random field

1. **Choose a spatial mesh** with nodes $(\hat{x}_j,\hat{y}_j)$ for $j = 1 \ldots n$ and a corresponding nodal basis $\{\Psi_j(\hat{x},\hat{y})\}_{j=1}^{n}$ and collect the functions in a row vector $\mathbf{\Psi}$.
2. **Compute the** $n \times n$ **Gramian matrix G** of the nodal basis with elements $[\mathbf{G}]_{ij} = \int_{\mathcal{D}} \Psi_i(\hat{x},\hat{y})\Psi_j(\hat{x},\hat{y}) \mathrm{d}x\, \mathrm{d}y$.

3. **Compute the $n \times n$ covariance matrix C** from the covariance function, $[\mathbf{C}]_{ij} = \mathrm{Cov}_{\tilde{\Theta}}(\hat{x}_i, \hat{y}_i, \hat{x}_j, \hat{y}_j) = \Sigma(\hat{x}_i, \hat{y}_i, \hat{x}_j, \hat{y}_j) = C_{\nu_c}(\sqrt{(\hat{x}_i - \hat{x}_j)^2 + (\hat{y}_i - \hat{y}_j)^2})$.

4. **Solve the generalized eigenvalue problem $\mathbf{GCGv}_i = \lambda_i \mathbf{Gv}_i$** for $i = 1 \ldots n$ for the eigenvectors $\mathbf{v}_i$ and the eigenvalues $\lambda_i = \sigma_i^2$. If n is very high, one may directly compute a truncated number of eigenvectors and eigenvalues.

5. **Truncate the expansion** by checking the captured variance with different number of eigenfunctions. The relative captured covariance with L eigenfunctions can be computed by

$$\rho_L = \frac{\sum_{i=1}^{L} \lambda_i}{\sum_{j=1}^{n} \lambda_j}.$$

Collect the L eigenvectors in the columns of the matrix $\mathbf{V}$ and the L values σ_i into the diagonal of the matrix $\mathbf{S}$.

The separated representation of $\theta(x, y, \omega)$ evaluated at the KLE mesh nodes is given by $\boldsymbol{\mu}_\theta + \mathbf{VSX}(\omega)$, with $\mathbf{X}$ being a vector of L independent standard Gaussian random variables. The complete field of θ is given by

$$\theta(\hat{x}, \hat{y}, \omega) = \mu_\theta(\hat{x}, \hat{y}) + \boldsymbol{\Psi}(\hat{x}, \hat{y})\mathbf{VSX}(\omega).$$

8.4.2 Proper Orthogonal Decomposition (POD)

Another approach to represent the random field is by the so-called Proper Orthogonal decomposition (POD). Let $\mathbf{T}$ be the random vector whose i^{th} element corresponds to the random field Θ taken at the ith node of the spatial mesh, that is, $[\mathbf{T}(\omega)]_i = \Theta(\hat{x}_i, \hat{y}_i, \omega)$ for all nodes $(\hat{x}_i, \hat{y}_i)$. Then this random vector can be written in the form

$$\begin{aligned} \mathbf{T}(\omega) &= \mu_{\mathbf{T}} + \tilde{\mathbf{T}}(\omega) \\ &= \mu_{\mathbf{T}} + \mathbf{VSW}(\omega) \\ &= \mu_{\mathbf{T}} + \sum_i \sigma_i \mathbf{v}_i W_i(\omega), \end{aligned} \tag{8.57}$$

where $\mathbf{S}$ is a diagonal matrix with $[\mathbf{S}]_{ii} = \sigma_i$, $\mathbf{V}$ an orthogonal matrix and $\mathbf{W}(\omega)$ a vector of uncorrelated random variables, that is, $\mathbb{E}[W_i(\omega)W_j(\omega)] = \delta_{ij}$. The orthogonal matrix $\mathbf{V}$ can be computed from the eigenvalue decomposition of the covariance matrix of $\mathbf{T}$, namely

$$\mathbf{C}_T = \mathbb{E}[\tilde{\mathbf{T}}\tilde{\mathbf{T}}^T] = \mathbf{VS}\underbrace{\mathbb{E}[\mathbf{W}(\omega)^T\mathbf{W}(\omega)]}_{=\mathbf{I}}\mathbf{SV}^T = \mathbf{VS}^2\mathbf{V}^T = \sum_i \sigma_i^2 \mathbf{v}_i \mathbf{v}_i^T. \tag{8.58}$$

The POD then can be computed by solving the eigenvector problem

$$\mathbf{C}_T = \mathbf{V}\mathbf{S}^2\mathbf{V}^T \tag{8.59}$$

$$\mathbf{C}_T\mathbf{V} = \mathbf{V}\mathbf{S}^2 \underbrace{\mathbf{V}^T\mathbf{V}}_{=\mathbf{I}} \tag{8.60}$$

or in a different form

$$\mathbf{C}_T\mathbf{v}_i = \sigma_i^2\mathbf{v}_i = \lambda_i\mathbf{v}_i \tag{8.61}$$

Note, that the POD in expression (8.57) can be regarded as the discretised form of the KLE Equation (8.52). The vectors $\mathbf{v}_i$ represent the nodal values of some field-typical spatial functions and the W_i represent independent Gaussian random variables, given that Θ is a Gaussian random field. If the same spatial nodes with coordinates $\{\hat{x}_i, \hat{y}_i\}_{i=1}^n$ are chosen as the one in the KLE, then the covariance matrix $\mathbf{C}_T$ matches the matrix $\mathbf{C}$ introduced in the KLE discretization. In this way, Equations (8.53) and (8.61) are almost the same, and are identical when the chosen spatial functions are orthonormal, so when the Gramian $\mathbf{G}$ is the identity matrix. In this case, the eigenvalues λ_i and eigenvectors $\mathbf{v_i}$ are the same as the one computed from the POD approach. In general, the nodal basis is not orthogonal, but if the nodal points are distributed uniformly, then the results from the two computations give very similar results.

Algorithm 13: Computation of the POD of a discretised random field

1. **Choose a spatial mesh** with nodes $(\hat{x}_j, \hat{y}_j)$ for $j = 1 \dots n$.
2. **Compute the** $n \times n$ **covariance matrix C** from the covariance function, $[\mathbf{C}]_{ij} = \mathrm{Cov}_{\tilde{\Theta}}(\hat{x}_i, \hat{y}_i, \hat{x}_j, \hat{y}_j) = \Sigma(\hat{x}_i, \hat{y}_i, \hat{x}_j, \hat{y}_j) = C_{\nu_c}(\sqrt{(\hat{x}_i - \hat{x}_j)^2 + (\hat{y}_i - \hat{y}_j)^2})$
3. **Solve the eigenvalue problem** $\mathbf{C}\mathbf{v}_i = \lambda_i\mathbf{v}_i$ for $i = 1 \dots n$ for the eigenvectors $\mathbf{v}_i$ and the eigenvalues $\lambda_i = \sigma_i^2$. If n is very high, it may be more efficient to compute only a limited number of eigenvectors and eigenvalues.
4. **Truncate the expansion** by checking the captured variance with a different number of eigenfunctions. The relative captured energy with L eigenfunctions can be computed by

$$\rho_L = \frac{\sum_{i=1}^L \lambda_i}{\sum_{j=1}^n \lambda_j}.$$

The POD of the field is written in the form $\Theta(\hat{x}_j, \hat{y}_j, \omega) = \mu_\Theta(\hat{x}_j, \hat{y}_j) + [\mathbf{V}\mathbf{S}\mathbf{X}(\omega)]_j$ with $\mathbf{X}$ being a vector of L independent standard Gaussian random variables.

Example 3: Separated representation of the scaling factor. Let the prior distribution of F_{cM} be defined by Equation (8.9) and (8.12). First, we write a separated representation of the underlying Gaussian field $\Theta \sim \mathcal{N}(0, \Sigma)$, which is later mapped to the lognormal field F_{cM} by Equation (8.13). The representation is computed by the following steps:

1. **Define the spatial mesh.**

   The FEM mesh with quadratic elements of the domain $\mathcal{D}_s$ is shown with a light grey color in Figure 8.5. For the KLE mesh, we used the center points of the quadratic elements of the FEM mesh. We extended the mesh with some additional outer nodes, because the KL Expansion has sometimes poor accuracy at the boundary. The resulting 192 nodes were connected with triangular elements, which is shown with black lines in Figure 8.5. On the mesh, we defined piecewise linear basis functions $\{\Psi_j(\hat{x}, \hat{y})\}_{j=1}^{192}$, taking the value one on the jth node and zero on every other node. These functions are collected in a horizontal vector $\boldsymbol{\Psi}$.

2. **Compute the Gramian.**

   The Gramian matrix $\mathbf{G}$ has size 192×192, but only seven diagonals have non-zero elements due to the nodal basis used.

3. **Compute covariance matrix.**

   The covariance matrix $\mathbf{C}$ has also size 192×192. The element at row i and column j is computed by

   $$[\mathbf{C}]_{ij} = \text{Cov}_\Theta(\hat{x}_i, \hat{y}_i, \hat{x}_j, \hat{y}_j) = C_{\nu_c}\left(\sqrt{(\hat{x}_i - \hat{x}_j)^2 + (\hat{y}_i - \hat{y}_j)^2}/l_c\right) \tag{8.62}$$

   where the function C_ν is given in Equation (8.11).

4. **Solve the generalized eigenvalue problem.**

   After solving Equation (8.54) for λ_i and $\mathbf{v}_i$, the eigenfunctions can be computed by $r_i(\hat{x}, \hat{y}) = \boldsymbol{\Psi}(\hat{x}, \hat{y})\mathbf{v}_i$. The first nine eigenfunctions are shown in the right panel of Figure 8.6.

5. **Truncate the expansion.**

   The left panel of Figure 8.6 shows the relative captured variance ρ_L using different truncations of the expansion. We decided to keep $L = 11$ eigenfunctions with which more then 93% of the total variance is captured.

   Now we can write the field Θ with the expansion $\Theta = \sum_{i=1}^{11} \sigma_i X_i r_i(\hat{x}, \hat{y})$, where the X_i are 11 independent standard Gaussian random variables and $\sigma_i = \sqrt{\lambda_i}$.

6. **Map the random field Θ to F_{cM}.**

   Following Equation (8.13) and using the separated representation of Θ, the map from the reference Gaussian random variable $\mathbf{X}$ to F_{c_M} is given by

   $$F_{c_M}(\hat{x}, \hat{y}, \mathbf{X}) = e^{1.562+0.534\cdot\left(\sum_{i=1}^{L} \sigma_i r_i(\hat{x}, \hat{y}) X_i\right)} \tag{8.63}$$

Example 4: Surrogate modelling of seabed displacement with random fields. Once we have the separated representation of the input field Equation (8.63), the forward model and the derivation of a surrogate model can be done the same way as in the scalar case. The only difference is that we have more then one germ now, meaning that the dimension of problem grew to $L = 11$. The steps to build the surrogate model (following Algorithm 11 to generate a gPCE model) are described below.

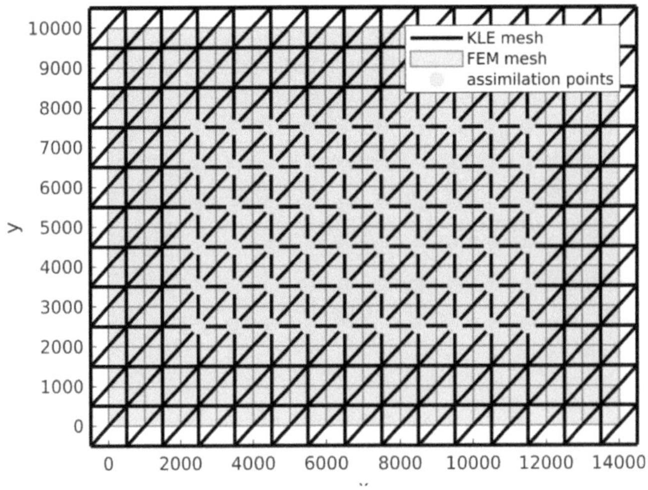

Figure 8.5: Triangular KLE mesh used for the separated representation of the scaling factor (with bold lines), the central part of the quadrilateral FEM mesh (with grey lines, see B panel of Figure 8.1), and the assimilation points (with gray dots).

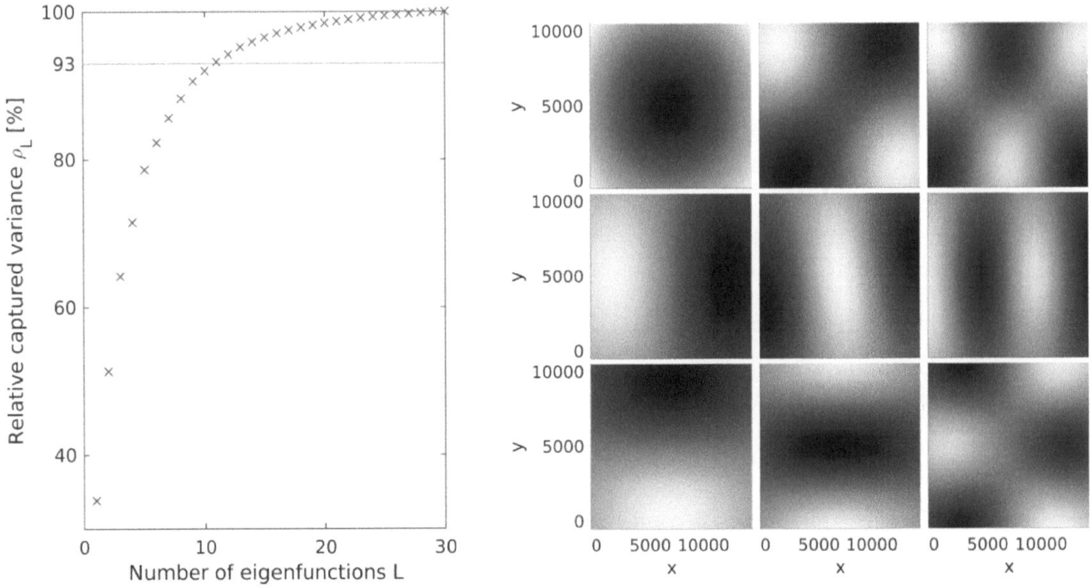

Figure 8.6: Relative captured variance ρ_L for different numbers of eigenfunctions L and the desired 93% level (to the left); the first nine eigenfunctions r_1 to r_9 (to the right).

1. **Define the prior distribution of the random field F_{cM} and determine the map from $\boldsymbol{\xi}$ to f_{cM}.**
   This was mainly done in the previous example, see Equation (8.63) for the map from $\mathbf{X}$ to F_{cM}. Choosing the $\mathbf{Q}$ parameters to be $\mathbf{X}$ and using the identity map (8.29) connecting the reference parameters to the input parameters $\mathbf{Q}$ the map becomes[7]

$$f_{c_M}(\hat{x}, \hat{y}, \boldsymbol{\xi}) = e^{1.562+0.534 \cdot \left(\sum_{i=1}^{L} \sigma_i r_i(\hat{x},\hat{y})\xi_i\right)}. \quad (8.64)$$

2. **Specify the orthogonal basis polynomials used for the expansion of y.**
   As we chose the number of eigenfunctions to be $L = 11$, we need eleven independent Gaussian random variables to describe the random input and thus Ξ is a random vector of eleven variables. Accordingly, the basis functions Φ_i are multivariate polynomials expressed as products of the univariate Hermite polynomials. The multivariate polynomials can be defined by the so-called multi-indices (see more in Appendix B.1 and the table below). For polynomials of maximum degree $d = 1$, the multi-index set and the corresponding basis are shown below.

	multi-index				basis functions
	ξ_1	ξ_2	$\cdots$	ξ_L	
i	$\alpha = i_1,$	i_2	...	$i_L)$	Φ_i
0	0	0	$\cdots$	0	$\Phi_0 = H_0(\xi_1)H_0(\xi_2)\cdots H_0(\xi_L) = 1 \cdot 1 \cdots 1 = 1$
1	1	0	$\cdots$	0	$\Phi_1 = H_1(\xi_1)H_0(\xi_2)\cdots H_0(\xi_L) = \xi_1 \cdot 1 \cdots 1 = \xi_1$
2	0	1	$\cdots$	0	$\Phi_2 = H_0(\xi_1)H_1(\xi_2)\cdots H_0(\xi_L) = 1 \cdot \xi_2 \cdots 1 = \xi_2$
$\vdots$			$\ddots$		$\vdots$
$L+1$	0	0	$\cdots$	1	$\Phi_{L+1} = H_0(\xi_1)H_0(\xi_2)\cdots H_1(\xi_L) = 1 \cdot 1 \cdots \xi_L = \xi_L$

(8.65)

3. **Compute the squared norms γ_i of the basis polynomials.**
   In our case, we have effectively only univariate polynomials, for which we have defined the norms already. As the norm of $H_0(\xi_i)$ and also of $H_1(\xi_i)$ is one, all twelve polynomials have norm one (see Appendix B.2).

4. **Get the integration points and weights.**
   The full tensor integration rule would be too expensive here to compute because of the high dimension of the stochastic space $L = 11$. If we took the tensor product of the two-point univariate rules, that would give $N = 2^L = 2048$ integration points, which means the deterministic solver would have to be run 2,048 times, which is prohibitive in our case. Instead, we use the Smolyak sparse integration grid (see e.g. [179]), which adds up to only twenty-three points.

5. **Map the integration points $\boldsymbol{\xi}_j$ to $f_{cM,j}(\hat{x}, \hat{y})$.**
   We have twenty-three samples of the reference random vector, which results in twenty-three field realisations of the scaling factor F_{cM} (see Figure 8.7). F_{cM} scales the compressibility field in the FEM code. As the FEM computation supposes a constant compressibility field within the FEM elements, we need to extract one value of the scaling factor f_{cM} for each quadrilateral FEM element. Since the centers of the elements coincide with the KLE nodes, and it is enough to possess the values of f_{cM} at these nodes, we do not need the basis functions Ψ_j for the

[7] When coding, it is practical to write Equation (8.29), $\mathbf{q} = \mathcal{I}(\boldsymbol{\xi})$, in a PCE form; this will also be convenient later in the update process.

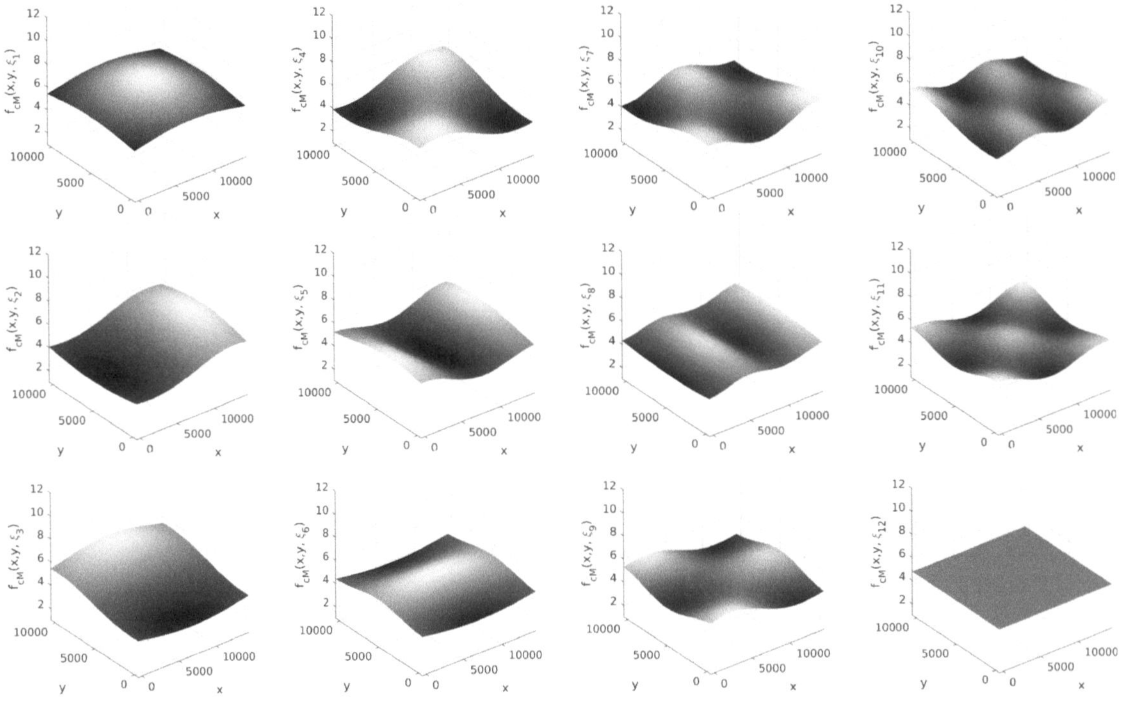

Figure 8.7: Realisations of the random field $f_{cM,j}$ computed at the first twelve integrations points $\boldsymbol{\xi}_j$.

computation. The compressibility field at the kth KLE node with coordinates $(\hat{x}_k, \hat{y}_k)$ is given by

$$f_{c_M}(\hat{x}_k, \hat{y}_k, \boldsymbol{\xi}) = e^{1.562+0.534\left(\sum_{i=1}^{L} \sigma_i [\mathbf{v}_i]_k [\boldsymbol{\xi}]_i\right)}, \tag{8.66}$$

or the values of the scaling factor for all KLE nodes at the jth integration point

$$\mathbf{f}_{c_M}(\boldsymbol{\xi}_j) = e^{1.562+0.534\mathbf{VS}\boldsymbol{\xi}_j}. \tag{8.67}$$

6. **Compute the measurable response.**
For the computation, we only need the $\mathbf{f}_{cM}$ values that correspond to the FEM center points (see Figure 8.5) and thus we drop the ones corresponding to the extended nodes. These selected values are used for setting the scaling factor scaling the compressibility value for each FEM element. We compute the measurable responses $\mathbf{y}(f_{cM}(\boldsymbol{\xi}_j))$ (by the map given in point 1) for all integration points $\boldsymbol{\xi}_j$. This step involves running the FEM solver with the twenty-three different compressibility fields.

7. **Evaluate the basis functions at all integration points.**
Here the twelve basis functions are evaluated at the twenty-three integration points.

8. **Compute the PCE coefficients.**
Similar to the scalar case, we compute the PCE coefficients from $\boldsymbol{v}_i \approx \frac{1}{\gamma_i}\sum_{j=1}^{N} w_j \mathbf{y}(\boldsymbol{\xi}_j)\Phi_i(\boldsymbol{\xi}_j)$ and collect them in the matrix $\boldsymbol{\Upsilon} = [\boldsymbol{v}_0, \ldots, \boldsymbol{v}_{M-1}]$.

8.5 Identification of the compressibility field

8.5.1 Bayes' Theorem

Having samples from similar soil conditions, we possess some knowledge about what values the scaling factor of the compressibility field F_{c_M} could possibly take. Unfortunately, as we do not know the exact value of this scaling factor, we represent this 'belief' of ours with probabilities. So, we model the scaling factor with a random variable/field whose prior probability distribution mirrors our 'belief'. When we have the opportunity to learn more about the scaling factor of the compressibility field from new measurements—in this case from measuring the displacement field—then our task is to change our belief in the simplest way so that it does not contradict the newly received information. The mathematical formulation of 'updating' such a 'belief' based on some evidence is expressed with the conditional probabilities of Bayes' Theorem (refer to Chapter 1 for more insight about Bayes' Theorem). Bayes' Theorem states that the updated (or assimilated or posterior) probability distribution of the scaling factor given a certain measurement $\mathbf{z}_m$ of the displacements at the assimilation points is

$$\underbrace{\pi_{F|\mathbf{z_m}}(f_{c_M})}_{\text{posterior}} = \frac{\overbrace{\pi_{\mathbf{z_m}|F}(f_{c_M})}^{\text{likelihood}}\,\overbrace{\pi_F(f_{c_M})}^{\text{prior}}}{\underbrace{\int_{I_F} \pi_{\mathbf{z_m}|F}(f_{c_M})\pi_F(f_{c_M})\mathrm{d}f_{c_M}}_{\text{evidence}}} = \frac{L(f_{cM})\pi_F(f_{cM})}{\zeta}, \tag{8.68}$$

where

- $\pi_{F|\mathbf{z_m}}$ is the posterior distribution function of the scaling factor, representing our 'changed belief' about the scaling factor after assimilating the measurement $\mathbf{z_m}$,
- $L(f_{cM}) = \pi_{\mathbf{z_m}|F}$ is the likelihood, expressing how likely it is to observe the measured value $\mathbf{z}_m$ given a certain value of the scaling factor,
- π_F is the prior distribution function of the scaling factor representing our 'belief' prior to learning about the new evidence, and
- $\zeta = \int_{I_F} \pi_{\mathbf{z_m}|F}\pi_F df_{c_M}$ is the evidence, the normalising factor assuring that the new posterior distribution function integrates to one.

For simplicity's sake, we use here the above derived separated representation of the random field and we update in this case the distribution of the random variables $\mathbf{Q}$ instead of updating directly the random field. For the input random vector $\mathbf{Q}$, Bayes' Theorem becomes

$$\pi_{\mathbf{Q}|\mathbf{z_m}}(\mathbf{q}) = \frac{\pi_{\mathbf{z_m}|\mathbf{Q}}(\mathbf{q})\pi_{\mathbf{Q}}(\mathbf{q})}{\int_{I_{\mathbf{Q}}} \pi_{\mathbf{z_m}|\mathbf{Q}}(\mathbf{q})\pi_{\mathbf{Q}}(\mathbf{q})d\mathbf{q}} = \frac{L(\mathbf{q})\pi_{\mathbf{Q}}(\mathbf{q})}{\eta}. \tag{8.69}$$

In Bayes' Theorem, the challenging part is to compute the evidence and the likelihood. The likelihood is a function of $\mathbf{q}$ and it expresses how likely it is to measure the observed displacement values $\mathbf{z}_m$ given a certain $\mathbf{q}$ value of the input variables. Our measurements are usually contaminated by measurement errors. Supposing that the error is additive, the measurement becomes

$$\mathbf{z}_m = \mathbf{y}(\mathbf{q}_m) + \boldsymbol{\epsilon}, \tag{8.70}$$

where $\boldsymbol{\epsilon}$ is one realisation of the random variables $\mathbf{E}$ of the measurement noise with probability density π_E, and $\mathbf{q}_m$ is the 'true' value of the input variable(s). The likelihood can be computed from evaluating the probability density of the measurement error π_E at $\boldsymbol{\epsilon} = \mathbf{z}_m - \mathbf{y}(\mathbf{q})$, that is,

$$L(\mathbf{q}) = \pi_E(\mathbf{z}_m - \mathbf{y}(\mathbf{q})). \tag{8.71}$$

8.5.2 Sampling-based procedures—the MCMC method

Unfortunately, the posterior can not be written in general in an explicit, closed form. That is why the simplest way is to go with some sampling-based method: Instead of having a closed form of the updated density, we generate samples of the updated scaling factor and compute statistics of the posterior density via those samples.

The most commonly known algorithm to sample from the posterior distribution is probably the Metropolis-Hastings random walk algorithm, presented as Algorithm 1 in Chapter 1, which belongs to the group of Markov-Chain Monte Carlo (MCMC) methods. The basic idea of the Metropolis-Hastings algorithm is to generate samples by a random walk in the parameter space, constructed such that the stationary distribution of this process equals the target distribution (in this case the Bayes posterior)[150]. That is, after an initial transient phase, taking samples from the random walk is equivalent to directly sampling from the posterior.

The random walk is governed by the so-called proposal density. The proposal density is a conditional distribution that selects a new point for consideration depending on which point of the parameter space we are actually standing at. Theoretically, the algorithm converges for almost any possible proposal density; however, in practice the speed of convergence depends crucially on how well and problem-adapted it was chosen. There are many sophisticated and adaptive ways of choosing this density [83, 7], the simplest (and most-frequently used) being to choose a multivariate Gaussian density $\mathcal{N}(\mathbf{q}, \sigma_{\mathbf{p}}^{\mathbf{2}}\mathbf{I})$ centered around the current point $\mathbf{q}$.

To assure that we really sample from the posterior distribution, not all the proposed steps are accepted. For symmetric proposal densities, the probability with which we accept the new point is governed by the ratio of the posterior density at the new point and that of the current point. If the posterior probability at this new value of the parameter is higher than at the previous point, the step will be accepted. If the posterior probability is lower, the new step may be rejected. The lower the ratio of the two probabilities is, the new step as a sample is accepted with the lower probability.

At every point only the numerator of Equation (8.69) has to be evaluated, since the denominator is a constant value and only the ratio of the probabilities matters in determining whether the new point will be accepted or not. This is a great advantage of the method, since the denomiator, that is, the evidence, is usually hard to compute due to the integral involved.

In a nutshell, the basic algorithm to generate N samples from the posterior is formulated below:

Algorithm 14: Sampling from posterior distribution by MCMC

1. **Choose the starting point** $\mathbf{q}_0$ of the random walk, for example, the mean values of the prior distribution of $\mathbf{Q}$, $\mathbf{q}_0 = \mathbb{E}[\mathbf{Q}]$.
2. **Choose proposal density** $\pi_{\mathbf{Q}'|\mathbf{q}}(\mathbf{q}')$, where $\mathbf{q}'$ is the proposed new value of the reference parameter, and $\pi_{\mathbf{Q}'|\mathbf{q}}(\mathbf{q}')$ is the conditional probability of proposing a new value of the parameter $\mathbf{q}'$ given the actual value of the parameter $\mathbf{q}$.
3. For $i = 0 \dots N$ repeat:
   (a) **Draw one** $\mathbf{q}_i'$ **sample from the proposal density** $\pi_{\mathbf{Q}'|\mathbf{q}_i}(\mathbf{q}')$.
   (b) Decide whether the proposed new point $\mathbf{q}_i'$ can be accepted:
      i. **Compute the acceptance ratio**

$$\alpha = \min\left\{1, \frac{\pi_{\mathbf{Q}|\mathbf{z_m}}(\mathbf{q}_i')\ \pi_{\mathbf{Q}'|\mathbf{q}_i}(\mathbf{q}_i')}{\pi_{\mathbf{Q}|\mathbf{z_m}}(\mathbf{q}_i)\ \pi_{\mathbf{Q}|\mathbf{q}_i'}(\mathbf{q}_i)}\right\}$$
$$= \min\left\{1, \frac{L(\mathbf{q}_i')\pi_{\mathbf{Q}}(\mathbf{q}_i')\ \pi_{\mathbf{Q}'|\mathbf{q}_i}(\mathbf{q}_i')}{L(\mathbf{q}_i)\pi_{\mathbf{Q}}(\mathbf{q}_i)\ \pi_{\mathbf{Q}|\mathbf{q}_i'}(\mathbf{q}_i)}\right\}.$$

      Where $\pi_{\mathbf{Q}|\mathbf{q}_i'}(\mathbf{q}_i)$ is the probability of proposing the actual value of the parameter $\mathbf{q}_i$ when we stand at the proposed value $\mathbf{q}_i'$. When the proposal distribution is symmetric, the ratio simplifies to the ratio of the posterior density at the proposed point and the one at the actual point:

$$\alpha = \min\left\{1, \frac{L(\mathbf{q}_i')\pi_{\mathbf{Q}}(\mathbf{q}_i')}{L(\mathbf{q}_i)\pi_{\mathbf{Q}}(\mathbf{q}_i)}\right\} \tag{8.72}$$

      ii. **Draw a sample** u **from the uniform distribution** $\mathcal{U}[0,1]$.
      iii. **If** $u \leq \alpha$, **accept** the new point $\mathbf{q}_i'$, make the step to a new point $\mathbf{q}_{i+1} = \mathbf{q}_i'$. **Otherwise reject the new point**, stay at the previous location $\mathbf{q}_{i+1} = \mathbf{q}_i$.

The classical method creates one chain from the random walk. The coordinates of the chain are the samples. It is common to drop samples from the beginning of the chain, the so-called burn-in period, because those depend too much on the chosen starting point $\mathbf{q}_0$. The algorithm can be slightly modified by computing several chains in parallel.

As described above, the time-consuming part of the procedure is the computation of the likelihood. In the MCMC computation, the likelihood has to be evaluated at every step of the random walk to compute the acceptance ratio α from Equation (8.72). For that, at each new value of $\mathbf{q}$, Equation (8.71) has to be evaluated, which involves the computation of a prognosis of nodal displacements $\mathbf{y}$ via the deterministic simulation. For a computationally expensive model, this may not be affordable. However, with the already determined PCE surrogate model of the nodal displacements, the evaluation of the likelihood is computationally very cheap.

Example 5: Updating the homogeneous scaling factor by MCMC. For testing the Bayesian inversion scheme, a synthetic measurement $\mathbf{z}_m$ was used. First, the deterministic code was run with a chosen 'true' value of the scaling factor $f_{c_M,m} = 1.89$ to generate the measurement. At this value of the scaling factor we have $\theta_m = (\ln(f_{c_M}) - \mu)/\sigma = (\ln(1.89) - 1.562)/0.534 = -1.733$, and thus also $\xi_m = q_m = -1.733$. We assumed the measurement error model given by Equation (8.24). The synthetic measurements are generated by computing $\mathbf{y}(f_{cM,m})$ and adding a randomly drawn sample of some 'true' $\mathbf{E}_m$ measurement error model with slightly lower standard deviation $\sigma_{\epsilon m,j} = 0.1 \cdot [\mathbf{y}_m]_j$ than that of $\mathbf{E}$. The random walk could be done directly in the space of the scaling factor F_{cM}, but for keeping the general framework, here we carry out the random walk for the input parameter Q.

1. As starting point of the random walk we chose the prior mean, that is,

$$q_0 = \mathbb{E}[Q] = 0. \tag{8.73}$$

2. A Gaussian proposal density $\mathcal{N}(q, \sigma_p^2)$ centered at the last step q was used with $\sigma_p = 0.2$, which has 30% of the variance of the prior parameter Q.

3. For the random walk, we need to define the likelihood function. Instead of the computationally expensive FEM solver, we evaluate $\mathbf{y}$ using its PCE approximation in Equation (8.33) computed as in Example 1 but with polynomial degree five

$$\begin{aligned} L(q) &= \det(2\pi \mathbf{C}_E)^{-1/2} e^{-\frac{1}{2}(\mathbf{z}_m - \mathbf{Y}(q))^T \mathbf{C}_E^{-1}(\mathbf{z}_m - \mathbf{Y}(q))} \\ &\approx \det(2\pi \mathbf{C}_E)^{-1/2} e^{-\frac{1}{2}(\mathbf{z}_m - \boldsymbol{\Upsilon}\boldsymbol{\Phi}(\mathcal{I}^{-1}(q)))^T \mathbf{C}_E^{-1}(\mathbf{z}_m - \boldsymbol{\Upsilon}\boldsymbol{\Phi}(\mathcal{I}^{-1}(q)))}. \end{aligned} \tag{8.74}$$

   The posterior distribution of Q is then defined by the product of the likelihood and the prior

$$\pi_Q(q) = \frac{1}{\sqrt{2\pi\sigma^2}} e^{-\frac{q^2}{2\sigma^2}}. \tag{8.75}$$

4. We ran Algorithm 14 with the preceding inputs which produced the MCMC chain as shown in the left panel of Figure 8.8. In the zoomed out beginning part shown in the right panel of the figure, one can see that at the beginning of the chain, the walk is stuck for a while at the starting point because of the many non-accepted proposals. This is due to the prior mean being far from the posterior mean. As the walk goes towards the posterior zone, the chain starts to look like a white noise. These samples after the transitional burn-in period were stored and mapped to the scaling factor. The posterior density and its mean were estimated by the probability density of these samples (see Figure 8.9).

Example 6: Updating the random field by MCMC. For the second scenario, another synthetic measurement $\mathbf{z}_m$ was created by adding random samples of $\mathbf{E}_m$ to the displacements $\mathbf{y}$ computed by the FEM solver with a 'true' spatially varying f_{c_M} field (see left upper and left bottom panels of Figure 8.12).

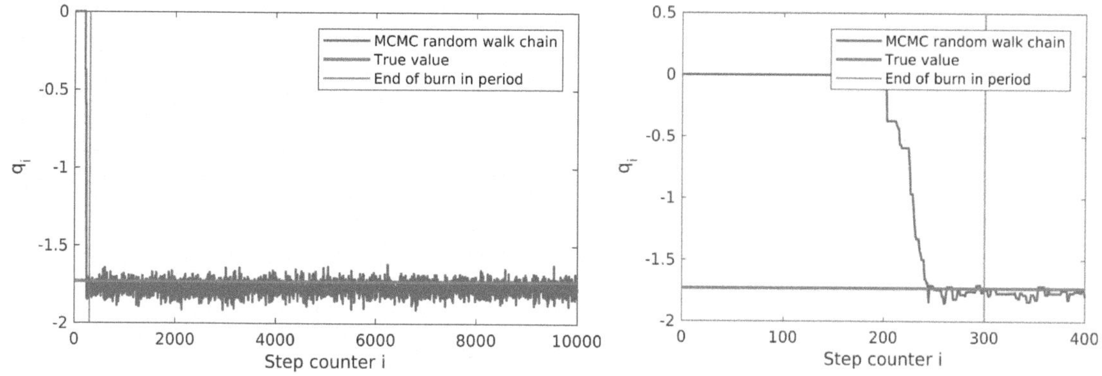

Figure 8.8: The posterior samples $\{q_i\}_{i=1}^{10000}$ from the MCMC random walk chain (to the left) and zoomed out part of it (to the right) to show the selected transitional burn-in period.

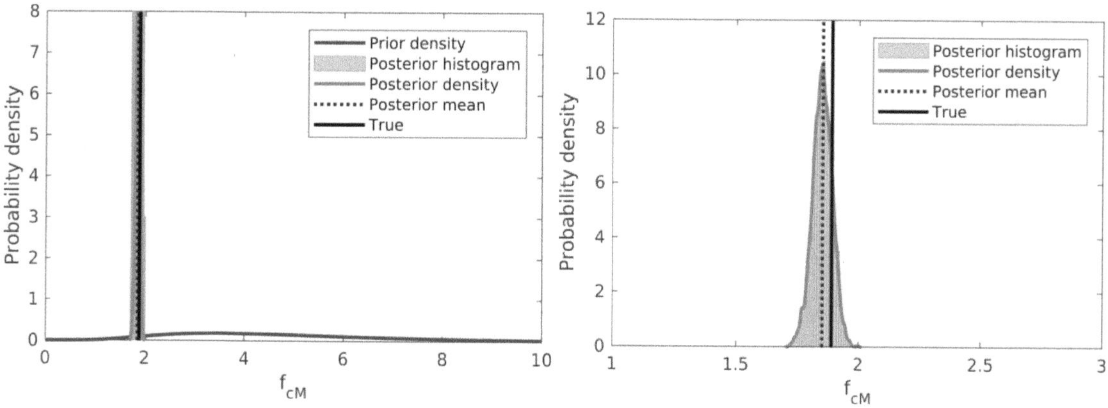

Figure 8.9: MCMC update results: the prior probability density and the posterior density estimate of F_{cM} with the mean of the posterior and the 'true' value.

The measurement error model and the likelihood are the same as in the scalar case, the only difference being that the random walk is now taking place in a higher dimensional space ($L = 11$), the number of eigenfunctions we chose to represent the random field with, and thus also the number of independent standard Gaussian random variables. For evaluating the likelihood, the PCE derived in Example 4 was used. The random chain of the first four input parameters are shown on Figure 8.10. The posterior domain shrinks as the scatter plot of the posterior samples of the left panel of Figure 8.11 shows. The right panel of Figure 8.11 shows the 90% confidence region and the mean of the prior and the posterior distribution. At the points where no measurement data was available (at the boundaries of the domain, specially at the corner points), the uncertainties stay pretty high. However, in the critical region, the updated mean gives a good estimated value of the scaling field. Figure 8.12 shows together the 'true' field $f_{cM,m}$ and the mean posterior estimate as

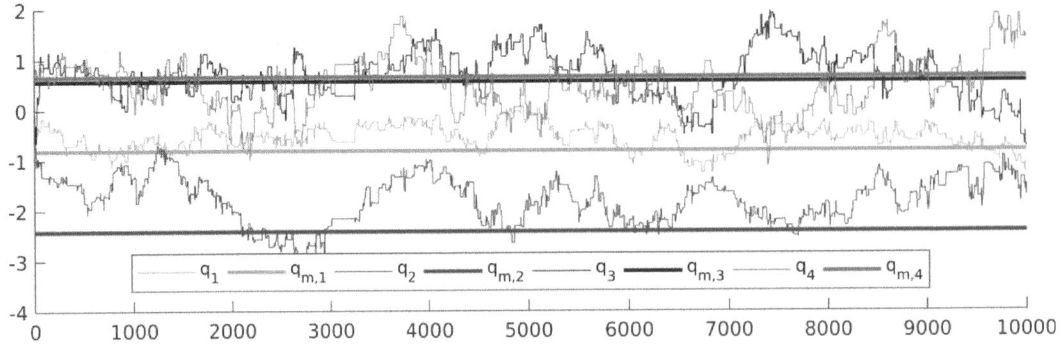

Figure 8.10: Random chain of the first four elements of $\mathbf{q}$ (right).

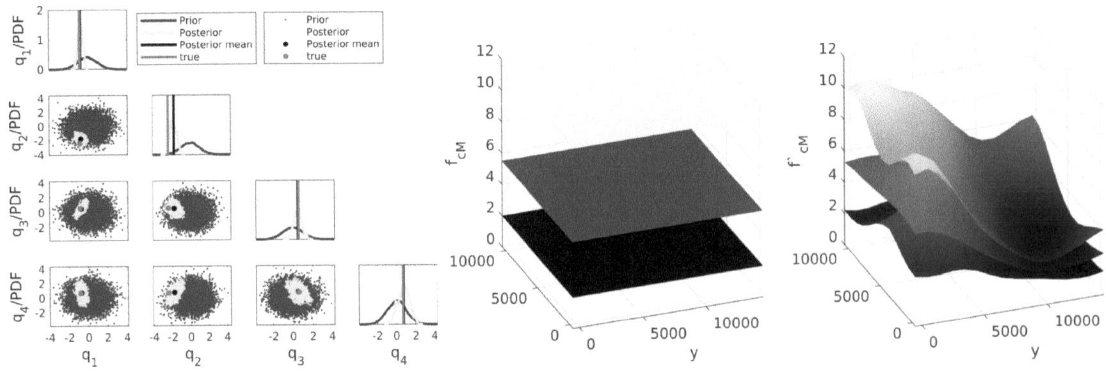

Figure 8.11: MCMC update results: scatter plot of the samples of the prior and the posterior first four input parameter $\mathbf{q}$ and their PDFs (to the left), mean and 90% confidence region of the prior (in the middle), and the posterior (to the right).

well as the standard deviation of the posterior and the deviation of the mean estimate from $f_{cM,m}$. It can be seen, that in the area of the assimilation points, the standard deviation of the posterior reduced significantly compared to the standard deviation of the prior ($\sigma_f = 3.16$). An exception is the region where no significant displacements were measured. Here the posterior standard deviation stays somewhat higher as these measurements could not contribute much to the assimilation process. Though only few eigenmodes were used, and we also kept the PCE representation of $\mathbf{y}$ linear, the Bayesian posterior mean captures pretty well the 'true' field at the assimilation area. It is important to highlight that for this result, the inversion cost only twenty-three deterministic solver calls.

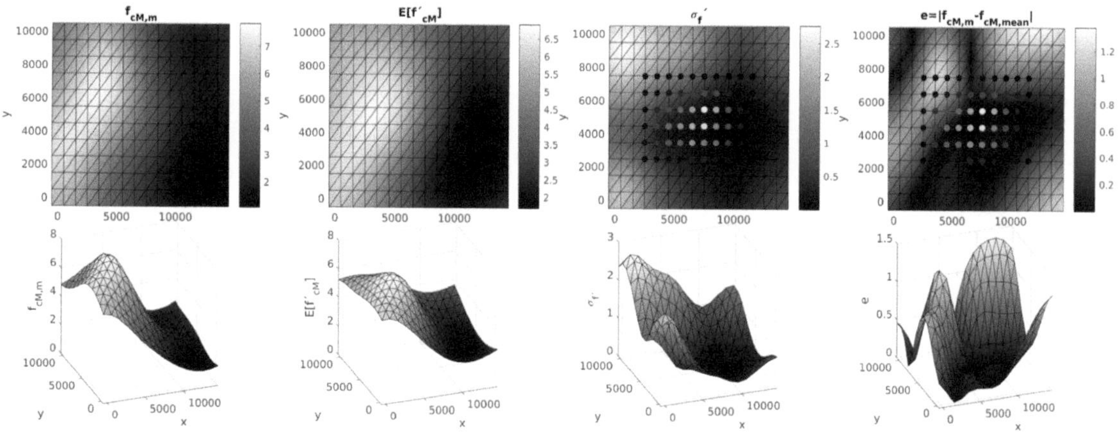

Figure 8.12: Result of the update by MCMC: 2D (on the top) and 3D view (on the bottom) of the 'true' f_{cM}' field, from which the measurement was generated from (to the left), mean of the posterior of f_{cM} computed from the MCMC samples (second column), standard deviation of the posterior f_{c_M} field computed from the MCMC samples (third column) and the deviation of the mean posterior from the 'true' field (to the right). The upper right figures also show the sixty assimilation points used for the update, with colors showing the magnitude of the measured displacement.

8.5.3 The Kalman filter and its modified versions

8.5.3.1 The Kalman filter

The Kalman filter was first proposed as a state estimation technique, but it can be also used for parameter estimation when the parameter is Gaussian and the estimate can be given with a linear function of the data. The filter is in detail explained in [64]. The main assumption of the Kalman filtering is that the map from the parameter $\mathbf{Q}$ to the measurable can be written in a linear form

$$\mathbf{Y}(\mathbf{Q}) = \mathbf{H}\mathbf{Q}, \tag{8.76}$$

where $\mathbf{H}$ is a matrix representing the linear map. Assuming Gaussian distribution for the measurement error $\mathbf{E}$, the likelihood reads

$$L(\mathbf{q}) = \det(2\pi\mathbf{C}_E)^{-1/2} e^{-\frac{1}{2}(\mathbf{z}_m - \mathbf{H}\mathbf{q})^T \mathbf{C}_E^{-1}(\mathbf{z}_m - \mathbf{H}\mathbf{q})}. \tag{8.77}$$

The Bayes' Theorem states that the posterior distribution of $\mathbf{Q}$ is

$$\pi_{Q'}(\mathbf{q}) = \frac{L(\mathbf{q})\pi_Q(\mathbf{q})}{\int_{I_Q} L(\mathbf{q})\pi_Q(\mathbf{q})}. \tag{8.78}$$

If we assume a Gaussian prior distribution for the parameter $\mathbf{Q}$

$$\pi_Q(\mathbf{q}) = \det(2\pi\mathbf{C}_Q)^{-1/2} e^{-\frac{1}{2}(\mathbf{q}-\boldsymbol{\mu}_Q)^T \mathbf{C}_Q^{-1}(\mathbf{q}-\boldsymbol{\mu}_Q)} \tag{8.79}$$

with mean $\boldsymbol{\mu}_Q$ and covariance $\mathbf{C}_Q$, then it can be shown that this posterior distribution corresponds to a Gaussian distribution

$$\pi_{Q'}(\mathbf{q}) = \det(2\pi\mathbf{C}_{Q'})^{-1/2} e^{-\frac{1}{2}(\mathbf{q}-\boldsymbol{\mu}_{Q'})^T \mathbf{C}_{Q'}^{-1}(\mathbf{q}-\boldsymbol{\mu}_{Q'})}, \tag{8.80}$$

with

$$\boldsymbol{\mu}_{Q'} = \mathbb{E}[\mathbf{Q}'] = \boldsymbol{\mu}_Q + \mathbf{K}(\mathbf{z}_m - \mathbf{H}\boldsymbol{\mu}_Q), \quad \mathbf{C}_{Q'} = \mathrm{cov}[\mathbf{Q}', \mathbf{Q}'] = (\mathbf{I} - \mathbf{KH})\mathbf{C}_Q, \tag{8.81}$$

where $\mathbf{C}_{Q'}$ is the covariance of the posterior random variable and $\mathbf{K}$ is the Kalman gain

$$\mathbf{K} = \mathbf{C}_Q\mathbf{H}^T(\mathbf{H}\mathbf{C}_Q\mathbf{H}^T + \mathbf{C}_E)^{-1} \tag{8.82}$$

$$\mathbf{K} = \mathbf{C}_{QY}(\mathbf{C}_Y + \mathbf{C}_E)^{-1} \tag{8.83}$$

with $\mathbf{C}_{QY}$ and $\mathbf{C}_Y$ being the covariances $\mathrm{cov}[\mathbf{Q}, \mathbf{Y}]$ and $\mathrm{cov}[\mathbf{Y}, \mathbf{Y}]$. The random variable

$$\mathbf{Q}' = \mathbf{Q} + \mathbf{K}(\mathbf{z}_m - \mathbf{Z}) \tag{8.84}$$

can also be shown to be Gaussian and has mean $\boldsymbol{\mu}_{Q'}$ and covariance $\mathbf{C}_{Q'}$. Gaussian posteriors are completely described by their first two moments, the mean and the variance, thus Equation (8.84) gives the exact posterior random variable when $\mathbf{Q}$ is Gaussian and the map from $\mathbf{Q}$ to $\mathbf{Y}$ is linear.

8.5.3.2 The ensemble Kalman filter

The main problem is that in our case—and usually in many engineering problems—the measurable $\mathbf{Y}$ is not linearly depending on $\mathbf{Q}$, so the map can not be given simply by a transformation matrix $\mathbf{H}$. Actually, in our case the operator

$$\mathbf{y}(\mathbf{q}) = \mathcal{G}\left(f_{cM}\left(\mathcal{F}\left(\mathbf{q}\right)\right)\right) = \mathcal{H}(\mathbf{q}) \tag{8.85}$$

was only introduced in an abstract way, and is not even explicitly given. However, many variations of the Kalman filter were proposed that are applicable for non-linear problems. An example is the Ensemble Kalman filter (EnKF), which keeps the form of the filter given in Equation (8.84), but represents the random variables $\mathbf{Q}'$, $\mathbf{Q}$, and $\mathbf{Z}$ by MC samples

$$\mathbf{q}'(\boldsymbol{\xi}_j) = \mathbf{q}(\boldsymbol{\xi}_j) + \mathbf{K}(\mathbf{z}_m - \mathbf{z}(\boldsymbol{\xi}_j, \boldsymbol{\eta}_j)) \quad \forall j = 1 \dots N, \tag{8.86}$$

where N is the number of samples, $\mathbf{z}(\boldsymbol{\xi}_j, \boldsymbol{\eta}_j) = \mathcal{H}(\mathbf{q}(\boldsymbol{\xi}_j)) + \boldsymbol{\epsilon}(\boldsymbol{\eta}_j)$, with $\boldsymbol{\epsilon}(\boldsymbol{\eta}_j)$ a sample from the measurement noise model $\mathbf{E}$, and $\boldsymbol{\eta}$ the germ of the measurement error model. The Kalman gain is computed using (8.83), but approximating the covariances from MC samples

$$\mathbf{C}_Y = \frac{1}{N-1}\sum_{j=1}^{N} (\mathbf{y}(\boldsymbol{\xi}_j) - \bar{\mathbf{y}})(\mathbf{y}(\boldsymbol{\xi}_j) - \bar{\mathbf{y}})^T, \tag{8.87}$$

$$\mathbf{C}_{QY} = \frac{1}{N-1}\sum_{j=1}^{N} (\mathbf{q}(\boldsymbol{\xi}_j) - \bar{\mathbf{q}})(\mathbf{y}(\boldsymbol{\xi}_j) - \bar{\mathbf{y}})^T, \tag{8.88}$$

$$\mathbf{C}_E = \frac{1}{N-1}\sum_{j=1}^{N} (\boldsymbol{\epsilon}(\boldsymbol{\eta}_j) - \bar{\boldsymbol{\epsilon}})(\boldsymbol{\epsilon}(\boldsymbol{\eta}_j) - \bar{\boldsymbol{\epsilon}})^T, \tag{8.89}$$

where the bar above a variable means mean of the samples, for example

$$\bar{\mathbf{q}} = \frac{1}{N}\sum_{j=1}^{N}\mathbf{q}(\boldsymbol{\xi}_j). \tag{8.90}$$

The ensemble Kalman filter is also applicable for (slightly) non-linear models, but as all MC procedures, necessitates a very large number of samples N, involving N evaluations of the deterministic solver to compute all responses $\mathbf{y}(\boldsymbol{\xi}_j)$ from the samples $\mathbf{q}(\boldsymbol{\xi}_j)$ of the input random variable. However, with the help of our surrogate model, the evaluation can be done with much less computational effort.

The algorithm of the EnKF using a PCE surrogate model is as follows:

Algorithm 15: Parameter update by the ensemble Kalman filter

1. Generate N samples of the random variables $\mathbf{Q}$ and $\mathbf{E}$: $\mathbf{q}(\boldsymbol{\xi}_j)$, $\boldsymbol{\epsilon}(\boldsymbol{\eta}_j)$ $j = 1 \ldots N$.
2. Compute the $\mathbf{y}_h$ displacements using the surrogate model $\mathbf{y}_h(\boldsymbol{\xi}_j) = \sum_{j=0}^{M-1} \boldsymbol{v}_i\Phi_i(\boldsymbol{\xi}_j)$ for all $j = 1 \ldots N$.
3. Compute the Kalman-gain:
   (a) Compute the sample covariances $\mathbf{C}_Y$, $\mathbf{C}_{QY}$, and $\mathbf{C}_E$ using Equations:

$$\mathbf{C}_Y = \frac{1}{N-1}\sum_{j=1}^{N}\left(\mathbf{y}(\boldsymbol{\xi}_j) - \bar{\mathbf{y}}\right)\left(\mathbf{y}(\boldsymbol{\xi}_j) - \bar{\mathbf{y}}\right)^T,$$

$$\mathbf{C}_{QY} = \frac{1}{N-1}\sum_{j=1}^{N}\left(\mathbf{q}(\boldsymbol{\xi}_j) - \bar{\mathbf{q}}\right)\left(\mathbf{y}(\boldsymbol{\xi}_j) - \bar{\mathbf{y}}\right)^T,$$

$$\mathbf{C}_E = \frac{1}{N-1}\sum_{j=1}^{N}\left(\boldsymbol{\epsilon}(\boldsymbol{\eta}_j) - \bar{\boldsymbol{\epsilon}}\right)\left(\boldsymbol{\epsilon}(\boldsymbol{\eta}_j) - \bar{\boldsymbol{\epsilon}}\right)^T.$$

   (b) Compute samples of the measurement model by adding to the predicted $\mathbf{y}_i$ samples the samples of the measurement error; $\mathbf{z}(\boldsymbol{\xi}_j, \boldsymbol{\eta}_j) = \mathbf{y}(\boldsymbol{\xi}_j) + \boldsymbol{\epsilon}(\boldsymbol{\eta}_j)$.
   (c) Compute the Kalman gain $\mathbf{K} = \mathbf{C}_{QY}(\mathbf{C}_Y + \mathbf{C}_E)^{-1}$.
4. Compute samples of the posterior using

$$\mathbf{q}'(\boldsymbol{\xi}_j) = \mathbf{q}(\boldsymbol{\xi}_j) + \mathbf{K}(\mathbf{z}_m - \mathbf{z}(\boldsymbol{\xi}_j, \boldsymbol{\eta}_j)) \quad \forall j = 1 \ldots N.$$

5. Map the samples to the scaling factor f_{cM}.

The advantage of this approach is that it is much faster than doing the random walk of the MCMC, and we do not have the problem of getting stuck with the random walk, or with decisions such as how to choose the proposal density or how long the burn-in period should be. However, as later shown in the examples, the linear filter still has its limits when used for nonlinear problems.

8.5.3.3 The PCE-based Kalman filter

With the help of the PCE approximation, one can compute the (linear) Kalman filter in an even more straightforward way. This is done purely using the PCE algebra for the updating. The method is explained in detail in [153]. The main idea is that we can represent all the random variables used for the Kalman filtering in a PCE form. Then the filtering is done by updating the PCE coefficients of $\mathbf{Q}$ – the map from the germ to the input parameter – without any sampling. This method is a complete sample-free update procedure. Let us write the input parameter, the error and the measurable response in a PCE format

$$\mathrm{q} = \sum_{\alpha=0}^{L} \mathrm{q}_\alpha \Phi_\alpha^{(Q)}(\boldsymbol{\xi}) \quad \epsilon = \sum_{\alpha=0}^{n_y} \mathrm{e}_\alpha \Phi_\alpha^{(E)}(\boldsymbol{\eta}) \quad \mathrm{y} = \sum_{\alpha=0}^{M-1} \upsilon_\alpha \Phi_\alpha(\boldsymbol{\xi}). \tag{8.91}$$

The first map is just the PCE of the map $\mathcal{I}(\boldsymbol{\xi})$, which is now the identity map, but in general could be any one-to-one map.

The given PCE representations of the three different random vectors are defined each in their own suitable PCE basis. Our first task is to bring the above given PCEs in one unified basis, combining all the Polynomial Chaos basis functions together into one joint basis (see Example 7). Furthermore, we suppose that the measurement errors $\mathbf{E}$ are independent from the random input parameters $\mathbf{Q}$ and so are $\boldsymbol{\xi}$ and $\boldsymbol{\eta}$. The new PCE representations in the extended basis then take the form

$$\mathrm{q} = \sum_{\alpha=0}^{n_c} \hat{\mathrm{q}}_\alpha \hat{\Phi}_\alpha(\boldsymbol{\xi}, \boldsymbol{\eta}) = \hat{\mathbf{Q}}\hat{\boldsymbol{\Phi}} \quad \epsilon = \sum_{\alpha=0}^{n_c} \hat{\mathrm{e}}_\alpha \hat{\Phi}_\alpha(\boldsymbol{\xi}, \boldsymbol{\eta}) = \hat{\mathbf{E}}\hat{\boldsymbol{\Phi}} \quad \mathrm{y} = \sum_{\alpha=0}^{n_c} \hat{\upsilon}_\alpha \hat{\Phi}_\alpha(\boldsymbol{\xi}, \boldsymbol{\eta}) = \hat{\boldsymbol{\Upsilon}}\hat{\boldsymbol{\Phi}}. \tag{8.92}$$

The new set of basis functions $\hat{\Phi}_\alpha$ have germs $\boldsymbol{\xi}$ and $\boldsymbol{\eta}$. Similarly, we can also write the deterministic values $\mathbf{z}_m$ in a PCE format

$$\mathbf{z}_m = \sum_{\alpha=0}^{n_c} \hat{\mathbf{z}}_\alpha \hat{\Phi}_\alpha(\boldsymbol{\xi}, \boldsymbol{\eta}) = \hat{\mathbf{Z}}\hat{\boldsymbol{\Phi}}, \tag{8.93}$$

where the first coefficient $\hat{\mathbf{z}}_0$ corresponding to the polynomial $\hat{\Phi}_0(\boldsymbol{\xi}, \boldsymbol{\eta}) = 1$ is directly the vector $\mathbf{z}_m$ and the rest of the coefficients are zero.

With the help of the new, extended PCE expansions, we can rewrite the filtering step (8.84) as

$$\hat{\mathbf{Q}}'\hat{\boldsymbol{\Phi}} = \hat{\mathbf{Q}}\hat{\boldsymbol{\Phi}} + \mathbf{K}\left(\hat{\mathbf{Z}}\hat{\boldsymbol{\Phi}} - \left(\hat{\boldsymbol{\Upsilon}}\hat{\boldsymbol{\Phi}} + \hat{\mathbf{E}}\hat{\boldsymbol{\Phi}}\right)\right), \tag{8.94}$$

where $\hat{\mathbf{Q}}'$ is the matrix collecting the new PCE coefficients of the updated random variable $\mathbf{Q}'$. It can be easily seen from the above equation that these new coefficients can be determined from

$$\hat{\mathbf{Q}}' = \hat{\mathbf{Q}} + \mathbf{K}(\hat{\mathbf{Z}} - \hat{\boldsymbol{\Upsilon}} - \hat{\mathbf{E}}). \tag{8.95}$$

The covariances for the Kalman gain can be computed without any samples from the unified PCE coefficients following Equation (8.45).

Algorithm 16: Parameter identification by the PCE-based Kalman filter

1. Write input parameter $\mathbf{Q}$, measurement error model $\mathbf{E}$, and the response $\mathbf{Y}$ in a PCE form:

$$\mathbf{q} = \sum_{\alpha=0}^{L} \mathbf{q}_\alpha \Phi_\alpha^{(Q)}(\boldsymbol{\xi}) \quad \boldsymbol{\epsilon} = \sum_{\alpha=0}^{n_y} \mathbf{e}_\alpha \Phi_\alpha^{(E)}(\boldsymbol{\eta}) \quad \mathbf{y} = \sum_{\alpha=0}^{M-1} \boldsymbol{v}_\alpha \Phi_\alpha(\boldsymbol{\xi}).$$

2. Generate a combined PCE basis $\hat{\Phi}_\alpha(\boldsymbol{\xi}, \boldsymbol{\eta})$ and compute the $\hat{\gamma}_\alpha$ squared norm of each basis function.

3. Rewrite the PCE coefficients in this extended basis; write the measured displacements $\mathbf{z}_m$ also in a PCE using the extended basis:

$$\mathbf{q} = \sum_{\alpha=0}^{n_c} \hat{\mathbf{q}}_\alpha \hat{\Phi}_\alpha(\boldsymbol{\xi}, \boldsymbol{\eta}) \quad \boldsymbol{\epsilon} = \sum_{\alpha=0}^{n_c} \hat{\mathbf{e}}_\alpha \hat{\Phi}_\alpha(\boldsymbol{\xi}, \boldsymbol{\eta}) \quad \mathbf{y} = \sum_{\alpha=0}^{n_c} \hat{\boldsymbol{v}}_\alpha \hat{\Phi}_\alpha(\boldsymbol{\xi}, \boldsymbol{\eta}) \quad \mathbf{z}_m = \sum_{\alpha=0}^{n_c} \hat{\mathbf{z}}_\alpha \hat{\Phi}_\alpha(\boldsymbol{\xi}, \boldsymbol{\eta}).$$

4. Compute the Kalman gain.

   (a) Compute the covariances $\mathbf{C}_{Y,}$, $\mathbf{C}_{QY}$, and $\mathbf{C}_E$ from the PCE coefficients:

$$\mathbf{C}_Y = \sum_{\alpha=1}^{n_c} \boldsymbol{v}_\alpha \boldsymbol{v}_\alpha^T \gamma_\alpha \quad \mathbf{C}_{QY} = \sum_{\alpha=1}^{n_c} \hat{\mathbf{q}}_\alpha \hat{\boldsymbol{v}}_\alpha^T \hat{\gamma}_\alpha \quad \mathbf{C}_E = \sum_{\alpha=1}^{n_c} \hat{\mathbf{e}}_\alpha \hat{\mathbf{e}}_\alpha^T \hat{\gamma}_\alpha.$$

   (b) Compute the Kalman gain:

$$\mathbf{K} = \mathbf{C}_{QY}(\mathbf{C}_Y + \mathbf{C}_E)^{-1}.$$

5. Compute the PCE coefficients $\hat{\mathbf{q}}'_\alpha$ of the updated input random variables $\mathbf{Q}'$:

$$\hat{\mathbf{q}}'_\alpha = \hat{\mathbf{q}}_\alpha + \mathbf{K}(\hat{\mathbf{z}}_\alpha - \hat{\boldsymbol{v}}_\alpha - \hat{\mathbf{e}}_\alpha).$$

The PCE of the updated reference random is

$$\mathbf{q}' = \sum_{\alpha=0}^{n_c} \hat{\mathbf{q}}'_\alpha \hat{\Phi}_\alpha(\boldsymbol{\xi}, \boldsymbol{\eta}).$$

Example 7: Updating the scalar scaling factor by the PCE-based Kalman filter. Here we present how to do the Kalman filtering for the scalar scaling factor of the compressibility field.

1. **Write Q, E, and Y in a PCE form.**

   For scenario I, Q is a scalar valued random variable and so is the Gaussian germ Ξ. The map $\mathcal{I}$ from the germ to the parameter is the identity map (see Equation (8.29)), whose PCE is

$$q = \begin{bmatrix} 0 & 1 \end{bmatrix} \begin{bmatrix} H_0(\xi) \\ H_1(\xi) \end{bmatrix} = \begin{bmatrix} 0 & 1 \end{bmatrix} \begin{bmatrix} 1 \\ \xi \end{bmatrix}. \tag{8.96}$$

   The measurement error model **E** is a vector valued random variable with $n_y = 60$ random elements – corresponding to the sixty assimilation points. The distribution of **E** is given in Equation (8.24). For representing the sixty independent Gaussian random variables given in this error model, we need a germ $\boldsymbol{\eta}$ with sixty independent standard Gaussians. The map from the germ $\boldsymbol{\eta}$ to **E** is given by scaling the germ $\boldsymbol{\eta}$ by the standard deviation of the measurement errors. The scaling in a PCE form is given by

$$\boldsymbol{\epsilon} = \begin{bmatrix} 0 & \sigma_{\epsilon,1} & 0 & \cdots & 0 \\ 0 & 0 & \sigma_{\epsilon,2} & \cdots & 0 \\ \vdots & & & \ddots & \vdots \\ 0 & 0 & 0 & \cdots & \sigma_{\epsilon,60} \end{bmatrix} \begin{bmatrix} H_0 \\ H_1(\eta_1) \\ H_1(\eta_2) \\ \vdots \\ H_1(\eta_{60}) \end{bmatrix} = \begin{bmatrix} 0 & \sigma_{\epsilon,1} & 0 & \cdots & 0 \\ 0 & 0 & \sigma_{\epsilon,2} & \cdots & 0 \\ \vdots & & & \ddots & \vdots \\ 0 & 0 & 0 & \cdots & \sigma_{\epsilon,60} \end{bmatrix} \begin{bmatrix} 1 \\ \eta_1 \\ \eta_2 \\ \vdots \\ \eta_{60} \end{bmatrix}. \tag{8.97}$$

   We use the PCE approximation of **y** computed in Example 1, but with polynomial degree five, which reads

$$\mathbf{y} = \boldsymbol{\Upsilon} \begin{bmatrix} 1 & \xi & \xi^2 - 1 & \xi^3 - 3\xi & \xi^4 - 6\xi^2 + 3 & \xi^5 - 10\xi^3 + 15\xi \end{bmatrix}^T \quad \boldsymbol{\Upsilon} \in \mathbb{R}^{60\times 6}. \tag{8.98}$$

2. **Generate a combined PCE basis.**

   The new, combined basis containing all PCE polynomials used for the expansion of Q, **E**, and **Y** becomes

$$\hat{\boldsymbol{\Phi}} = \begin{bmatrix} 1 & \xi & \xi^2 - 1 & \xi^3 - 3\xi & \xi^4 - 6\xi^2 + 3 & \xi^5 - 10\xi^3 + 15\xi & \eta_1 & \eta_2 & \cdots & \eta_{60} \end{bmatrix}^T. \tag{8.99}$$

3. **Rewrite the PCE coefficients in the extended basis.**

   The coefficients collected in a matrix reads

$$\begin{aligned} \hat{Q} &= \begin{bmatrix} 0 & 1 & 0 & \dots & 0 \end{bmatrix}, & \hat{Q} &\in \mathbb{R}^{1\times 66}, \\ \hat{\mathbf{E}} &= \begin{bmatrix} 0 & 0 & \dots & 0 & \sigma_{\epsilon,1} & 0 & \cdots & 0 \\ 0 & 0 & \dots & 0 & 0 & \sigma_{\epsilon,2} & \cdots & 0 \\ \vdots & \vdots & \ddots & \vdots & \vdots & \vdots & \ddots & \vdots \\ 0 & 0 & \dots & 0 & 0 & 0 & \cdots & \sigma_{\epsilon,60} \end{bmatrix} & \hat{\mathbf{E}} &\in \mathbb{R}^{60\times 66}, \\ \hat{\boldsymbol{\Upsilon}} &= \begin{bmatrix} \nu_{1,0} & \nu_{1,1} & \cdots & \nu_{1,5} & 0 & 0 & \cdots & 0 \\ \nu_{2,0} & \nu_{2,1} & \cdots & \nu_{2,5} & 0 & 0 & \cdots & 0 \\ \vdots & \vdots & \ddots & \vdots & \vdots & \vdots & \ddots & \vdots \\ \nu_{60,0} & \nu_{60,1} & \cdots & \nu_{60,5} & 0 & 0 & \cdots & 0 \end{bmatrix} & \hat{\boldsymbol{\Upsilon}} &\in \mathbb{R}^{60\times 66}, \end{aligned} \tag{8.100}$$

with $\sigma_{\epsilon,j}$ the standard deviation of the measurement error at the jth assimilation point and $\upsilon_{i,j}$ the elements $[\boldsymbol{\Upsilon}]_{i,j}$. The matrix of PCE coefficients of the measurement $\mathbf{z}_m$ reads

$$\hat{\mathbf{Z}} = \begin{bmatrix} z_{m,1} & 0 & \dots & 0 & 0 & 0 & \cdots & 0 \\ z_{m,2} & 0 & \dots & 0 & 0 & 0 & \cdots & 0 \\ \vdots & \vdots & \ddots & \vdots & \vdots & \vdots & \ddots & \vdots \\ z_{m,60} & 0 & \dots & 0 & 0 & 0 & \cdots & 0 \end{bmatrix}, \qquad \hat{\mathbf{Z}} \in \mathbb{R}^{60\times 66}. \tag{8.101}$$

4. **Compute the Kalman gain.**

   The covariance matrices $\mathbf{C}_Y$ and $\mathbf{C}_E$ can be computed from the original PCE expansions, but for the computation of $\mathbf{C}_{QY}$, Q and $\mathbf{Y}$ needs to be described in the same basis.

5. **Compute the coefficients of the updated input variables $\mathbf{Q}'$.**

   The PCE of the updated reference random variable is given in the extended basis, which means that its dimension is the sum of the size L (in Example 1) of the input random variables Q and the dimension n_y of the germ $\boldsymbol{\eta}$ (in the example 60) needed for the description of the measurement error $\mathbf{E}$, that is, the updated input parameters are given in stochastic dimension $L{+}n_y = 61$. If new measurements are available, the update can be repeated, but the new updated variable is then given in a new space extended with further stochastic dimensions. For example, when we receive a new data with measured displacements at the same sixty assimilation points, the stochastic dimension of the updated input random variable will be $L + n_y + n_y = 121$.

At this point, we have a PCE expansion of the posterior input parameter Q'. What we are interested in though is the posterior distribution of the scaling factor f_{cM}. The posterior scaling factor can be given now in an analytical form because we have a PCE of the updated input parameter Q' and an explicit map from the parameter to the scaling factor. If a probability density plot is desired, we can sample from Q' and map all the samples to the scaling factor. Figure 8.13 shows the updated density of the scaling factor computed by the PCE-based Kalman filter. One can see that the filter somewhat overestimates the scaling factor, which is due to the filter being linear.

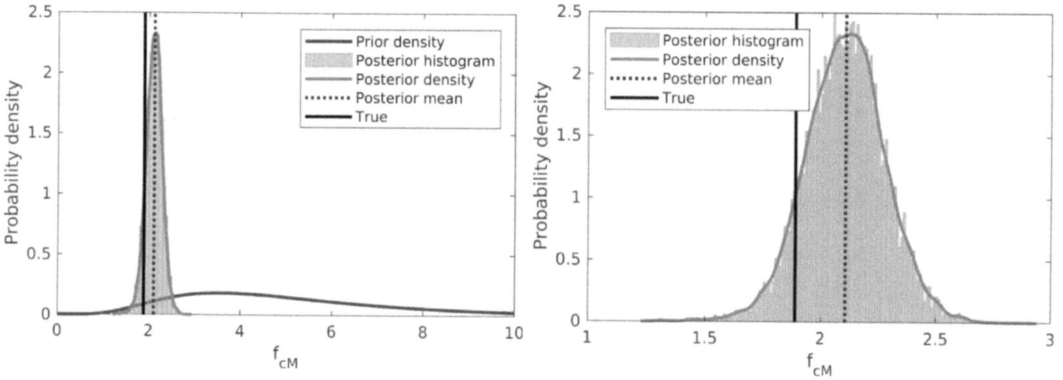

Figure 8.13: PCE based Kalman filter update results: the prior probability density and the posterior density estimate of F_{cM} with the mean of the posterior and the 'true' value.

Example 8: Updating the spatially varying random field by the PCE-based Kalman filter. Here we present how to do the Kalman filtering for the scaling factor of the compressibility field when no homogeneity conditions are assumed.

1. **Represent Q, E, and Y by a PCE.**

   We use the truncated KLE expansion of the field θ computed in Example 3, which represented the field by $L = 11$ independent, standard Gaussian random variables, which we use for $\mathbf{Q}$. The PCE expansion of $\mathbf{Q}$ is

$$\mathbf{q} = \begin{bmatrix} 0 & 1 & 0 & \dots & 0 \\ 0 & 0 & 1 & \dots & 0 \\ \vdots & \vdots & \vdots & \ddots & \vdots \\ 0 & 0 & 0 & \dots & 1 \end{bmatrix} \begin{bmatrix} H_0 \\ H_1(\xi_1) \\ H_1(\xi_2) \\ \vdots \\ H_1(\xi_L) \end{bmatrix} = \begin{bmatrix} 0 & 1 & 0 & \dots & 0 \\ 0 & 0 & 1 & \dots & 0 \\ \vdots & \vdots & \vdots & \ddots & \vdots \\ 0 & 0 & 0 & \dots & 1 \end{bmatrix} \begin{bmatrix} 1 \\ \xi_1 \\ \xi_2 \\ \vdots \\ \xi_L \end{bmatrix}. \tag{8.102}$$

   The PCE of the measurement error $\mathbf{E}$ is the same as in the scalar case in the previous example. For the PCE expansion of $\mathbf{Y}$ we use the results in Example 4

$$\mathbf{y} = \mathbf{\Upsilon} \begin{bmatrix} 1 & \xi_1 & \xi_2 & \dots & \xi_L \end{bmatrix}^T, \tag{8.103}$$

   which is given in a simple linear PCE basis.

2. **Generate a combined PCE basis.**

   The extended basis is

$$\hat{\mathbf{\Phi}} = \begin{bmatrix} 1 & \xi_1 & \xi_2 & \dots & \xi_L & \eta_1 & \eta_2 & \dots & \eta_{60} \end{bmatrix}^T. \tag{8.104}$$

3. **Rewrite the PCE coefficients in the extended basis.**

   The coefficients in the extended basis become

$$\hat{\mathbf{Q}} = \begin{bmatrix} 0 & 1 & 0 & \dots & 0 & 0 & 0 & \dots & 0 \\ 0 & 0 & 1 & \dots & 0 & 0 & 0 & \dots & 0 \\ \vdots & \vdots & \vdots & \ddots & \vdots & \vdots & \vdots & \ddots & \vdots \\ 0 & 0 & 0 & \dots & 1 & 0 & 0 & \dots & 0 \end{bmatrix} \quad \hat{\mathbf{E}} = \begin{bmatrix} 0 & 0 & 0 & \dots & 0 & \sigma_{\epsilon,1} & 0 & \dots & 0 \\ 0 & 0 & 0 & \dots & 0 & 0 & \sigma_{\epsilon,2} & \dots & 0 \\ \vdots & \vdots & \vdots & \ddots & \vdots & \vdots & \vdots & \ddots & \vdots \\ 0 & 0 & 0 & \dots & 0 & 0 & 0 & \dots & \sigma_{\epsilon,60} \end{bmatrix}$$

$$\hat{\mathbf{\Upsilon}} = \begin{bmatrix} \nu_{1,0} & \nu_{1,1} & \nu_{1,2} & \dots & \nu_{1,L} & 0 & 0 & \dots & 0 \\ \nu_{2,0} & \nu_{2,1} & \nu_{2,2} & \dots & \nu_{2,L} & 0 & 0 & \dots & 0 \\ \vdots & \vdots & \vdots & \ddots & \vdots & \vdots & \vdots & \ddots & \vdots \\ \nu_{60,0} & \nu_{60,1} & \nu_{60,2} & \dots & \nu_{60,L} & 0 & 0 & \dots & 0 \end{bmatrix}$$

$$\hat{\mathbf{Z}} = \begin{bmatrix} z_{m,1} & 0 & 0 & \dots & 0 & 0 & 0 & \dots & 0 \\ z_{m,2} & 0 & 0 & \dots & 0 & 0 & 0 & \dots & 0 \\ \vdots & \vdots & \vdots & \ddots & \vdots & \vdots & \vdots & \ddots & \vdots \\ z_{m,60} & 0 & 0 & \dots & 0 & 0 & 0 & \dots & 0 \end{bmatrix}. \tag{8.105}$$

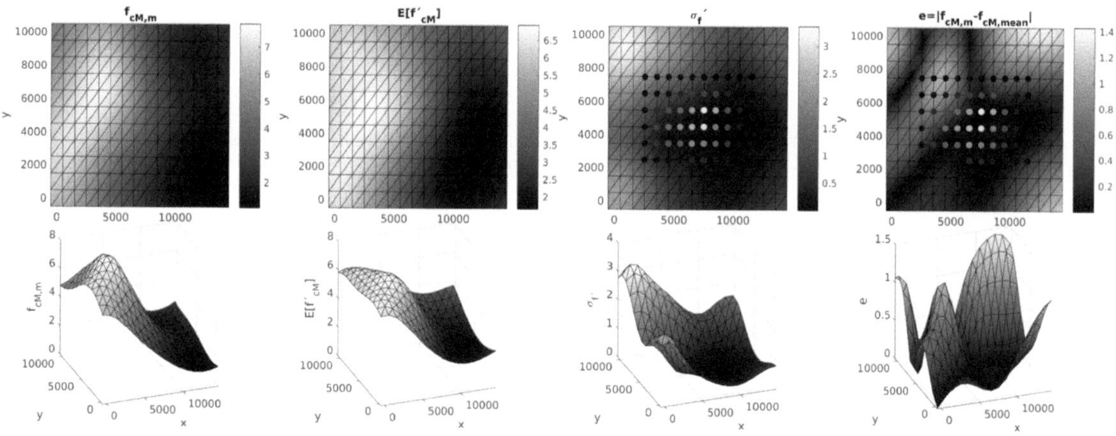

Figure 8.14: Result of the PCE-based Kalman filter field update: 2D and 3D view of the 'true' scaling factor (to the left),the mean (second column from left), and the standard deviation (third column from left) of the posterior scaling factor and the deviation of the mean from the 'true' (to the right). See explanation of the dots in Figure 8.12.

4. **Compute the coefficients of the updated input variables $\mathbf{Q}'$.**

   The Kalman gain and the coefficients are then computed in the same way as in the previous example.

To have samples of the scaling factor, one first needs to sample from the updated PCE of $\mathbf{Q}'$ (first by sampling from both germs, $\boldsymbol{\xi}$ and $\boldsymbol{\eta}$, that is, from the $11 + 60$ independent standard Gaussians, then by evaluating the polynomials at those sample points and multiplying it with the coefficients of $\hat{\mathbf{Q}}'$). These samples are then mapped to the scaling factor by

$$\mathbf{f}_{c_M,j} = e^{1.562+0.534\mathbf{V}\mathbf{S}\mathbf{q}'_j} \tag{8.106}$$

where $\mathbf{V}$ and $\mathbf{S}$ are computed in Example 3. The updated mean field of the scaling factor together with the 'true' field, the standard deviation of the posterior, and the deviation of the mean from the 'true' are shown in Figure 8.14. The result is very similar to the one with the MCMC, but the computation was much less demanding in terms of runtime here.

8.5.4 Non-linear filters

It is shown for the PCE-based Kalman filter that the use of a functional approximation of the measurement model allows a functional approximation of the posterior. This method is much easier to compute than the sampling based MCMC approach. The Kalman filter applies a linear estimator to obtain the posterior. As presented before, the linearity of the estimator has its limitations. In an attempt to overcome this problem, it is shown in Chapter 4, how a non-linear filter can be built based on the conditional expectation, which can be seen as an extension to the Kalman filter. Here, we draft only the main ideas, to be able to implement it for the reservoir characterization task. The idea is to find for all L components Q_i of the parameter $\mathbf{Q}$ an optimal estimator $\varphi_i : \mathbb{R}^{n_y} \to \mathbb{R}$

minimizing the mean square error

$$\mathbb{E}[||Q_i(\boldsymbol{\xi}) - \varphi_i\left(\mathbf{Z}(\boldsymbol{\xi},\boldsymbol{\eta})\right)||_2^2]. \tag{8.107}$$

Furthermore, we restrict this estimator to be in some finite dimensional space by writing it as a linear combination of some given basis functions Ψ_j

$$\varphi_i = \sum_j \hat{\varphi}_{ij}\Psi_j(\mathbf{Z}). \tag{8.108}$$

The task is then to find the coefficients $\hat{\varphi}_j$ from the minimisation of (8.107). As described in Chapter 4, the minimisation leads to the linear system

$$\mathbf{A}\hat{\boldsymbol{\varphi}}_i = \mathbf{b}_i, \tag{8.109}$$

where $\hat{\boldsymbol{\varphi}}$ is the vector of coefficients, and the matrix $\mathbf{A}$ and the vector $\mathbf{b}_i$ are the integrals

$$[\mathbf{A}]_{ij} = \mathbb{E}[\psi_i(\mathbf{Z})\psi_j(\mathbf{Z})] = \int \,_i(\mathbf{Z}(\boldsymbol{\xi},\boldsymbol{\eta}))\psi_j(\mathbf{Z}(\boldsymbol{\xi},\boldsymbol{\eta}))\pi_\Xi(\boldsymbol{\xi})\pi_E(\boldsymbol{\eta})\mathrm{d}\boldsymbol{\xi}\mathrm{d}\boldsymbol{\eta}, \tag{8.110}$$

$$[\mathbf{b}_i]_j = \mathbb{E}[Q_i\psi_j] = \int Q_i(\boldsymbol{\xi})\psi_j(\mathbf{Z}(\boldsymbol{\xi},\boldsymbol{\eta}))\pi_\Xi(\boldsymbol{\xi})\pi_E(\boldsymbol{\eta})\mathrm{d}\boldsymbol{\xi}\mathrm{d}\boldsymbol{\eta}. \tag{8.111}$$

which can be evaluated with the help of numerical integration. Once the coefficients are computed from the system of Equations (8.109), the updated components of $\mathbf{Q}$ can be written with the help of the MMSE filter by

$$Q_i' = Q_i(\boldsymbol{\xi}) - \varphi_i\left(\mathbf{Z}(\boldsymbol{\xi},\boldsymbol{\eta})\right) + \varphi_i(\mathbf{z}_m). \tag{8.112}$$

The updated Q_i' would have the correct conditional mean if in Equation (8.108) an infinite series were taken, so that φ_i could be any measurable function. With the finite expansion, this is only approximately true.

The algorithm to compute an update of the L dimensional $\mathbf{Q}$ parameter using the n_y dimensional measurement $\mathbf{z}_m$ is as follows:

Algorithm 17: Parameter identification by the MMSE

1. Write input parameter $\mathbf{Q}$, measurement error model $\mathbf{E}$, and the response $\mathbf{Y}$ in a PCE form

$$\mathbf{q} = \sum_{\alpha=0}^{L} \mathbf{q}_\alpha \Phi_\alpha^{(Q)}(\boldsymbol{\xi}) \quad \boldsymbol{\epsilon} = \sum_{\alpha=0}^{n_y} \mathbf{e}_\alpha \Phi_\alpha^{(E)}(\boldsymbol{\eta}) \quad \mathbf{y} = \sum_{\alpha=0}^{M-1} \boldsymbol{\upsilon}_\alpha \Phi_\alpha(\boldsymbol{\xi}).$$

2. Generate a combined PCE basis $\hat{\Phi}_\alpha(\boldsymbol{\xi},\boldsymbol{\eta})$, and introduce the combined germ vector

$$\boldsymbol{\zeta} = \begin{bmatrix}\boldsymbol{\xi} & \boldsymbol{\eta}\end{bmatrix}^T.$$

3. Rewrite the PCE coefficients in this extended basis

$$\mathbf{q} = \sum_{\alpha=0}^{n_c} \hat{\mathbf{q}}_\alpha \hat{\Phi}_\alpha(\boldsymbol{\zeta}) \quad \boldsymbol{\epsilon} = \sum_{\alpha=0}^{n_c} \hat{\mathbf{e}}_\alpha \hat{\Phi}_\alpha(\boldsymbol{\zeta}) \quad \mathbf{y} = \sum_{\alpha=0}^{n_c} \hat{\boldsymbol{\upsilon}}_\alpha \hat{\Phi}_\alpha(\boldsymbol{\zeta}).$$

   The measurement model then becomes

$$\mathbf{z}(\boldsymbol{\zeta}) = \boldsymbol{\epsilon} + \mathbf{y} = \sum_{\alpha=0}^{n_c} \hat{\mathbf{e}}_\alpha \hat{\Phi}_\alpha(\boldsymbol{\zeta}) + \sum_{\alpha=0}^{n_c} \hat{\boldsymbol{\upsilon}}_\alpha \hat{\Phi}_\alpha(\boldsymbol{\zeta}) = \sum_{\alpha=0}^{n_c} \left(\hat{\mathbf{e}}_\alpha + \hat{\boldsymbol{\upsilon}}_\alpha\right) \hat{\Phi}_\alpha(\boldsymbol{\zeta}).$$

4. Determine the basis for the approximation of the estimators φ_i (e.g. polynomials of maximum total degree p_φ) by determining the basis functions $\Psi_j(\mathbf{z})$ collected in a vector $\boldsymbol{\Psi}$.

5. Compute the elements of the matrix $\mathbf{A}$:

   (a) Determine the integration points $\boldsymbol{\zeta}_k$ and weights w_k;

   (b) Evaluate the measurement model at the integration points $\mathbf{z}_k = \mathbf{z}(\boldsymbol{\zeta}_k)$, and compute the elements of the matrix $\mathbf{A}$ by

$$[\mathbf{A}]_{ij} \approx \sum_k \quad _i(\mathbf{z}_k)\psi_j(\mathbf{z}_k) w_k.$$

6. Compute the elements of vectors $\mathbf{b}_i$:

   (a) Determine integration points $\boldsymbol{\zeta}_l$ and weights w_l (which could be different from the ones we used in the computation of $\mathbf{A}$);

   (b) Evaluate the measurement model and all input parameters Q_i at the integration points, that is, $\mathbf{z}_l = \mathbf{z}(\boldsymbol{\zeta}_l)$, $q_{i,l} = q_i(\boldsymbol{\zeta}_l)$, and compute the elements of vectors $\mathbf{b}_i$ for all $i = 1 \dots L$ by

$$[\mathbf{b}_i]_j \approx \sum_l q_{i,l} \psi_j(\mathbf{z}_l) w_l.$$

7. Solve the system of equations $\mathbf{A}\begin{bmatrix}\hat{\boldsymbol{\varphi}}_1 & \cdots & \hat{\boldsymbol{\varphi}}_L\end{bmatrix} = \begin{bmatrix}\mathbf{b}_1 & \cdots & \mathbf{b}_L\end{bmatrix}$ for all coefficients $\hat{\boldsymbol{\varphi}}_i$ of the estimator.

The estimators are then given by

$$\varphi_i = \sum_j \hat{\varphi}_{ij} \Psi_j(\mathbf{Z}), \quad i = 1 \dots L.$$

The best estimate of $\mathbf{Q}$ for all components of the input parameter is

$$q_i' = \varphi_i(\mathbf{z}_m) = \sum_j \hat{\varphi}_{ij} \Psi_j(\mathbf{z}_m) = \hat{\boldsymbol{\varphi}}_i^T \boldsymbol{\Psi}(\mathbf{z}_m).$$

The components of the updated reference parameter reads $Q_i'(\boldsymbol{\xi}, \boldsymbol{\eta}) = Q_i(\boldsymbol{\xi}) - \varphi_i\left(\mathbf{Z}(\boldsymbol{\xi}, \boldsymbol{\eta})\right) + q_i'$.

One of the problematic parts of the method is the computation of integrals (8.110) and (8.111). For Gaussian $\boldsymbol{\xi}$ and $\boldsymbol{\eta}$, the Gauss-Hermite quadrature can be used for the integration. As explained in Chapter 4 for the computation of matrix $\mathbf{A}$, an integration rule of order $p_y p_\varphi + 1$ will suffice. Here, p_y is the maximum total degree of the PCE of $\mathbf{y}$, and p_φ is the degree of the estimator. For the computation of $\mathbf{b}_i$, a minimum order $(p_x + p_y p_\varphi + 1)/2$ should be used. When the dimension $L + n_y$ of the integration is too high, a sparse grid can be used to keep the number of points in a computationally affordable regime. If such a rule makes the computations too expensive, a MC or QMC integration rule could also be applied.

A big problem is that the dimension even for small problems can get very high. This results in a high dimensional integration demanding high computational cost or questionable accuracy. To release the burden of dimensionality, we propose here a low-rank approximation of the measurement model $\mathbf{Z}$ by using the POD basis of the response $\mathbf{Y}(\omega)$. The POD approximation of the fluctuating part of $\mathbf{Y}$ following Section 8.4.2 is given by

$$\tilde{\mathbf{Y}}(\omega) \approx \mathbf{V}_Y \mathbf{S}_Y \mathbf{Y}_r(\omega), \tag{8.113}$$

where

$$\mathbf{Y}_r = \mathbf{S}_Y^{-1} \mathbf{V}_Y^T \tilde{\mathbf{Y}} = \mathbf{P} \tilde{\mathbf{Y}}, \tag{8.114}$$

is the reduced model of the response field, $\mathbf{P} = \mathbf{S}_Y^{-1} \mathbf{V}_Y^T$ is the projection matrix, $\mathbf{V}_Y$ is the matrix collecting the eigenvectors $\mathbf{v}_{y,i}$ of the covariance matrix $\mathbf{C}_Y$ and $\mathbf{S}_y$ is a diagonal matrix with the squared eigenvalues $\lambda_{Y,i}$. The eigenvalues and the eigenvectors are computed from

$$\mathbf{C}_Y \mathbf{v}_{y,i} = \lambda_{Y,i} \mathbf{v}_{y,i}. \tag{8.115}$$

$\mathbf{C}_Y$ is an $n_y \times n_y$ matrix, thus we have at most n_y eigenvectors. We ignore the vectors with indices which are not contributing to the covariance of $\mathbf{Y}$ more than some percentage δ_σ. We use this eigenbasis to represent the measurement model, so we project the fluctuating part of the measurement model $\tilde{\mathbf{Z}}$ onto this reduced basis spanned by the spatial vectors typical for the response, that is

$$\mathbf{Z}_r = \mathbf{P}\tilde{\mathbf{Z}} = \mathbf{P}(\mathbf{Z} - \mathbb{E}[\mathbf{Z}]) = \mathbf{P}\underbrace{(\mathbf{Y} - \mathbb{E}[\mathbf{Y}])}_{\tilde{\mathbf{Y}}} + \underbrace{\mathbf{PE}}_{=\mathbf{E}_r} = \mathbf{Y}_r + \mathbf{E}_r, \tag{8.116}$$

where in the last but one equality we supposed a mean-free measurement error model $\mathbf{E}$. The transformed error $\mathbf{E}_r = \mathbf{PE}$ is not uncorrelated anymore, and its covariance matrix is

$$\mathbf{C}_{E_r} = \mathrm{Cov}(\mathbf{PE}) = \mathbf{P}\mathbf{C}_E\mathbf{P}^T. \tag{8.117}$$

This correlated error model $\mathbf{E}_r$ can be represented in a smaller stochastic space—by a germ $\boldsymbol{\eta}_r$ wich has much lower stochastic dimension R than n_y—namely, the projected error can be written by a simple linear transformation of the germ,

$$\boldsymbol{\epsilon}_r = \mathbf{L}\boldsymbol{\eta}_r, \tag{8.118}$$

where $\mathbf{L}$ is the lower triangular matrix from the Cholesky decomposition of the covariance matrix $\mathbf{C}_{E_r}$, that is

$$\mathbf{C}_{E_r} = \mathbf{L}\mathbf{L}^T. \tag{8.119}$$

The projected error can be also given in a PCE form

$$\boldsymbol{\epsilon}_r = \sum_{\beta=1}^{R} \mathbf{l}_\beta \Phi^{(E_r)}(\boldsymbol{\eta}_r) = \mathbf{L}\boldsymbol{\Phi}^{(E_r)}, \tag{8.120}$$

where $\mathbf{l}_\beta$ are the columns of the matrix $\mathbf{L}$ and $\Phi^{(E_r)}$ are the linear univariate Hermite polynomials. Plugging in the PCE approximations of $\mathbf{E}_r$ and $\mathbf{Y}$ into Equation (8.116), the fluctuating part of the measurement model is approximated by

$$\tilde{\mathbf{Z}}_r \approx \mathbf{P}\left(\sum_{\alpha=0}^{M-1} \boldsymbol{v}_\alpha \Phi_\alpha(\boldsymbol{\xi}) - \mathbb{E}\left[\mathbf{Y}\right]\right) + \sum_{\beta=1}^{R} \mathbf{l}_\beta \Phi^{(E_r)}(\boldsymbol{\eta}_r) = \mathbf{P}\tilde{\boldsymbol{\Upsilon}}\boldsymbol{\Phi} + \mathbf{L}\boldsymbol{\Phi}^{(E_r)}, \tag{8.121}$$

where the $\tilde{\boldsymbol{\Upsilon}}$ is the matrix $\boldsymbol{\Upsilon}$ with the zeroed out first column for subtraction of the mean. This latest description has the advantage that the dimension of the germ shrunk from $L + n_y$ to $L + R$. If the response $\mathbf{Y}(\boldsymbol{\xi})$ is smooth, only a few number of eigenmodes R already give a good approximation. The optimal estimator will then be a function of the projected error model and can be computed by solving the linear system in Equation (8.109), where in the computation of matrix $\mathbf{A}$ and the right hand side $\mathbf{b}$ (see Equations (8.110) and (8.111)) the measurement model $\mathbf{Z}(\boldsymbol{\xi}, \boldsymbol{\eta})$ must be replaced by the projected one $\mathbf{Z}_r(\boldsymbol{\xi}, \boldsymbol{\eta}_r)$. Once the optimal maps ϕ_i are computed, the best estimate of Q_i can be computed by plugging in the projected measurement $\mathbf{z}_m$ into the optimal estimator

$$q'_m = \phi_i(\mathbf{P}(\mathbf{z}_m - \mathbb{E}[\mathbf{Y}])). \tag{8.122}$$

Using the MMSE filter, the posterior input parameter becomes

$$Q'_i = Q_i(\boldsymbol{\xi}) - \phi_i\left(\mathbf{Z}_r(\boldsymbol{\xi}, \boldsymbol{\eta}_r)\right) + q'_m. \tag{8.123}$$

We only present here examples of updating the spatially not varying scaling factor. When we drop the assumption of the scalar scaling factor, the field update by MMSE is done the same way, and the update gives very similar result to the one computed by MCMC or the Kalman filter (see Figure 8.17). This is normal due to the fact that we used a linear PCE of the response $\mathbf{Y}$, and so the limitation is more in the PCE approximation of the response. As the response surface is slightly non-linear, we could improve the accuracy of the MCMC update and also the one of a non-linear estimator by a higher degree approximation of the response, but that would greatly increase the cost of the surrogate modelling.

Example 9: Computation of the low rank representation of the measurement model for the MMSE update. Consider the first scenario, where Q is a scalar valued random variable, that is, $L = 1$ and a PCE of $\mathbf{Y}$ of maximum polynomial degree $p_y = 5$. When we aim at finding a quadratic estimator, that is, $p_\varphi = 2$ by the MMSE estimator, for the computation of the matrix $\mathbf{A}$ by (8.110) we would need to generate integration points from the $p_y p_\varphi + 1 = 11$-point univariate rule, which results in $11^{(L+n_y)} = 11^{61} = 3.349 \cdot 10^{63}$ points using a full tensor grid. Instead of replacing the integration rule with a Smolyak sparse grid or a QMC grid, we use the low-rank approximation of the measurement model as explained above.

1. **Compute the covariance matrix $\mathbf{C}_Y$.**

   The 60×60 covariance matrix was determined from the PCE coefficients using Equation (8.43).

2. **Compute the eigenvectors and eigenvalues of $\mathbf{C}_Y$ and truncate.**

   From the eigenvalues $\{\lambda_i\}_{i=1}^{n_y}$, it can be shown that more than 99% of the total variance of $\mathbf{Y}$ can be represented by using only $R = 3$ modes out of the 60, that is

$$\frac{\sum_{i=1}^{3} \lambda_i}{\sum_{j=1}^{n_y} \lambda_j} > 0.99. \tag{8.124}$$

   We collect the first three computed eigenvectors $\{\mathbf{v}_{y,i}\}_{i=1}^{3}$ in the columns of the matrix $\mathbf{V}_y$, and the first three eigenvectors in the diagonal of the matrix $\mathbf{S}$ and set their product to be the projection matrix $\mathbf{P} = \mathbf{S}_y^{-1}\mathbf{V}_y^T$.

3. **Compute the PCE of the projected measurement error $\mathbf{E}_r$.**

   The covariance matrix of the measurement error model $\mathbf{C}_E$ is a 60×60 diagonal matrix whose diagonal entries are the variances of the error at the sixty assimilation points. The 3×3 sized covariance matrix of the projected measurement error $\mathbf{C}_{E_r}$ is computed by Equation (8.117). For the PCE of the projected measurement error $\mathbf{E}_r$, we first carry out the Cholesky decomposition (8.119) of $\mathbf{C}_{E_r}$ which gives us a lower triangular matrix $\mathbf{L}$ of size $R \times R$. The new, projected measurement error can be represented in PCE form by

$$\mathbf{E}_r = \mathbf{L}\mathbf{\Phi}^{(E_r)} = \begin{bmatrix} l_{11} & 0 & 0 \\ l_{21} & l_{22} & 0 \\ l_{31} & l_{32} & l_{33} \end{bmatrix} \begin{bmatrix} \eta_{r,1} \\ \eta_{r,2} \\ \eta_{r,3} \end{bmatrix}. \tag{8.125}$$

4. **Determine the PCE of the projected measurable response $\mathbf{Y}_r$.**

   The computation of the PCE coefficients $\mathbf{\Upsilon}$ was performed as in Example 1, but here we used polynomial degree $p_y = 5$. In order to get only the fluctuating part $\tilde{\mathbf{Y}}$ of $\mathbf{Y}$, the PCE coefficients corresponding to the basis function $\Phi_0 = 1$ are set to zero. The coefficients $\tilde{\mathbf{\Upsilon}}$ of $\tilde{\mathbf{Y}}$ are then projected onto the main three eigenvectors by

$$\mathbf{\Upsilon}_r = \mathbf{P}\tilde{\mathbf{\Upsilon}}. \tag{8.126}$$

   The PCE of the projected model $\mathbf{Y}_r$ becomes

$$\mathbf{Y}_r = \mathbf{P}\tilde{\mathbf{\Upsilon}}\mathbf{\Phi}(\boldsymbol{\xi}) = \mathbf{\Upsilon}_r\mathbf{\Phi}(\boldsymbol{\xi}). \tag{8.127}$$

5. **Compute the projected measurement $\mathbf{z}_{r,m}$**

$$\mathbf{z}_{r,m} = \mathbf{P}(\mathbf{z}_m - \mathbb{E}[\mathbf{Y}]). \tag{8.128}$$

Example 10: Test low-rank approximation on the MCMC update. The low-rank approximation was first tested by using it in an MCMC computation. The procedure is the same as in Example 5, the only difference being that we use $\mathbf{z}_{r,m}$ instead of $\mathbf{z}_m$, and $\mathbf{Y}_r$ instead of the response $\mathbf{Y}$. The likelihood is computed from the multivariate Gaussian distribution of the projected error by

$$L(q) = \det(2\pi\mathbf{C}_{E_r})^{-1/2} e^{-\frac{1}{2}(\mathbf{z}_{r,m} - \mathbf{Y}_r(\mathcal{I}^{-1}q))^T \mathbf{C}_{E_r}^{-1}(\mathbf{z}_{r,m} - \mathbf{Y}_r(\mathcal{I}^{-1}q))}. \tag{8.129}$$

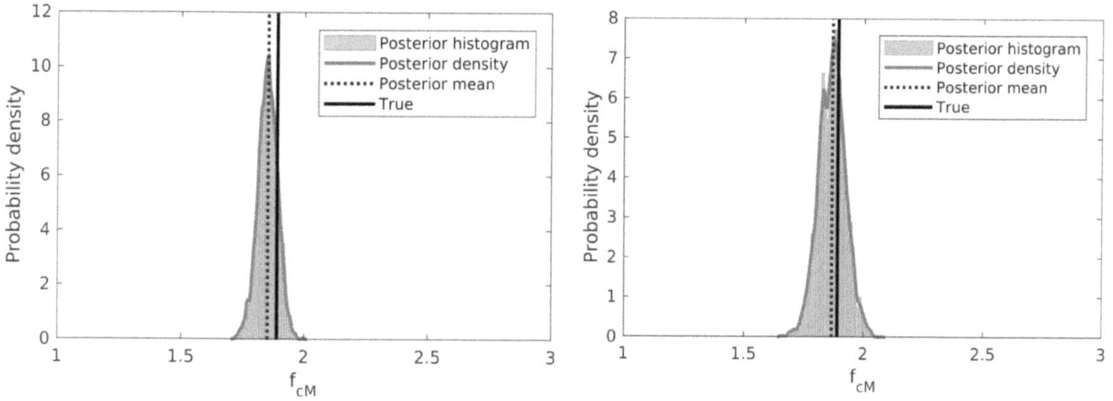

Figure 8.15: Testing low-rank approximation on MCMC update: the posterior density estimate of F_{cM} without (to the left) and with low-rank approximation (to the right).

Results of the update are shown in Figure 8.15. Cutting the stochastic dimension of the measurement model from $n_y = 60$ to $R = 3$ did not influence significantly the results of the MCMC update.

Example 10: Updating the scalar scaling factor by a quadratic MMSE filter. Here we present how to apply the MMSE filter to the reservoir characterization.

1. **Write Q, $\mathbf{E}$, and $\mathbf{Y}$ in a PCE form.**

   The PCE expansions would be the same as the ones given in the first point of Example 7, but we use here the projected response $\mathbf{Y}_r$ and the projected error $\mathbf{E}_r$ whose PCE representation are explained in Example 9 and given in Equations (8.127) and (8.125).

2. **Generate a combined PCE basis.**

   The PCE basis becomes

$$\hat{\mathbf{\Phi}} = \begin{bmatrix} 1 & \xi & \xi^2 - 1 & \xi^3 - 3\xi & \xi^4 - 6\xi^2 + 3 & \xi^5 - 10\xi^3 + 15\xi & \eta_{r,1} & \eta_{r,2} & \eta_{r,3} \end{bmatrix}^T \tag{8.130}$$

   with the combined germ

$$\boldsymbol{\zeta} = \begin{bmatrix} \xi & \eta_{r,1} & \eta_{r,2} & \eta_{r,3} \end{bmatrix}. \tag{8.131}$$

3. **Rewrite the coefficients in the extended basis.**

   The coefficients in the extended basis are assembled in the same manner as in Example 7. Due to the low-rank representation, the modified sizes of the coefficient matrices of Q, $\mathbf{Y}_r$, and $\mathbf{E}_r$ are

$$\hat{\mathbf{Q}} \in \mathbb{R}^{1\times 9}, \quad \hat{\mathbf{\Upsilon}}_r \in \mathbb{R}^{3\times 9}, \quad \hat{\mathbf{E}}_r \in \mathbb{R}^{3\times 9}. \tag{8.132}$$

   The PCE coefficients of the measurement model $\mathbf{Z}_r$ in this extended basis are simply given by adding the coefficients of the response and the error, that is,

$$\hat{\mathbf{Z}}_r = \hat{\mathbf{\Upsilon}}_r + \hat{\mathbf{E}}_r. \tag{8.133}$$

4. **Chose approximating basis for the estimator.**

   We restrict the estimator φ to linear combinations of products of the monomials. We tried different degree estimators. The basis of maximum total degree two example is

$$\mathbf{\Psi} = [1 \quad z_{r,1} \quad z_{r,2} \quad z_{r,3} \quad {z_{r,1}}^2 \quad z_{r,1}z_{r,2} \quad z_{r,1}z_{r,3} \quad {z_{r,2}}^2 \quad z_{r,2}z_{r,3} \quad {z_{r,3}}^2]^T, \tag{8.134}$$

   thus, 10 basis functions of the projected measurement model $\mathbf{z}_r$[8]. Without the projection, there would be 1,891 basis functions.

5. **Compute matrix A and vector b.**

   We used the same integration rule for computing $\mathbf{A}$ and $\mathbf{b}$. The tensor product of the one dimensional $n = \max\{(p_y p_\varphi + 1), (1 + p_y p_\varphi + 1)/2\} = \max\{(5 \cdot 2 + 1), (1 + 5 \cdot 2 + 1)/2\} = 11$ point Gauss quadrature rule was used. This sums up to $11^4 = 14,641$ integration points $\boldsymbol{\zeta}_k$.

   For the computation of the matrix $\mathbf{A}$ and vector $\mathbf{b}$, we evaluate the measurement model $\mathbf{Z}$ and the Q parameter at these integration points $\boldsymbol{\zeta}_k$ by using their PCE approximation in the extended basis.

6. **Solve the system of equation to determine the estimator.**

   The coefficients of the estimator $\hat{\varphi}$ can be computed by solving the system of equations $\mathbf{A}\hat{\varphi} = \mathbf{b}$, where φ is a vector of 10 coefficients corresponding to the 10 Ψ_i basis functions given above. The estimator is given by

$$\varphi = \sum_j \hat{\varphi}_j \Psi_j(\mathbf{z}_r). \tag{8.135}$$

The best estimate of Q is computed by plugging in the projected data $\mathbf{z}_{r,m}$ computed in Equation (8.128) into the estimator

$$q'_m = \varphi(\mathbf{z}_{r,m}). \tag{8.136}$$

The updated Q' is

$$Q'(\boldsymbol{\zeta}) = Q(\boldsymbol{\zeta}) - \varphi(\mathbf{Z}_r(\boldsymbol{\zeta})) + q'_m. \tag{8.137}$$

To generate samples of Q', we first sample from the germ of four identical and independent standard Gaussians ($\boldsymbol{\zeta} = [\xi, \eta_{r,1}, \eta_{r,2}, \eta_{r,3}]^T$), then we compute the input parameter and the projected measurement model from their PCE representation given in the extended basis. Then the above expression can be evaluated for any samples of the germ $\boldsymbol{\zeta}$. By further mapping the values of q' to f_{cM}, we can estimate the posterior distribution, which is shown in Figure 8.16 using different degree estimators ($p_\varphi = 1, 2, 3$).

[8]Please note, monomials are typically bad for approximation, but with such a low degree estimator it is applicable.

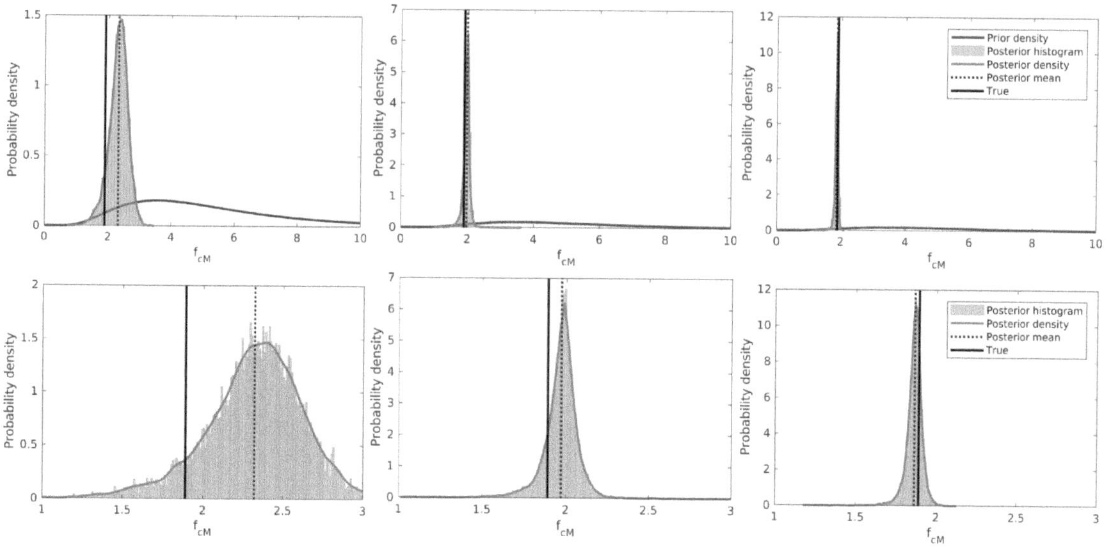

Figure 8.16: Update results of the MMSE: the prior probability density and the posterior density estimate of F_{cM} using different degree estimators (linear estimator to the left, degree two in the middle column and degree three to the right). In the bottom row the density is zoomed to the posterior.

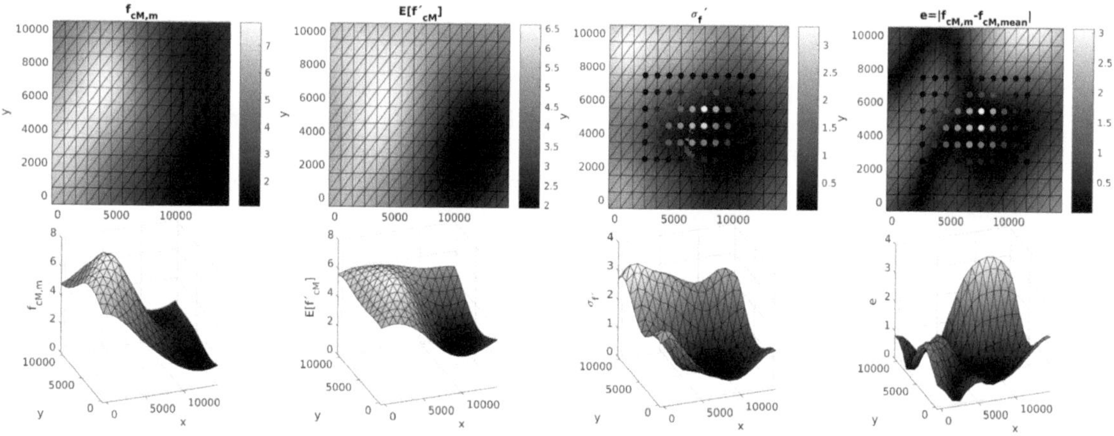

Figure 8.17: Result of the MMSE field update using a linear estimator and a low rank approximation of the measurement model, projecting to $R = 10$ eigenvectors: 2D and 3D view of the 'true' scaling factor (to the left), the mean (second column from left), and the standard deviation (third column from left) of the posterior scaling factor and the deviation of the mean from the 'true' (to the right). See the explanation of the dots in Figure 8.12.

8.6 Summary, conclusion, and outlook

In this chapter, we presented a general framework for parameter and field identification using different techniques, but all of which are based on a PCE surrogate model of the response. The identification is presented on a gas reservoir problem, where we wished to identify the scaling factor F_{c_M} of the compressibility field by inference from measurement of seabed displacements. We provided examples for two different scenarios, one assuming a spatially not varying, scalar scaling factor and one dropping the homogeneity condition. A summary of the used identification methods is as follows:

- The Markov Chain Monte Carlo (MCMC) is based on a Metropolis-Hastings random walk, which generates samples of the posterior random field of the input parameter F_{c_M}. Though the method is converging very slowly, the computational expense was largely reduced by computing the likelihood at every random step from the PCE surrogate model instead of the deterministic solver.

- The Ensemble Kalman filter (EnKF) generates samples of the posterior F_{c_M} with the help of a linear estimator. First, an ensemble of the measurable response is created by sampling from the PCE surrogate model. The ensemble is then mapped by the linear estimator of the Kalman filter. The matrix of the Kalman gain is computed from the ensemble.

- The sample-free PCE-based Kalman-Filter (PCEKF) is similar to the EnKF method, but the Kalman-gain is computed from the PCE coefficients and the updated parameter F_{c_M} is determined in a PCE form, leading to an analytical form of the posterior scaling factor. Sampling was used here only for plotting the posterior density.

- The Minimum Mean Square Estimator (MMSE) also gives a functional representation of the updated parameter, but not necessarily in a PCE form. Here, the map from the measurement model to the updated parameter is not restricted to linear maps but to a finite dimensional space of polynomials and is determined by minimizing the minimum mean squared error. Here, different degree polynomials were tested ($p = 1, 2, 3$).

The results for the posterior mean and variance and the deviation of the posterior from its 'true' value $f_{cM,m}$—from which the synthetic measurement was generated—for the first scenario are shown in Table 8.2. The MCMC update method can be seen as a reference solution representing the Bayesian posterior distribution. We used 10,000 samples to get sufficiently accurate statistics of the posterior. The MCMC update method was also carried out using the low-rank representation of the measurement model to be able to see the error originating from projecting to the POD basis. This was important, because the MMSE method was only feasible with the low-rank approximation. It is shown in the examples that the Kalman filter can largely speed up the computation of the inverse method, but it only gives the exact posterior distribution if the input variable is Gaussian and the map to the measurement model is linear. For slightly non-linear problems, the PCE-based Kalman filter is a fast and efficient method to get an approximate posterior of the input variables. The MMSE method gives a good extension of the Kalman filter, and it is shown in the last example how the increased non-linearity of the filter can improve the update.

Table 8.2: Performance of the different PCE-based update methods: posterior mean $\mathbb{E}[f'_{cM}]$, the deviation of the mean from the true value $f_{cM,m}$, and the posterior variance $\text{var}(f'_{cM})$.

	MCMC	EnKF	PCEKF	MCMC LR	MMSE $p = 1$	MMSE $p = 2$	MMSE $p = 3$
low-rank	no	no	no	yes	yes	yes	yes
$\mathbb{E}[f'_{cM}]$	1.8521	2.1159	2.1066	1.8679	2.3235	1.9715	1.8647
$\lvert\mathbb{E}[f'_{cM}] - f_{cM,m}\rvert$	0.0379	0.2259	0.2161	0.0221	0.4336	0.0814	0.0253
$\text{var}(f'_{cM})$	0.0016	0.0384	0.0303	0.0033	0.0907	0.0109	0.0026

For the identification of the spatially varying field, the same framework is used by separating the random field's spatial structure from the stochastic one. This way the random field can be represented by a set of random variables. For the surrogate modelling in this scenario, we limited the PCE to a linear one to keep the computational cost affordable. This limitation constraints somewhat the update procedure and does not emphasize that much the difference between the different update methods. Even this linear PCE of the response resulted in a sufficiently accurate estimation of the spatially varying scaling factor, capturing well the spatial structure and the magnitude of the scaling factor. For the quality of the identification the location of the measurement points plays a crucial role. Obviously, in the region where no measurements are available, the scaling factor can not be updated, and also assimilation points with a measured response that are not sensitive to the scaling factor do not contribute much to the identification.

To conclude, we have shown how a PCE surrogate model of the response can enable different Bayesian techniques to update the model's input parameter. The methods were all tested on the slightly non-linear model of the gas reservoir characterization, where the sample-free PCEKF gave a good rough estimate of the posterior with very low computational cost. The performance of the update could be further improved by using the non-linear filter of the MMSE method. It should be mentioned though that the computational cost of the MMSE method can easily grow very high, and we kept it in a manageable scale here by using a low-rank approximation of the measurement model. Another problem with the PCEKF and the MMSE methods could be that the conditional expectation is not so informative when the Bayesian posterior is multimodal, which can happen for strongly non-linear problems.

Appendices

Appendix A: FEM computation of seabed displacements

The governing equilibrium equations to compute the seabed displacements read

$$\nabla \cdot \boldsymbol{\sigma}' - \alpha \nabla p = \rho \mathbf{g} + \mathbf{b}. \tag{A1}$$

For a saturated porous volume $\mathcal{D}$ with surface Σ, the equilibrium is prescribed minimizing the total potential energy by the virtual work theorem. By using the hypothesis of quasi-static conditions and neglecting the inertial forces, the work theorem formulation reads

$$\int_D \delta\boldsymbol{\epsilon} : (\boldsymbol{\sigma}' - \alpha p \mathbf{1}) \mathrm{d}D = \int_D \delta\mathbf{u} \cdot \mathbf{b} \mathrm{d}D + \int_{\Gamma_N} \delta\mathbf{u} \cdot \mathbf{t} \mathrm{d}\Sigma, \tag{A2}$$

where $\boldsymbol{\epsilon}$ is the (small) strain tensor, $\mathbf{u} = [\; u_x(\hat{x},\hat{y},\hat{z}) \quad u_y(\hat{x},\hat{y},\hat{z}) \quad u_z(\hat{x},\hat{y},\hat{z}) \;]^T$ is the displacement vector, $\mathbf{1}$ the rank-2 identity tensor, and $\mathbf{t}$ represents the total external forces acting per unit surface. The symbol δ indicates the virtual variables. The integral formulation of Equation (A2) must be supplemented with appropriate Dirichlet and Neumann boundary conditions

$$\begin{aligned} \mathbf{u} &= \mathbf{u}_0 \quad \text{on} \quad \Gamma_D, \\ \boldsymbol{\sigma} \cdot \mathbf{n} &= \mathbf{t} \quad \text{on} \quad \Gamma_N, \quad \Gamma_N \cup \Gamma_D = \Gamma, \end{aligned} \tag{A3}$$

in which $\boldsymbol{\sigma}$ is the total stress. The FEM discretisation of Equation (A2) is only briefly explained here. Interested readers are directed to [98, 173] for further details. Here, we introduce a set of basis functions collected in a row vector

$$\mathbf{N_u} = \begin{bmatrix} N_{u1} & 0 & 0 & N_{u2} & 0 & 0 & \cdots & N_{un} & 0 & 0 \\ 0 & N_{u1} & 0 & 0 & N_{u2} & 0 & \cdots & 0 & N_{un} & 0 \\ 0 & 0 & N_{u1} & 0 & 0 & N_{u2} & \cdots & 0 & 0 & N_{un} \end{bmatrix}, \tag{A4}$$

and restrict our solution space and the space of virtual displacements to the span of these basis functions. Accordingly, the virtual and real displacements and strains are approximated by

$$\begin{aligned} \delta\mathbf{u} &\approx \mathbf{N}_u \delta\mathbf{a} \quad & \delta\boldsymbol{\epsilon} \approx \mathbf{B}\delta\mathbf{a} \\ \mathbf{u} &\approx \mathbf{N}_u \mathbf{a} \quad & \boldsymbol{\epsilon} \approx \mathbf{B}\mathbf{a}, \end{aligned} \tag{A5}$$

where the elements of $\mathbf{a}$ are the unknown nodal displacements and $\mathbf{B} = L\mathbf{N}_u$ with L, the differential operator defining the kinematic equation

$$\boldsymbol{\epsilon} \approx L\mathbf{u} \quad \text{or} \quad \delta\boldsymbol{\epsilon} \approx L\delta\mathbf{u}. \tag{A6}$$

The fluid pore pressure p is also discretised by approximating it with the ansatz

$$p \approx \mathbf{N}_p \mathbf{p}, \tag{A7}$$

where $\mathbf{p}$ are the nodal component vectors of the pore pressure and $\mathbf{N}_p$ is the vector of another set of basis functions. First, rearranging Equation (A2) and then plugging in Equations (A5) through (A7) into Equation (A2) and differentiating with respect to the virtual variables, yields the equation:

$$\underbrace{\int_{\mathcal{D}} \mathbf{B}^T \hat{\boldsymbol{\sigma}}' \mathrm{d}D}_{\Pi} = \underbrace{\int_{\mathcal{D}} \alpha \mathbf{B}^T \hat{\mathbf{1}} \mathbf{N}_p^T \mathbf{p} \mathrm{d}D \int_{\mathcal{D}} \mathbf{N}_u^T \mathbf{b} \mathrm{d}D + \int_{\Gamma_N} \mathbf{N}_u^T \mathbf{t} \mathrm{d}\Gamma}_{\mathbf{f}}, \tag{A8}$$

where $\hat{(\cdot)}$ is the Voigt notation of $(\cdot)$. In this case, a non-linear constitutive model is implemented leading to a nonlinear form of Π. Thus, the system of equations in (A8) can be solved by a Newton-like method with the Jacobian matrix $\mathbf{K}_T$ computed as

$$\mathbf{K}_T = \frac{\mathrm{d}\Pi}{\mathrm{d}\mathbf{a}} = \frac{\mathrm{d}}{\mathrm{d}\mathbf{a}} \int_{\mathcal{D}} \mathbf{B}^T \hat{\boldsymbol{\sigma}}'(\mathbf{a}) \mathrm{d}D = \int_{\mathcal{D}} \mathbf{B}^T \frac{\mathrm{d}\hat{\boldsymbol{\sigma}}'(\mathbf{a})}{\mathrm{d}\mathbf{a}} \mathrm{d}D = \int_{\mathcal{D}} \mathbf{B}^T \mathbf{C} \mathbf{B} \mathrm{d}D. \tag{A9}$$

The tangent constitutive tensor $\mathbf{C}$ governs stress-strain relationship

$$\mathrm{d}\hat{\boldsymbol{\sigma}}' = \mathbf{C}\mathrm{d}\hat{\boldsymbol{\epsilon}}. \tag{A10}$$

For an isotropic porous material, the tangent operator $\mathbf{C}$ reads

$$\mathbf{C} = c_M^{-1} \begin{bmatrix} 1 & \frac{\nu}{1-\nu} & \frac{\nu}{1-\nu} & 0 & 0 & 0 \\ \frac{\nu}{1-\nu} & 1 & \frac{\nu}{1-\nu} & 0 & 0 & 0 \\ \frac{\nu}{1-\nu} & \frac{\nu}{1-\nu} & 1 & 0 & 0 & 0 \\ 0 & 0 & 0 & \frac{1-2\nu}{2(1-\nu)} & 0 & 0 \\ 0 & 0 & 0 & 0 & \frac{1-2\nu}{2(1-\nu)} & 0 \\ 0 & 0 & 0 & 0 & 0 & \frac{1-2\nu}{2(1-\nu)} \end{bmatrix}, \tag{A11}$$

where ν is the Poisson's moduli, $c_M = [(1+\nu)(1-2\nu)]/[E(1-\nu)]$ is the oedometric compressibility and E is Young's modulus, and the tangent stiffness $\mathbf{K}_T$ in Equation (A9) depends linearly on $c_M{-}1$. Following some experimental results, the constitutive model depends on the stress state of the seabed. The Poisson's moduli can be fixed to a constant value but the oedometric compressibility and thus the Young's modulus was observed to depend on the vertical component of the effective stress, σ'_z by a power relation

$$c_M(\sigma_z) = c_{M0} \left(\frac{\sigma'_z}{f_0} \right)^{-\lambda}, \tag{A12}$$

where f_0 is a fixed value of stress enabling a dimensionless form, and λ and c_{M0} are material coefficients estimated via measurements. For more detail about this experimental constitutive law, the reader is directed to [173].

Appendix B: Hermite polynomials

B.1 Generation of Hermite Polynomials

The univariate Hermite polynomials H are orthogonal polynomials with respect to the standard Gaussian distribution [71], that is they satisfy the orthogonality condition

$$\mathbb{E}[H_i(\xi)H_j(\xi)] = \int_{\mathbb{R}} H_i(\xi)H_j(\boldsymbol{\xi})\pi_{\Xi}(\xi)\mathrm{d}\xi = h_i\delta_{ij}, \tag{B1}$$

where δ_{ij} is the Kroenecker delta, $\pi_{\Xi}(\xi)$ is the standard Gaussian distribution

$$\pi_{\Xi}(\xi) = \frac{1}{\sqrt{2\pi}}e^{-\frac{\xi^2}{2}}, \tag{B2}$$

and h_i is the squared norm of the polynomials, that is

$$h_i = \mathbb{E}[H_i(\xi)H_i(\xi)]) = i!. \tag{B3}$$

The Hermite polynomials can be generated by starting with the monomial basis $\{1, \xi, \xi^2, \xi^3 \dots\}$ and determine the Hermite polynomials by the Gram-Schmidt orthogonalization. However, for numerical purposes it is more convenient to use the three term recurrence formula [71]

$$H_{n+1}(\xi) = (A_n\xi + B_n)H_n(\xi) - C_nH_{n-1}(\xi), \quad n \geq 0, \tag{B4}$$

where the first two polynomials are set to

$$\begin{aligned} H_{-1} &= 0 \\ H_0 &= 1. \end{aligned}$$

For Hermite polynomials, the generating sequences are $A_n = 1$, $B_n = 0$ and $C_n = n$. For example, the first few Hermite polynomials are

$$H_0 = 1 \quad H_1 = \xi \quad H_2 = \xi^2 - 1 \quad H_3 = \xi^3 - 3\xi \quad H_4 = \xi^4 - 6\xi^2 + 3. \tag{B5}$$

The generation of multivariate polynomials from the univariate Hermite polynomials is straightforward. The polynomials are generated by taking products of the univariate Hermite polynomials. The easiest is to generate the products by the so-called multi-index set, tuples of indices $\boldsymbol{\alpha} = (i_1, i_2, ...i_j, ...)$ defining which indexed univariate polynomials are to be multiplied together. As an example, the multi-index set for polynomials with two variables of max total degree 3 and the corresponding Hermite polynomials read

index	multi–	index	Deg. of polyn.	The αth multiv. polyn.
i	$\boldsymbol{\alpha} = (i_1$	$i_2)$	$\lvert\boldsymbol{\alpha}\rvert = \sum_j i_j$	$\Phi_i = \Phi_{\boldsymbol{\alpha}} = H_{i_1}(\xi_1)H_{i_2}(\xi_2)$
0	0	0	0	$\Phi_0 = H_0(\xi_1)H_0(\xi_2) = 1 \cdot 1 = 1$
1	1	0	1	$\Phi_1 = H_1(\xi_1)H_0(\xi_2) = \xi_1 \cdot 1 = \xi_1$
2	0	1	1	$\Phi_2 = H_0(\xi_1)H_1(\xi_2) = 1 \cdot \xi_2 = \xi_2$
3	2	0	2	$\Phi_3 = H_2(\xi_1)H_0(\xi_2) = (\xi_1^2 - 1) \cdot 1 = \xi_1^2 - 1$
4	1	1	2	$\Phi_4 = H_1(\xi_1)H_1(\xi_2) = \xi_1\xi_2$
5	0	2	2	$\Phi_5 = H_0(\xi_1)H_2(\xi_2) = 1 \cdot (\xi_2^2 - 1) = \xi_2^2 - 1$
6	3	0	3	$\Phi_6 = H_3(\xi_1)H_0(\xi_2) = (\xi_1^3 - 3\xi_1) \cdot 1 = \xi_1^3 - 3\xi_1$
7	2	1	3	$\Phi_7 = H_2(\xi_1)H_1(\xi_2) = (\xi_1^2 - 1) \cdot \xi_2 = \xi_1^2\xi_2 - \xi_2$
8	1	2	3	$\Phi_8 = H_1(\xi_1)H_2(\xi_2) = \xi_1 \cdot (\xi_2^2 - 1) = \xi_2^2\xi_1 - \xi_1$
9	0	3	3	$\Phi_9 = H_0(\xi_1)H_3(\xi_2) = 1 \cdot (\xi_2^3 - 3\xi_2) = \xi_2^3 - 3\xi_2$

(B6)

The above tuples were generated in such a way that their sum is less or equal to three, corresponding to a maximal polynomial degree of three. The multivariate polynomials are still orthogonal with respect to the underlying multivariate Gaussian measure, that is, the orthogonality condition

$$\mathbb{E}[\Phi_i(\boldsymbol{\xi})\Phi_j(\boldsymbol{\xi})] = \int_{\mathbb{R}^n} \Phi_i(\boldsymbol{\xi})\Phi_j(\boldsymbol{\xi})\pi_{\Xi}(\boldsymbol{\xi})\mathrm{d}\boldsymbol{\xi} = \gamma_i\delta_{ij}, \tag{B7}$$

holds for multivariate polynomials. In the above equation, γ_i is the squared norm of the multivariate polynomial with linear index i. To see why the orthogonality still holds, it is easier to index the multivariate polynomials with the multi-index set. As an example to show that, let us express the expected value of the product of two second variate Hermite polynomials with multi-indices $\boldsymbol{\alpha} = (k, \ell)$ and $\boldsymbol{\beta} = (m, n)$:

$$\begin{aligned}
\mathbb{E}[\Phi_{\boldsymbol{\alpha}}(\boldsymbol{\xi})\Phi_{\boldsymbol{\beta}}(\boldsymbol{\xi})] &= \int_{\mathbb{R}}\int_{\mathbb{R}} \underbrace{\Phi_{(k,\ell)}(\xi_1,\xi_2)}_{H_k(\xi_1)H_\ell(\xi_2)}\,\underbrace{\Phi_{(m,n)}(\xi_1,\xi_2)}_{H_m(\xi_1)H_n(\xi_2)}\,\pi_{\Xi_1}(\xi_1)\pi_{\Xi_2}(\xi_2)\mathrm{d}\xi_1\mathrm{d}\xi_2 \\
&= \int_{\mathbb{R}}\int_{\mathbb{R}} H_k(\xi_1)H_\ell(\xi_2)H_m(\xi_1)H_n(\xi_2)\pi_{\Xi_1}(\xi_1)\pi_{\Xi_2}(\xi_2)\mathrm{d}\xi_1\mathrm{d}\xi_2 \\
&= \underbrace{\int_{\mathbb{R}} H_k(\xi_1)H_m(\xi_1)\pi_{\Xi_1}(\xi_1)\mathrm{d}\xi_1}_{h_k\delta_{km}}\,\underbrace{\int_{\mathbb{R}} H_l(\xi_2)H_n(\xi_2)\pi_{\Xi_2}(\xi_2)\mathrm{d}\xi_2}_{h_\ell\delta_{\ell n}} \\
&= h_k h_\ell \delta_{km}\delta_{\ell n}.
\end{aligned} \tag{B8}$$

The squared norm γ_i of the multivariate polynomial Φ_i with linear index i corresponding to the multi-index $\boldsymbol{\alpha} = (i_1, i_2, ...)$ can be computed from the product of the squared norms of the univariate polynomials that generated the multivariate polynomial, that is

$$\gamma_i = \mathbb{E}[\Phi_{\alpha}(\boldsymbol{\xi})\Phi_{\boldsymbol{\alpha}}(\boldsymbol{\xi})] = h_{\boldsymbol{\alpha}} = \prod_k h_{i_k} = i_1!i_2!.... \tag{B9}$$

To conclude, the multivariate polynomials are generated from the product of univariate polynomials. When these univariate polynomials are orthogonal, the multivariate polynomials are also orthogonal and their squared norms can be computed from the product of the squared norms of the univariate polynomials.

B.2 Calculation of the norms

The squared norms of orthogonal univariate polynomials can be computed from the A_n and C_n sequences [71]

$$h_n = \mathbb{E}[H_n^2(\xi)] = h_0 C_1 C_2 \dots C_n \frac{A_n}{A_0} = n!, \tag{B10}$$

where

$$h_0 = \mathbb{E}[H_0^2(\xi)] = \mathbb{E}[1] = 1. \tag{B11}$$

For the underlying Gaussian measure, for the Hermite polynomials this is simply $h_n = n!$. As shown above in Equation (B8), the squared norms of the multivariate Hermite polynomials can be computed from the product of squared norms of the univariate polynomials they were generated from.

B.3 Quadrature points and weights

The integration points and the weights can be also computed from the recurrence formula in Equation (B4) but by first rewriting it to a form, which generates the monic version of the polynomials

$$\phi_{n+1}(\xi) = \left(1\xi + \underbrace{\frac{B_n}{A_n}}_{\alpha_n} \right) \phi_n(\xi) - \underbrace{\frac{C_n}{A_n A_{n-1}}}_{\beta_n} \phi_{n-1}(\xi), \quad n \geq 0. \tag{B12}$$

According to the Golub-Welsch algorithm [71], the integration points and integration weights can be calculated from the eigenvalues and the eigenvectors of the Jacobi matrix

$$J = \begin{bmatrix} \alpha_0 & \sqrt{\beta_1} & \sqrt{\beta_2} & & \\ \sqrt{\beta_1} & \alpha_1 & \sqrt{\beta_1} & & \\ & \sqrt{\beta_2} & \alpha_2 & \sqrt{\beta_2} & \\ & & \ddots & \ddots & \ddots \end{bmatrix} \tag{B13}$$

from the eigenvalue problem

$$J v_i = \lambda_i v_i \quad ||v_i|| = 1. \tag{B14}$$

The integration weights are given by the first component squared of this eigenvector

$$w_i = (v_{i,1})^2 \tag{B15}$$

and the integration points by:

$$x_i = \lambda_i. \tag{B16}$$

Appendix C: Galerkin solution of the Karhunen Loève eigenfunction problem

Here the target is to find an optimal expansion for the mean-free random field $\tilde{\Theta}$ in the form

$$\tilde{\Theta}(\hat{x}, \hat{y}, \omega) = \sum_{i=1}^{\infty} \sigma_i X_i(\omega) r_i(\hat{x}, \hat{y}). \tag{C1}$$

Let's suppose, that we already have an orthogonal set of basis functions,

$$\int_{\mathcal{D}} r_i(\hat{x}, \hat{y}) r_j(\hat{x}, \hat{y}) \mathrm{d}\hat{x}\mathrm{d}\hat{y} = \gamma_i \delta_{ij} \tag{C2}$$

with squared norms γ_i

$$\gamma_i = \int_{\mathcal{D}} r_i(\hat{x}, \hat{y}) r_i(\hat{x}, \hat{y}) \mathrm{d}\hat{x}\mathrm{d}\hat{y}. \tag{C3}$$

The $\sigma_i X_i$ coefficients of the expansion (C1) can be computed by minimizing the squared error

$$\int_{\mathcal{D}} \left(\tilde{\Theta}(\hat{x}, \hat{y}, \omega) - \sum_{i=1}^{\infty} \sigma_i X_i(\omega) r_i(\hat{x}, \hat{y}) \right)^2 \mathrm{d}\hat{x}\mathrm{d}\hat{y}. \tag{C4}$$

As in the orthogonal projection used in the stochastic space for computing the PCE coefficients, we can proceed here similarly, but in the spatial domain using the inner product

$$\langle a(\hat{x}, \hat{y}), b(\hat{x}, \hat{y}) \rangle_{L_2(\mathcal{D})} = \int_{\mathcal{D}} a(\hat{x}, \hat{y}) b(\hat{x}, \hat{y}) \mathrm{d}\hat{x}\mathrm{d}\hat{y}. \tag{C5}$$

Then, the coefficient of the expansion can be computed from

$$\sigma_i X_i = \frac{\int_{\mathcal{D}} \Theta(\hat{x}, \hat{y}, \omega) r_i(\hat{x}, \hat{y}) \mathrm{d}\hat{x}\mathrm{d}\hat{y}}{\gamma_i}. \tag{C6}$$

However, we would like that we use here a special set of r_i basis functions, the one that maximises the captured variance for a certain truncation of the expansion. Thus, we determine the spatial functions by the maximisation problem

$$\sum_i \mathbb{E}[(\sigma_i X_i)^2] = \sum_i \mathbb{E}\left[\frac{\left(\int_{\mathcal{D}} \Theta(\hat{x}, \hat{y}, \omega) r_i(\hat{x}, \hat{y}) \mathrm{d}\hat{x}\mathrm{d}\hat{y}\right)^2}{\left(\int_{\mathcal{D}} r_i(\hat{x}, \hat{y}) r_i(\hat{x}, \hat{y}) \mathrm{d}\hat{x}\mathrm{d}\hat{y}\right)^2} \right] = \sum_i \sigma_i^2 \to \max. \tag{C7}$$

In the above term, the expected value of the nominator can be rewritten in terms of the covariance function

$$\mathbb{E}\left[\left(\int_{\mathcal{D}} \Theta(\hat{x},\hat{y},\omega) r_i(\hat{x},\hat{y}) \mathrm{d}\hat{x}\mathrm{d}\hat{y}\right)^2\right] = \int_{\mathcal{D}}\int_{\mathcal{D}} \underbrace{\mathbb{E}[\theta(\hat{x},\hat{y},\omega)\theta(\hat{x}',\hat{y}',\omega)]}_{=\mathrm{Cov}_{\tilde{\Theta}}} r_i(\hat{x}',\hat{y}') r_i(\hat{x},\hat{y}) \mathrm{d}\hat{x}'\mathrm{d}\hat{y}'\mathrm{d}\hat{x}\mathrm{d}\hat{y}, \tag{C8}$$

which can be shown by the following equality

$$\int_{\mathcal{D}}\int_{\mathcal{D}} \underbrace{\mathbb{E}[\theta(\hat{x},\hat{y},\omega)\theta(\hat{x}',\hat{y}',\omega)]}_{=\mathrm{Cov}_{\tilde{\Theta}}} r_i(\hat{x}',\hat{y}') r_i(\hat{x},\hat{y}) \mathrm{d}\hat{x}'\mathrm{d}\hat{y}'\mathrm{d}\hat{x}\mathrm{d}\hat{y} = \tag{C9}$$

$$= \mathbb{E}\left[\int_{\mathcal{D}} \theta(\hat{x},\hat{y},\omega) r_i(\hat{x},\hat{y}) \mathrm{d}\hat{x}\mathrm{d}\hat{y} \int_{\mathcal{D}} \theta(\hat{x}',\hat{y}',\omega) r_i(\hat{x}',\hat{y}') \mathrm{d}\hat{x}'\mathrm{d}\hat{y}'\right]. \tag{C10}$$

Then the term to be maximised is

$$\frac{\langle \int_{\mathcal{D}} \mathrm{Cov}_{\tilde{\Theta}}(\hat{x},\hat{y},\hat{x}',\hat{y}') r_i(\hat{x}',\hat{y}') \mathrm{d}\hat{x}'\mathrm{d}\hat{y}', r_i(\hat{x},\hat{y})\rangle_{L2(\mathcal{D})}}{\langle r_i(\hat{x},\hat{y}), r_i(\hat{x},\hat{y})\rangle_{L2(\mathcal{D})}}.$$

It can be shown that the maximisation problem is equivalent to solving the eigenvalue problem

$$\int_{\mathcal{D}} \mathrm{Cov}_{\tilde{\Theta}}(\hat{x},\hat{y},\hat{x}',\hat{y}') r_i(\hat{x}',\hat{y}') \mathrm{d}\hat{x}'\mathrm{d}\hat{y}' = \sigma_i^2 r_i(\hat{x},\hat{y}) \tag{C11}$$

for the $r_i(\hat{x},\hat{y})$ eigenfunctions and for the $\lambda_i = \sigma_i^2$ eigenvalues. As the $r_i(\hat{x},\hat{y})$ eigenfunctions are in an infinite dimensional space, for a numerical solution one has to discretise this on some finite dimensional space by approximating $r_i(\hat{x},\hat{y})$ by a finite linear combination of some fixed spatial basis functions $\Psi_j(\hat{x},\hat{y})$, like, for example, the FEM piecewise linear basis functions used for the FEM computation. We restrict these basis function to be some nodal basis with the Kronecker property

$$\psi_i(x_j) = \delta_{ij}. \tag{C12}$$

The approximation of the eigenfunctions reads

$$r_i(x) \approx r_{h,i}(x) = \sum_{j=1}^{n} \Psi_j(\hat{x},\hat{y}) v_{ji} = \mathbf{\Psi}\mathbf{V}, \tag{C13}$$

where $\mathbf{\Psi}$ is a row vector with the fixed spatial $\Psi_i(\hat{x},\hat{y})$ basis functions and $\mathbf{V}$ is a matrix with the v_{ji} coefficients of the expansion. The covariance can be also discretised with the help of the Ψ_i nodal basis in the form

$$\mathrm{Cov}_{\tilde{\Theta}}(\hat{x},\hat{y},\hat{x}',\hat{y}') \approx \mathrm{Cov}_{h,\tilde{\Theta}}(\hat{x},\hat{y},\hat{x}',\hat{y}') = \sum_{l,m} \underbrace{\mathrm{Cov}_{\tilde{\Theta}}(x_l,y_l,x_m,y_m)}_{=C_{lm}} \Psi_l(\hat{x},\hat{y})\Psi_m(\hat{x}',\hat{y}') \tag{C14}$$

$$= C_{lm}\Psi_l(\hat{x},\hat{y})\Psi_m(\hat{x}',\hat{y}')), \tag{C15}$$

where C_{lm} are the elements of the covariance matrix computed from the discrete spatial points. To solve the eigenproblem (C11), we formulate it in a weak form, using the Galerkin method, that is by requiring the residual

$$\int_{\mathcal{D}} \mathrm{Cov}_{h,\tilde{\Theta}}(\hat{x},\hat{y},\hat{x}',\hat{y}') r_{h,i}(\hat{x}',\hat{y}') \mathrm{d}\hat{x}' \mathrm{d}\hat{y}' - \lambda_i r_i(\hat{x},\hat{y}) \tag{C16}$$

to be orthogonal to the approximating subspace, or in other words orthogonal to all the spatial basis functions Ψ_j. This orthogonality condition reads

$$\left\langle \Psi_j(\hat{x},\hat{y}), \int_{\mathcal{D}} \mathrm{Cov}_{h,\tilde{\Theta}}(\hat{x},\hat{y},\hat{x}',\hat{y}') r_{h,i}(\hat{x}',\hat{y}') \mathrm{d}\hat{x}' \mathrm{d}\hat{y}' - \lambda_i r_{h,i}(\hat{x},\hat{y}) \right\rangle_{L_2(\mathcal{D})} = 0 \quad j = 1 \ldots n \tag{C17}$$

$$\left\langle \Psi_j(\hat{x},\hat{y}), \int_{\mathcal{D}} \mathrm{Cov}_{h,\tilde{\Theta}}(\hat{x},\hat{y},\hat{x}',\hat{y}') r_{h,i}(\hat{x}',\hat{y}') \mathrm{d}\hat{x}' \mathrm{d}\hat{y}' \right\rangle_{L_2(\mathcal{D})} = \left\langle \Psi_j(\hat{x},\hat{y}), \lambda_i r_{h,i}(\hat{x},\hat{y}) \right\rangle_{L_2(\mathcal{D})} \quad j = 1 \ldots n. \tag{C18}$$

Using the discretised approximations (8.53) and (C14), the left side of the orthogonality condition (C18) modifies to

$$\begin{aligned}
&\left\langle \Psi_j(\hat{x},\hat{y}) \int_{\mathcal{D}} \mathrm{Cov}_{h,\tilde{\Theta}}(\hat{x},\hat{y},\hat{x}',\hat{y}') r_{h,i}(\hat{x}',\hat{y}') \mathrm{d}\hat{x}' \mathrm{d}\hat{y}' \right\rangle_{L_2(\mathcal{D})} = \\
&= \int_{\mathcal{D}} \Psi_j(\hat{x},\hat{y}) \int_{\mathcal{D}} \mathrm{Cov}_{h,\tilde{\Theta}}(\hat{x},\hat{y},\hat{x}',\hat{y}') r_{h,i}(\hat{x}',\hat{y}') \mathrm{d}\hat{x}' \mathrm{d}\hat{y}' \mathrm{d}\hat{x} \mathrm{d}\hat{y} \\
&= \int_{\mathcal{D}} \int_{\mathcal{D}} \Psi_j(\hat{x},\hat{y}) \mathrm{Cov}_{h,\tilde{\Theta}}(\hat{x},\hat{y},\hat{x}',\hat{y}') r_{h,i}(\hat{x}',\hat{y}') \mathrm{d}\hat{x}' \mathrm{d}\hat{y}' \mathrm{d}\hat{x} \mathrm{d}\hat{y} \\
&= \int_{\mathcal{D}} \int_{\mathcal{D}} \Psi_j(\hat{x},\hat{y}) \sum_{l,m} C_{lm} \Psi_l(\hat{x},\hat{y}) \Psi_m(\hat{x}',\hat{y}') \sum_k v_{ki} \Psi_k(\hat{x}',\hat{y}') \mathrm{d}\hat{x}' \mathrm{d}\hat{y}' \mathrm{d}\hat{x} \mathrm{d}\hat{y} \\
&= \sum_{l,m,k} C_{lm} v_{ki} \int_{\mathcal{D}} \int_{\mathcal{D}} \Psi_j(\hat{x},\hat{y}) \Psi_l(\hat{x},\hat{y}) \Psi_m(\hat{x}',\hat{y}')) \Psi_k(\hat{x}',\hat{y}') \mathrm{d}\hat{x}' \mathrm{d}\hat{y}' \mathrm{d}\hat{x} \mathrm{d}\hat{y} \\
&= \sum_{l,m,k} C_{lm} v_{ki} \underbrace{\int_{\mathcal{D}} \Psi_j(\hat{x},\hat{y}) \Psi_l(\hat{x},\hat{y}) \mathrm{d}\hat{x} \mathrm{d}\hat{y}}_{G_{jl}} \underbrace{\int_{\mathcal{D}} \Psi_m(\hat{x}',\hat{y}')) \Psi_k(\hat{x}',\hat{y}') \mathrm{d}\hat{x}' \mathrm{d}\hat{y}'}_{G_{mk}} \\
&= \sum_{l,m,k} C_{lm} v_{ki} G_{jl} G_{mk} = \sum G_{jl} C_{lm} G_{mk} v_{ki} \\
&= \mathbf{GCGv}_i,
\end{aligned} \tag{C19}$$

where $\mathbf{C}$ is the covariance matrix and $\mathbf{G}$ is the Gramian matrix of the Ψ_i nodal basis and $\mathbf{v}_i$ is the ith column of the $\mathbf{V}$ matrix. Using the approximation (8.53), the right hand side of the Equation (C18) reads

$$\begin{aligned}\left\langle \Psi_j(\hat{x},\hat{y}), \lambda_i r_{h,i}(\hat{x},\hat{y}) \right\rangle_{L_2(\mathcal{D})} &= \int_{\mathcal{D}} \Psi_j(\hat{x},\hat{y})\lambda_i \sum_k v_{ik}\Psi_k(\hat{x},\hat{y})\mathrm{d}\hat{x}\mathrm{d}\hat{y} \\ &= \lambda_i \sum_k v_{ik} \underbrace{\int_{\mathcal{D}} \Psi_j(\hat{x},\hat{y})\Psi_k(\hat{x},\hat{y})\mathrm{d}\hat{x}\mathrm{d}\hat{y}}_{G_{jk}} \\ &= \lambda_i \mathbf{G}\mathbf{v}_i. \end{aligned} \tag{C20}$$

Using the above two forms, the discretised weak form of the eigenvalue problem (8.52) boils down to the discrete generalized eigenvalue problem

$$\mathbf{GCG}\mathbf{v}_i = \underbrace{\lambda_i}_{\sigma_i^2} \mathbf{G}\mathbf{v}_i. \tag{C21}$$

Appendix D: Computation of the PCE Coefficients by Orthogonal projection

In the following, we show how to approximate the response $\mathbf{y}$, a function of some random variable $\boldsymbol{\xi}$ with a linear combination of the orthogonal polynomials $\Phi_1(\boldsymbol{\xi}), \Phi_2(\boldsymbol{\xi}), ...$, such that the approximation $\mathbf{y}_h$ minimizes

$$||\boldsymbol{\epsilon}_h||^2_{L_2(\Omega)} = \mathbb{E}\left[\boldsymbol{\epsilon}_h^2\right] = \mathbb{E}\left[(\mathbf{y} - \mathbf{y}_h)^2\right] = \int_{\mathbb{R}^s} (\mathbf{y}(\boldsymbol{\xi}) - \mathbf{y}_h(\boldsymbol{\xi}))^2 \pi_\Xi(\boldsymbol{\xi}) \mathrm{d}\boldsymbol{\xi}. \tag{D1}$$

The above quadratic form is induced by the norm generated from the inner product defined as

$$\langle u, v \rangle_{L_2(\Omega)} = \mathbb{E}[uv] = \int_{\mathbb{R}^s} uv\pi_\Xi \mathrm{d}\boldsymbol{\xi}, \tag{D2}$$

thus, the norm can be written with the help of this inner product

$$||u||_{L_2(\Omega)} = \sqrt{\langle u, u \rangle_{L_2(\Omega)}} \tag{D3}$$

and the value to be minimized is the squared induced norm, that is,

$$\mathbb{E}\left[\boldsymbol{\epsilon}_h^2\right] = \langle \boldsymbol{\epsilon}_h, \boldsymbol{\epsilon}_h \rangle. \tag{D4}$$

The task is to approximate the function $\mathbf{y}$, lying in some infinite dimensional space, in a subspace, a finite dimensional space spanned by the given orthogonal polynomials. This problem is similar to approximating let's say a three-dimensional vector in a plane. The solution to this later task is to project orthogonally the vector to the approximating plane, which will minimize the projection error, the distance between the original vector and the projected one. Similarly, in the functional space, one can do the same thing, only we need to define a proper inner product, with which we can define orthogonality. By orthogonally projecting to the subspace, the norm generated from the inner product of the error is minimized.

Returning to our original task, it is easy to show that the minimum of the quadratic form is attained when the approximation error is orthogonal to each of the stochastic polynomials. This orthogonality is written by

$$\langle (\mathbf{y} - \mathbf{y}_h), \Phi_j \rangle = 0 \quad \forall j = 0 \ldots (M-1). \tag{D5}$$

Plugging in the approximation (8.28) into this orthogonality equation, rearranging it and using the linearity of inner products and the orthogonality of the PCE polynomials in Equation (D5), the $\boldsymbol{v}_i$ coefficients can be expressed

$$\langle \left(\mathbf{y} - \sum_{i=0}^{M-1} \boldsymbol{v}_i \Phi_i \right), \Phi_j \rangle = 0 \quad \forall j = 0 \ldots M-1 \tag{D6}$$

$$\langle \mathbf{y}, \Phi_j \rangle = \sum_{i=0}^{M-1} \boldsymbol{v}_i \underbrace{\langle \Phi_i, \Phi_j \rangle}_{\gamma_i \delta_{ij}} \quad \forall j = 0 \ldots M-1 \tag{D7}$$

$$\boldsymbol{v}_i = \frac{1}{\gamma_i} \langle \mathbf{y}, \Phi_i \rangle = \frac{1}{\gamma_i} \mathbb{E}[\mathbf{y}\Phi_i] = \frac{1}{\gamma_i} \int_{\mathbb{R}^s} \mathbf{y}(\boldsymbol{\xi}) \Phi_i(\boldsymbol{\xi}) \pi_\Xi(\boldsymbol{\xi}) \mathrm{d}\boldsymbol{\xi}. \tag{D8}$$

Bibliography

[1] J.D. Achenbach. *Wave Propagation in Elastic Solids.* North-Holland Publishing Company/ American Elsevier, 1973.

[2] D.E. Acuña-Ureta and M.E. Orchard. Conditional predictive Bayesian Cramér-Rao lower bounds for prognostic algorithm design. *Applied Soft Computing*, 72: 647–665, 2018.

[3] D.E. Acuña-Ureta, M.E. Orchard and P. Wheeler. Computation of time probability distributions for the occurrence of uncertain future events. *Mechanical Systems and Signal Processing*, 150, 2021.

[4] J. Aldrich. R.A. Fisher and the making of maximum likelihood 1912–1922. *Statistical Science*, 12(3): 162–176, 1997.

[5] J. Aldrich. R.A. Fisher on Bayes and Bayes theorem. *Bayesian Analysis*, 3(1): 161–170, 2008.

[6] D. Alleyne and P. Cawley. A two-dimensional fourier transform method for the measurement of propagating multimode signals. *The Journal of the Acoustical Society of America*, 89(3): 1159–1168, 1991.

[7] C. Andrieu and J. Thoms. A tutorial on adaptive MCMC. *Statistics and Computing*, 18(4): 343–373, 2008.

[8] C. Argyris, S. Chowdhury, V. Zabel and C. Papadimitriou. Bayesian optimal sensor placement for crack identification in structures using strain measurements. *Structural Control and Health Monitoring*, 25(5): e2137, 2018.

[9] M.S. Arulampalam, S. Maskell, N. Gordon and T. Clapp. A tutorial on particle filters for online nonlinear/non-Gaussian Bayesian tracking. *IEEE Transactions on Signal Processing*, 50(2): 174–188, 2002.

[10] S.-K. Au. Assembling mode shapes by least squares. *Mechanical Systems and Signal Processing*, 25(1): 163–179, 2011.

[11] S.-K. Au. Fast Bayesian FFT method for ambient modal identification with separated modes. *Journal of Engineering Mechanics*, 137(3): 214–226, 2011.

[12] S.-K. Au. Fast Bayesian ambient modal identification in the frequency domain, Part i: Posterior most probable value. *Mechanical Systems and Signal Processing*, 26: 60–75, 2012.

[13] S.-K. Au. Fast Bayesian ambient modal identification in the frequency domain, Part ii: Posterior uncertainty. *Mechanical Systems and Signal Processing*, 26: 76–90, 2012.

[14] S.-K. Au and J.L. Beck. Estimation of small failure probabilities in high dimensions by subset simulation. *Probabilistic Engineering Mechanics*, 16(4): 263–277, 2001.

[15] S.-K. Au and F.-L. Zhang. Fundamental two-stage formulation for Bayesian system identification, Part i: General theory. *Mechanical Systems and Signal Processing*, 66: 31–42, 2016.

[16] S.-K. Au and J.L. Beck. Subset Simulation and its application to seismic risk based on dynamic analysis. *Journal of Engineering Mechanics*, 129(8): 901–917, 2003.

[17] S.-K. Au, J. Ching and J.L. Beck. Application of Subset Simulation methods to reliability benchmark problems. *Structural Safety*, 29(3): 183–193, 2007.

[18] H.R. Bae, R.V. Grandhi and R.A. Canfield. An approximation approach for uncertainty quantification using evidence theory. *Reliability Engineering & System Safety*, 86(3): 215–225, 2004.

[19] D. Balageas, C.-P. Fritzen and A. Güemes. *Structural Health Monitoring*, volume 90. John Wiley & Sons, 2010.

[20] C. Bao, H. Hao, Z.-X. Li and X. Zhu. Time-varying system identification using a newly improved HHT algorithm. *Computers & Structures*, 87(23-24): 1611–1623, 2009.

[21] T. Bayes and M. Price. An essay towards solving a problem in the Doctrine of Chances. by the late Rev. Mr. Bayes, FRS communicated by Mr. Price, in a letter to John Canton, AMFRS. *Philosophical Transactions (1683–1775)*, pages 370–418, 1763.

[22] J.L. Beck. Bayesian system identification based on probability logic. *Structural Control and Health Monitoring*, 17(7): 825–847, 2010.

[23] J.L. Beck, S.-K. Au and M.W. Vanik. Monitoring structural health using a probabilistic measure. *Computer-Aided Civil and Infrastructure Engineering*, 16(1): 1–11, 2001.

[24] J.L. Beck, C. Papadimitriou, S.-K. Au and M.W. Vanik. Entropy-based optimal sensor location for structural damage detection. In *Smart Structures and Materials 1998: Smart Systems for Bridges, Structures, and Highways*, volume 3325, pages 161–173. International Society for Optics and Photonics, 1998.

[25] J.L. Beck and K.M Zuev. Asymptotically independent Markov sampling: a new Markov chain Monte Carlo scheme for Bayesian inference. *International Journal for Uncertainty Quantification*, 3(5), 2013.

[26] J.L. Beck and L.S. Katafygiotis. Updating models and their uncertainties. I: Bayesian statistical framework. *Journal of Engineering Mechanics*, 124(4): 455–462, 1998.

[27] J.L. Beck and K.V. Yuen. Model selection using response measurements: Bayesian probabilistic approach. *Journal of Engineering Mechanics*, 130: 192, 2004.

[28] P. Bortot, S.G. Coles and S.A. Sisson. Inference for stereological extremes. *Journal of the American Statistical Association*, 102(477): 84–92, 2007.

[29] K.P. Burnham and D.R. Anderson. *Model Selection and Multi-Model Inference: A Practical Information-Theoretic Approach.* Springer, 2002.

[30] D. Calvetti and E. Somersalo. *Introduction to Bayesian Scientific Computing: Ten Lectures on Subjective Computing*, volume 2. Springer, 2007.

[31] J. Candy. *Bayesian Signal Processing: Classical, Modern and Particle Filtering Methods*. Wiley, 2009.

[32] S. Cantero-Chinchilla, G. Aranguren, M.K. Malik, J. Etxaniz and F.M. de la Escalera. An empirical study on transmission beamforming for ultrasonic guided-wave based structural health monitoring. *Sensors*, 20(5): 1445, 2020.

[33] S. Cantero-Chinchilla, J.L. Beck, M. Chiachío, J. Chiachío, D. Chronopoulos and A. Jones. Optimal sensor and actuator placement for structural health monitoring via an efficient convex cost-benefit optimization. *Mechanical Systems and Signal Processing*, 144: 106901, 2020.

[34] S. Cantero-Chinchilla, J. Chiachío, M. Chiachío, D. Chronopoulos and A. Jones. A robust Bayesian methodology for damage localization in plate-like structures using ultrasonic guided-waves. *Mechanical Systems and Signal Processing*, 122: 192–205, 2019.

[35] S. Cantero-Chinchilla, J. Chiachío, M. Chiachío, D. Chronopoulos and A. Jones. Optimal sensor configuration for ultrasonic guided-wave inspection based on value of information. *Mechanical Systems and Signal Processing*, 135: 106377, 2020.

[36] B.P. Carlin and T.A. Louis. *Bayesian Methods for Data Analysis*. Chapman and Hall/CRC, 3 Edition.

[37] J.R. Celaya, A. Saxena and K. Goebel. Uncertainty representation and interpretation in model-based prognostics algorithms based on Kalman filter estimation. In *Proceedings of the Annual Conference of the Prognostics and Health Management Society, 2012*, pages 23–27, 2012.

[38] C.C. Chang and C.W. Poon. Nonlinear identification of lumped-mass buildings using empirical mode decomposition and incomplete measurement. *Journal of Engineering Mechanics*, 136(3): 273–281, 2010.

[39] E.N. Chatzi and C. Papadimitriou. *Identification Methods for Structural Health Monitoring*, volume 567. Springer, 2016.

[40] S.H. Cheung and J.L. Beck. Calculation of posterior probabilities for Bayesian model class assessment and averaging from posterior samples based on dynamic system data. *Computer-Aided Civil and Infrastructure Engineering*, 25(5): 304–321, 2010.

[41] J. Chiachío, N. Bochud, M. Chiachío, S. Cantero and G. Rus. A multilevel Bayesian method for ultrasound-based damage identification in composite laminates. *Mechanical Systems and Signal Processing*, 88: 462–477, 2017.

[42] J. Chiachío, M. Chiachío, S. Sankararaman, A. Saxena and K. Goebel. Condition-based prediction of time-dependent reliability in composites. *Reliability Engineering & System Safety*, 142: 134–147, 2015.

[43] M. Chiachío, J. Chiachío, G. Rus and J.L. Beck. Predicting fatigue damage in composites: A Bayesian framework. *Structural Safety*, 51: 57–68, 2014.

[44] J. Chiachío, M. Chiachío, S. Sankararaman, A. Saxena and K. Goebel. Prognostics design for structural health management. In *Emerging Design Solutions in Structural Health Monitoring Systems*, Advances in Civil and Industrial Engineering Series (ACIE). IGI Global, 2014.

[45] J. Chiachío, M. Chiachío, A. Saxena, G. Rus and K. Goebel. An energy-based prognostics framework to predict fatigue damage evolution in composites. In *Proceedings of the Annual Conference of the Prognostics and Health Management Society, New Orleans LA, October 14–17, 2013*, volume 1, pages 363–371. Prognostics and Health Management Society, 2013.

[46] M. Chiachío, J.L. Beck, J. Chiachío and G. Rus. Approximate Bayesian computation by Subset Simulation. *SIAM Journal on Scientific Computing*, 36(3): 1339–1339, 2014.

[47] J. Ching, S.K. Au and J.L. Beck. Reliability estimation of dynamical systems subject to stochastic excitation using Subset Simulation with splitting. *Computer Methods in Applied Mechanics and Engineering*, 194(12-16): 1557–1579, 2005.

[48] J. Ching and J.L. Beck. New Bayesian model updating algorithm applied to a structural health monitoring benchmark. *Structural Health Monitoring*, 3(4): 313–332, 2004.

[49] J. Ching, M. Muto and J.L. Beck. Structural model updating and health monitoring with incomplete modal data using Gibbs sampler. *Computer-Aided Civil and Infrastructure Engineering*, 21(4): 242–257, 2006.

[50] C.K. Chui. *"An Introduction to Wavelets"*. Academic Press Professional, Inc., San Diego, CA, USA, 1992.

[51] L. Cohen. *"Time-frequency Analysis: Theory and Applications"*. Prentice-Hall, Inc., Upper Saddle River, NJ, USA, 1995.

[52] R.T. Cox. Probability, frequency and reasonable expectation. *American Journal of Physics*, (17): 1–13, 1946.

[53] M. Daigle, A. Saxena and K. Goebel. An efficient deterministic approach to model-based prediction uncertainty estimation. In *Proceedings of the Annual Conference of the Prognostics and Health Management Society, 2012*, pages 326–335, 2012.

[54] B. De Finetti. *Theory of Probability (Two Volumes)*. John Wiley & Sons, 1974.

[55] P. Del Moral, A. Doucet and A. Jasra. Sequential Monte Carlo samplers. *Journal of the Royal Statistical Society: Series B (Statistical Methodology)*, 68(3): 411–436, 2006.

[56] A. Der Kiureghian and T. Dakessian. Multiple design points in first and second-order reliability. *Structural Safety*, 20(1): 37–49, 1998.

[57] A. Der Kiureghian, H.Z. Lin and S.J. Hwang. Second-order reliability approximations. *Journal of Engineering Mechanics*, 113: 1208, 1987.

[58] A. Doucet, N. De Freitas and N. Gordon. *Sequential Monte Carlo Methods in Practice*. Springer Verlag, 2001.

[59] A. Doucet, S. J. Godsill and C. Andrieu. On sequential Monte Carlo sampling methods for Bayesian filtering. *Statistics and Computing*, 10(3): 197–208, 2000.

[60] A. Doucet, N. De Freitas and N. Gordon. An introduction to sequential Monte Carlo methods. In Arnaud Doucet, Nando De Freitas, and Neil Gordon, editors, *Sequential Monte Carlo Methods in Practice*, pages 3–14. Springer, 2001.

[61] D. Dubois, H. Prade and E.F. Harding. *Possibility Theory: An Approach to Computerized Processing of Uncertainty*, volume 2. Plenum Press New York, 1988.

[62] A.W.F. Edwards. *Likelihood*. Cambridge Univ. Pr., 1984.

[63] S.J. Engel, B.J. Gilmartin, K. Bongort and A. Hess. Prognostics, the real issues involved with predicting life remaining. In *Aerospace Conference Proceedings, 2000 IEEE*, volume 6, pages 457–469. IEEE, 2000.

[64] G. Evensen. *Data Assimilation: The Ensemble Kalman Filter*. Springer-Verlag, Berlin, Heidelberg, 2009.

[65] P. Fearnhead and D. Prangle. Constructing summary statistics for approximate Bayesian computation: semi-automatic approximate Bayesian computation. *Journal of the Royal Statistical Society, Series B*, 74(3): 419–474, 2012.

[66] S.E. Fienberg. When did Bayesian inference become Bayesian? *Bayesian Analysis*, 1(1): 1–40, 2006.

[67] R.A. Fisher. On an absolute criterion for fitting frequency curves. *Messenger of Mathematics*, 41(1): 155–160, 1912.

[68] E.B. Flynn, M.D. Todd, P.D. Wilcox, B.W. Drinkwater and A.J. Croxford. Maximum-likelihood estimation of damage location in guided-wave structural health monitoring. In *Proceedings of the Royal Society of London A: Mathematical, Physical and Engineering Sciences*, volume 467, pages 2575–2596. The Royal Society, 2011.

[69] A.A. Fraenkel, Y. Bar-Hillel and A. Levy. *Foundations of Set Theory*. Elsevier, 1973.

[70] G. Gambolati, M. Ferronato, P. Teatini, R. Deidda and G. Lecca. Finite element analysis of land subsidence above depleted reservoirs with pore pressure gradient and total stress formulations. *International Journal for Numerical and Analytical Methods in Geomechanics*, 25(4): 307–327, 2001.

[71] W. Gautschi. *Orthogonal Polynomials: Computation and Approximation*. Oxford University Press.

[72] A.E. Gelfand and D.K. Dey. Bayesian model choice: asymptotics and exact calculations. *Journal of the Royal Statistical Society. Series B (Methodological)*, 56(3): 501–514, 1994.

[73] A. Gelman, X.L. Meng and H. Stern. Posterior predictive assessment of model fitness via realized discrepancies. *Statistica Sinica*, 6: 733–759, 1996.

[74] A. Gelman, J.B. Carlin, H.S. Stern and D.B. Rubin. *Bayesian Data Analysis, Second Edition*. Chapman and Hall/CRC.

[75] R.G. Ghanem and P.D. Spanos. *Stochastic Finite Elements: A Spectral Approach*. Springer New York.

[76] W.R. Gilks, S. Richardson and D.J. Spiegelhalter. *Markov Chain Monte Carlo in Practice.* Chapman and Hall, 1996.

[77] D.E. Goldberg. *Genetic Algorithms.* Pearson Education India, 2006.

[78] N.R. Goodman. Statistical analysis based on a certain multivariate complex Gaussian distribution (an introduction). *The Annals of Mathematical Statistics*, 34(1): 152–177, 1963.

[79] N.J. Gordon, D.J. Salmond and A.F.M. Smith. Novel approach to nonlinear/non-Gaussian Bayesian state estimation. *IEEE-Proceedings-F*, 140: 107–113, 1993.

[80] M. Grigoriu. *Stochastic Calculus: Applications in Science and Engineering.* Birkhäuser, 2002 Edition.

[81] S.F. Gull. Bayesian inductive inference and maximum entropy. In *Maximum-entropy and Bayesian Methods in Science and Engineering*, pages 53–74. Springer, 1988.

[82] S.F. Gull. *Developments in Maximum Entropy Data Analysis*, pages 53–71. Springer Netherlands, Dordrecht, 1989.

[83] H.H., E. Saksman and J. Tamminen. An adaptive Metropolis algorithm. *Bernoulli*, 7(2): 223–242, 2001.

[84] J. Hadamard. Sur les problemes aux dérivées partielles et leur signification physique. *Princeton University Bulletin*, 13(49-52): 28, 1902.

[85] A. Haldar and S. Mahadevan. *Probability, Reliability, and Statistical Methods in Engineering Design.* John Wiley & Sons, Inc., 2000.

[86] A. Haldar and S. Mahadevan. *Reliability Assessment Using Stochastic Finite Element Analysis.* New York: John Wiley & Sons, 2000.

[87] U.D. Hanebeck, K. Briechle and A. Rauh. Progressive Bayes: A new framework for nonlinear state estimation. *Proc. of the SPIE AeroSense Symposium 2003, 5099, Orlando, USA*, 2003.

[88] U.D. Hanebeck, M.F. Huber and V. Klumpp. Dirac mixture approximation of multivariate Gaussian densities. *Proceedings of the 48th IEEE Conference on Decision and Control, 2009, held jointly with the 2009 28th Chinese Control Conference. CDC/CCC 2009.*, pages 3851–3858, 2009.

[89] N. Hansen, S.D. Müller and P. Koumoutsakos. Reducing the time complexity of the derandomized evolution strategy with covariance matrix adaptation (CMA-ES). *Evolutionary Computation*, 11(1): 1–18, 2003.

[90] W.K. Hastings. Monte Carlo sampling methods using Markov chains and their applications. *Biometrika*, 57(1): 97–109, 1970.

[91] R.A. Howard. Information value theory. *IEEE Transactions on Systems Science and Cybernetics*, 2(1): 22–26, 1966.

[92] X. Huan and Y.M. Marzouk. Simulation-based optimal Bayesian experimental design for nonlinear systems. *Journal of Computational Physics*, 232(1): 288–317, 2013.

[93] N.E. Huang, Z. Shen, S.R. Long, M.C. Wu, H.H. Shih, Q. Zheng, N.-C. Yen, C. Tung and H.H. Liu. The empirical mode decomposition and the Hilbert spectrum for nonlinear and non-stationary time series analysis. In *Proceedings of the Royal Society of London A: Mathematical, Physical and Engineering Sciences*, volume 454, pages 903–995. The Royal Society, 1998.

[94] Y. Huang and J.L. Beck. Hierarchical sparse Bayesian learning for strucutral health monitoring with incomplete modal data. *International Journal for Uncertainty Quantification*, 5(2): 139–169, 2015.

[95] Y. Huang, J.L. Beck and H. Li. Hierarchical sparse Bayesian learning for structural damage detection: theory, computation and application. *Structural Safety*, 64: 37–53, 2017.

[96] Y. Huang, J.L. Beck, S. Wu and H. Li. Robust Bayesian compressive sensing for signals in structural health monitoring. *Computer-Aided Civil and Infrastructure Engineering*, 29(3): 160–179, 2014.

[97] J.-B. Ihn and F.-K Chang. Pitch-catch active sensing methods in structural health monitoring for aircraft structures. *Structural Health Monitoring*, 7(1): 5–19, 2008.

[98] C. Janna, N. Castelletto, M. Ferronato, G. Gambolati and P. Teatini. A geomechanical transversely isotropic model of the Po River basin using PSInSAR derived horizontal displacement. *International Journal of Rock Mechanics and Mining Sciences*, 51: 105–118, 2012.

[99] E.T. Jaynes. *Papers on Probability, Statistics and Statistical Physics.* (Ed. R.D. Rosenkrantz) Kluwer Academic Publishers, 1983.

[100] E.T. Jaynes. Information theory and statistical mechanics. *Physical Review*, 106(4): 620–630, 1957.

[101] E.T. Jaynes. The well-posed problem. *Foundations of Physics*, 3(4): 477–492, 1973.

[102] E.T. Jaynes. *Probability Theory: The Logic of Science.* Cambridge University Press, 2003.

[103] H. Jeffreys. *Theory of Probability.* Oxford University Press, USA, 1998.

[104] H. Jo, S.-H. Sim, T. Nagayama and B.F. Spencer Jr. Development and application of high-sensitivity wireless smart sensors for decentralized stochastic modal identification. *Journal of Engineering Mechanics*, 138(6): 683–694, 2012.

[105] S.J. Julier. The scaled unscented transformation. *Proc. Am. Control Conf. (ACC 02)*, pages 4555–4559, 2002.

[106] S.J. Julier and J.K. Uhlmann. Unscented filtering and nonlinear estimation. *Proceedings of the IEEE*, 92(3): 401–422, 2004.

[107] R.E. Kalman. A new approach to linear filtering and prediction problems. *Journal of Basic Engineering*, 82(1): 35–45, 1960.

[108] L.S. Katafygiotis and J.L. Beck. Updating models and their uncertainties. II: Model identifiability. *Journal of Engineering Mechanics*, 124(4): 463–467, 1998.

[109] L.S. Katafygiotis and K.V. Yuen. Bayesian spectral density approach for modal updating using ambient data. *Earthquake Engineering & Structural Dynamics*, 30(8): 1103–1123, 2001.

[110] L.S. Katafygiotis and H.F. Lam. Tangential projection algorithm for manifold representation in unidentifiable model updating problems. *Earthquake Engineering & Structural Dynamics*, 31(4): 791–812, 2002.

[111] G. Kerschen, K. Worden, A.F. Vakakis and J.-C. Golinval. Past, present and future of nonlinear system identification in structural dynamics. *Mechanical Systems and Signal Processing*, 20(3): 505–592, 2006.

[112] A.N. Kolmogorov. *Foundations of the Theory of Probability.* Chelsea Pub. Co, 2 Edition.

[113] K. Konakli, B. Sudret and M.H. Faber. Numerical investigations into the value of information in lifecycle analysis of structural systems. *ASCE-ASME Journal of Risk and Uncertainty in Engineering Systems, Part A: Civil Engineering*, 2(3): B4015007, 2015.

[114] A. Kong, J.S. Liu and W.H Wong. Sequential imputations and Bayesian missing data problems. *Journal of the American Statistical Association*, 89(425): 278–288, 1994.

[115] A. Krause, C. Guestrin, A. Gupta and J. Kleinberg. Near-optimal sensor placements: Maximizing information while minimizing communication cost. In *Proceedings of the 5th International Conference on Information Processing in Sensor Networks*, pages 2–10. ACM, 2006.

[116] A. Krause, A. Singh and C. Guestrin. Near-optimal sensor placements in Gaussian processes: Theory, efficient algorithms and empirical studies. *Journal of Machine Learning Research*, 9: 235–284, 2008.

[117] P.M. Lee. *Bayesian Statistics.* Arnold London, UK, 2004.

[118] T. Leonard and J.S.J. Hsu. Bayesian Methods. *Cambridge Books*, 2001.

[119] M. Li and S. Azarm. Multiobjective collaborative robust optimization with interval uncertainty and interdisciplinary uncertainty propagation. *Journal of Mechanical Design*, 130, 2008.

[120] T. Li, S. Sun, T.P. Sattar and J.M. Corchado. Fight sample degeneracy and impoverishment in particle filters: A review of intelligent approaches. *Expert Systems With Applications*, 41(8): 3944–3954, 2014.

[121] F. Liang, C. Liu and J. Chuanhai. *Advanced Markov chain Monte Carlo methods.* Wiley Online Library, 2010.

[122] D.V. Lindley. On a measure of the information provided by an experiment. *The Annals of Mathematical Statistics*, pages 986–1005, 1956.

[123] D.G. Luenberger. *Optimization by Vector Space Methods.* Wiley-Interscience, 1969.

[124] B.R. Mace, D. Duhamel, M.J. Brennan and L. Hinke. Finite element prediction of wave motion in structural waveguides. *The Journal of the Acoustical Society of America*, 117(5): 2835–2843, 2005.

[125] D.J.C. MacKay. Bayesian interpolation. *Neural Computation*, 4(3): 415–447, 1992.

[126] Z. Mao. *Uncertainty Quantification in Vibration-Based Structural Health Monitoring for Enhanced Decision-Making Capability*. PhD thesis, UC San Diego, 2012.

[127] J.M. Marin, P. Pudlo, C.P. Robert and R.J. Ryder. Approximate Bayesian computational methods. *Statistics and Computing*, 22(6): 1167–1180, 2012.

[128] P. Marjoram, J. Molitor, V. Plagnol and S. Tavaré. Markov chain Monte Carlo without likelihoods. *Proceedings of the National Academy of Sciences of the United States of America*, 100(26): 15324–15328, 2003.

[129] A.M. Mathai. Moments of the trace of a noncentral Wishart matrix. *Communications in Statistics-Theory and Methods*, 9(8): 795–801, 1980.

[130] H.G. Matthies, E. Zander, B.V. Rosić, A. Litvinenko and O. Pajonk. Inverse Problems in a Bayesian Setting. In *Computational Methods for Solids and Fluids*, Computational Methods in Applied Sciences, pages 245–286. Springer, Cham.

[131] N. Metropolis, A.W. Rosenbluth, M.N. Rosenbluth, A.H. Teller and E. Teller. Equation of state calculations by fast computing machines. *The Journal of Chemical Physics*, 21: 1087–1092, 1953.

[132] M. Muto and J.L. Beck. Bayesian updating and model class selection for hysteretic structural models using stochastic simulation. *Journal of Vibration and Control*, 14(1-2): 7–34, 2008.

[133] E. Myötyri, U. Pulkkinen and K. Simola. Application of stochastic filtering for lifetime prediction. *Reliability Engineering & System Safety*, 91(2): 200–208, 2006.

[134] R.M. Neal. Probabilistic inference using Markov chain Monte Carlo methods. Technical Report CRG TR 93 1, Department of Computer Science, University of Toronto, 1993.

[135] M.A. Newton and A.E. Raftery. Approximate Bayesian inference with the weighted likelihood bootstrap. *Journal of the Royal Statistical Society. Series B (Methodological)*, 56(1): 3–48, 1994.

[136] M. Orchard and G. Vachtsevanos. A particle-filtering approach for on-line fault diagnosis and failure prognosis. *Transactions of the Institute of Measurement and Control*, 31: 221–246, 2009.

[137] M. Orchard, G. Kacprzynski, K. Goebel, B. Saha and G. Vachtsevanos. Advances in uncertainty representation and management for particle filtering applied to prognostics. In *Proceedings of the IEEE International Conference on Prognostics and Health Management*. IEEE, 2008.

[138] W. Ostachowicz, R. Soman and P. Malinowski. Optimization of sensor placement for structural health monitoring: a review. *Structural Health Monitoring*, 18(3): 963–988, 2019.

[139] C. Papadimitriou, J.L. Beck and S.K. Au. Entropy-based optimal sensor location for structural model updating. *Journal of Vibration and Control*, 6(12): 89–110, 2000.

[140] C. Papadimitriou, J.L. Beck and L.S Katafygiotis. Asymptotic expansions for reliability and moments of uncertain systems. *Journal of Engineering Mechanics*, 123(12): 89–110, 1997.

[141] C. Papadimitriou. Optimal sensor placement methodology for parametric identification of structural systems. *Journal of Sound and Vibration*, 278(4-5): 923–947, 2004.

[142] C. Papadimitriou, J.L. Beck and S.K. Au. Entropy-based optimal sensor location for structural model updating. *Journal of Vibration and Control*, 6(5): 781–800, 2000.

[143] C. Papadimitriou, J.L. Beck and L.S. Katafygiotis. Updating robust reliability using structural test data. *Probabilistic Engineering Mechanics*, 16(2): 103–113, 2001.

[144] C. Papadimitriou and G. Lombaert. The effect of prediction error correlation on optimal sensor placement in structural dynamics. *Mechanical Systems and Signal Processing*, 28: 105–127, 2012.

[145] C. Papadimitriou and D.-C. Papadioti. Component mode synthesis techniques for finite element model updating. *Computers & Structures*, 126: 15–28, 2013.

[146] K.R. Popper. The propensity interpretation of probability. *The British Journal for the Philosophy of Science*, 10(37): 25–42, 1959.

[147] M.M. Rao and R.J. Swift. *Probability Theory with Applications.* Mathematics and Its Applications. Springer, 2nd Edition.

[148] S.S. Rao and K.K. Annamdas. Dempster-Shafer theory in the analysis and design of uncertain engineering systems. *Product Research*, pages 135–160, 2009.

[149] A. Rauh, K. Briechle and U.D. Hanebeck. Nonlinear measurement update and prediction: Prior density splitting mixture estimator. *2009 IEEE Control Applications, (CCA) & Intelligent Control, (ISIC), St. Petersburg*, pages 1421–1426, 2009.

[150] C.P. Robert and G. Casella. *Introducing Monte Carlo Methods with R.* Springer, 2010.

[151] C.P. Robert and G. Casella. *Monte Carlo Statistical Methods, 2nd Ed.* Springer-Verlag, New York, 2004.

[152] G.O. Roberts and J.S. Rosenthal. Optimal scaling for various Metropolis-Hastings algorithms. *Statistical Science*, 16(4): 351–367, 2001.

[153] B.V. Rosić, A. Litvinenko, O. Pajonk and H.G. Matthies. Sampling-free linear Bayesian update of polynomial chaos representations. *Journal of Computational Physics*, 231(17): 5761–5787, 2012.

[154] G. Rus, J. Chiachío and M. Chiachío. Logical inference for inverse problems. *Inverse Problems in Science and Engineering*, 24(3): 448–464, 2016.

[155] K.J. Ryan. Estimating expected information gains for experimental designs with application to the random fatigue-limit model. *Journal of Computational and Graphical Statistics*, 12(3): 585–603, 2003.

[156] B. Saha and K. Goebel. Uncertainty management for diagnostics and prognostics of batteries using Bayesian techniques. In *Aerospace Conference, 2008 IEEE*, pages 1–8. IEEE, 2008.

[157] S. Sankararaman. *Uncertainty Quantification and Integration in Engineering Systems.* PhD thesis, Vanderbilt University, 2012.

[158] S. Sankararaman, M. Daigle and K. Goebel. Uncertainty quantification in remaining useful life prediction using first-order reliability methods. *IEEE Transactions on Reliability*, 63(2): 603–619, 2014.

[159] S. Sankararaman and K. Goebel. Remaining useful life estimation in prognosis: An uncertainty propagation problem. In *2013 AIAA Infotech@ Aerospace Conference*, pages 1–8, 2013.

[160] S. Sankararaman and S. Mahadevan. Integration of model verification, validation, and calibration for uncertainty quantification in engineering systems. *Reliability Engineering & System Safety*, 138: 194–209, 2015.

[161] A. Saxena, J. Celaya, E. Balaban, K. Goebel, B. Saha, S. Saha and M. Schwabacher. Metrics for evaluating performance of prognostic techniques. In *Proceedings of the International Conference on Prognostics and Health Management, PHM 2008*, pages 1–17. IEEE, 2008.

[162] A. Saxena, J. Celaya, B. Saha, B. Saha and K. Goebel. Metrics for offline evaluation of prognostic performance. *International Journal of the PHM Society*, 1(1): 20, 2010.

[163] A. Saxena, I. Roychoudhury, J. Celaya, B. Saha, S. Saha and K. Goebel. Requirement flowdown for prognostics health management. In *AIAA Infotech@Aerospace, Garden Grove, CA*. AIAA, 2012.

[164] A. Saxena, S. Shankararaman and K. Goebel. Performance evaluation for fleet-based and unit-based prognostic methods. In *Proceedings of the European Conference of the Prognostics and Health Management Society, Nantes, France 2014.*, pages 276–287. PHM Society, 2014.

[165] R. Schlaifer and H. Raiffa. *Applied Statistical Decision Theory.* 1961.

[166] J.Z. Sikorska, M. Hodkiewicz and L. Ma. Prognostic modelling options for remaining useful life estimation by industry. *Reliability Engineering & System Safety*, 25: 1803–1836, 2011.

[167] N.D. Singpurwalla. *Reliability and Risk: A Bayesian Perspective*, volume 637. Wiley, Hoboken, NJ, 2006.

[168] N.D. Singpurwalla. Betting on residual life: The caveats of conditioning. *Statistics & Probability Letters*, 77(12): 1354–1361, 2007.

[169] S.A. Sisson and Y. Fan. *Handbook of Markov chain Monte Carlo*, chapter Likelihood-free Markov chain Monte Carlo, pages 319–341. Chapman and Hall/CRC Press, 2011.

[170] R.C. Smith. *Uncertainty Quantification: Theory, Implementation, and Applications.* SIAM-Society for Industrial and Applied Mathematics.

[171] H. Sohn, D.W. Allen, K. Worden and C.R. Farrar. Structural damage classification using extreme value statistics. *Journal of Dynamic Systems, Measurement, and Control*, 127(1): 125–132, 2005.

[172] E. Somersalo and J. Kaipio. *Statistical and Computational Inverse Problems.* New York: Springer-Verlag, 2004.

[173] N. Spiezia, M. Ferronato, C. Janna and P. Teatini. A two-invariant pseudoelastic model for reservoir compaction. *International Journal for Numerical and Analytical Methods in Geomechanics*, 41(18): 1870–1893, 2017.

[174] W. Staszewski, C. Boller and G.R. Tomlinson. *Health Monitoring of Aerospace Structures: Smart Sensor Technologies and Signal Processing.* John Wiley & Sons, 2004.

[175] S.M. Stigler. Laplace's 1774 memoir on inverse probability. *Statistical Science*, 1(3): 359–363, 1986.

[176] S.M. Stigler. *Statistics on the Table: The History of Statistical Concepts and Methods.* Harvard Univ. Pr., 2002.

[177] Z. Su, L. Ye and Y. Lu. Guided Lamb waves for identification of damage in composite structures: A review. *Journal of Sound and Vibration*, 295(3-5): 753–780, 2006.

[178] Y. Suhov and M. Kelbert. *Markov Chains: A Primer in Random Processes and Their Applications.* Cambrigde, University Press, 2008.

[179] T.J. Sullivan. *Introduction to Uncertainty Quantification.* Springer International Publishing.

[180] L.E. Szabó. Objective probability-like things with and without objective indeterminism. *Studies In History and Philosophy of Science Part B: Studies in History and Philosophy of Modern Physics*, 38(3): 626–634, 2007.

[181] A. Tarantola. *Inverse Problem Theory and Methods for Model Parameter Estimation.* Society for Industrial Mathematics, 2005.

[182] S. Tavare, D.J. Balding, R.C. Griffiths and P. Donnelly. Inferring coalescence times from DNA sequence data. *Genetics*, 145(2): 505, 1997.

[183] M.E. Tipping. Sparse Bayesian learning and the relevance vector machine. *The Journal of Machine Learning Research*, 1: 211–244, 2001.

[184] M.E. Tipping and A.C. Faul. Fast marginal likelihood maximisation for sparse Bayesian models. In *AISTATS*, 2003.

[185] L. Tvedt. Distribution of quadratic forms in normal space: Application to structural reliability. *Journal of Engineering Mechanics*, 116: 1183, 1990.

[186] M.K. Vakilzadeh, J.L. Beck and T. Abrahamsson. Using approximate Bayesian computation by subset simulation for efficient posterior assessment of dynamic state-space model classes. *SIAM Journal on Scientific Computing*, 40(1): B168–B195, 2018.

[187] M.K. Vakilzadeh, Y. Huang, J.L. Beck and T. Abrahamsson. Approximate Bayesian Computation by Subset Simulation using hierarchical state-space models. *Mechanical Systems and Signal Processing*, 84: 2–20, 2017.

[188] M.K. Vakilzadeh. *Stochastic Model Updating and Model Selection: With Application to Structural Dynamics.* PhD thesis, Chalmers University of Technology, Gothenburg, Sweden, 2016.

[189] M.W. Vanik, J.L. Beck and S.K. Au. Bayesian probabilistic approach to structural health monitoring. *Journal of Engineering Mechanics*, 126(7): 738–745, 2000.

[190] R. Von Mises. *Probability, Statistics, and Truth.* Dover Pubns, 1981.

[191] J. Vondrejc and H.G. Matthies. Accurate computation of conditional expectation for highly nonlinear problems. *SIAM/ASA Journal on Uncertainty Quantification*, 7(4): 1349–1368, 2019.

[192] D. Xiu and G.E. Karniadakis. The Wiener–Askey Polynomial Chaos for Stochastic Differential Equations. *SIAM Journal on Scientific Computing*, 24(2): 619–644, 2002.

[193] W.-J. Yan and L.S. Katafygiotis. A novel Bayesian approach for structural model updating utilizing statistical modal information from multiple setups. *Structural Safety*, 52: 260–271, 2015.

[194] W.-J. Yan and L.S. Katafygiotis. A two-stage fast Bayesian spectral density approach for ambient modal analysis. Part i: Posterior most probable value and uncertainty. *Mechanical Systems and Signal Processing*, 54: 139–155, 2015.

[195] W.-J. Yan and L.S. Katafygiotis. A two-stage fast Bayesian spectral density approach for ambient modal analysis. Part ii: Mode shape assembly and case studies. *Mechanical Systems and Signal Processing*, 54: 156–171, 2015.

[196] W.-J. Yan. *Wireless Sensor Network Based Structural Health Monitoring Accommodating Multiple Uncertainties.* PhD thesis, Hong Kong University of Science and Technology, 2013.

[197] T.-H. Yi and H.-N. Li. Methodology developments in sensor placement for health monitoring of civil infrastructures. *International Journal of Distributed Sensor Networks*, 8(8): 612726, 2012.

[198] F.-G. Yuan. *Structural Health Monitoring (SHM) in Aerospace Structures.* Woodhead Publishing, 2016.

[199] K.V. Yuen. *Bayesian Methods for Structural Dynamics and Civil Engineering.* John Wiley & Sons, 2010.

[200] K.V. Yuen, J.L. Beck and L.S. Katafygiotis. Efficient model updating and health monitoring methodology using incomplete modal data without mode matching. *Structural Control and Health Monitoring*, 13(1): 91–107, 2006.

[201] K.V. Yuen, L.S. Katafygiotis and J.L. Beck. Spectral density estimation of stochastic vector processes. *Probabilistic Engineering Mechanics*, 17(3): 265–272, 2002.

[202] K.V. Yuen and J.L. Beck. Updating properties of nonlinear dynamical systems with uncertain input. *Journal of Engineering Mechanics*, 129(1): 9–20, 2003.

[203] X. Zhang and H.Z. Huang. Sequential optimization and reliability assessment for multidisciplinary design optimization under aleatory and epistemic uncertainties. *Structural and Multidisciplinary Optimization*, 40(1): 165–175, 2010.

[204] W.J. Zhou and M.N. Ichchou. Wave propagation in mechanical waveguide with curved members using wave finite element solution. *Computer Methods in Applied Mechanics and Engineering*, 199(33-36): 2099–2109, 2010.

[205] E. Zio and G. Peloni. Particle filtering prognostic estimation of the remaining useful life of nonlinear components. *Reliability Engineering & System Safety*, 96(3): 403–409, 2011.

[206] K.M. Zuev and J.L. Beck. Global optimization using the asymptotically independent Markov sampling method. *Computers & Structures*, 126: 107–119, 2013.

[207] K.M. Zuev, J.L. Beck, S.K. Au and L.S. Katafygiotis. Bayesian post-processor and other enhancements of Subset Simulation for estimating failure probabilities in high dimensions. *Computers & Structures*, 93: 283–296, 2011.

[208] K. Tron Golder and B. Plischke. History matched full field geomechanics model of the Valhall field including water weakening and re-pressurisation. *Paper Presented at the SPE EUROPEC/EAGE Annual Conference and Exhibition, Barcelona, Spain*, June, 2010.

[209] J. Li and D. Xiu. A generalized polynomial chaos based ensemble Kalman filter with high accuracy. *Journal of Computational Physics*, 228(15): 5454–5469, 2009.

[210] C. Zoccarato, D. Baù, F. Bottazzi, M. Ferronato, G. Gambolati, S. Mantica and P. Teatini. On the importance of the heterogeneity assumption in the characterization of reservoir geomechanical properties. *Geophysical Journal International*, 207(1): 47–58, 2016.

Index

R

S

T

U

V